Series on Complexity Science – Vol. 3

Hypernetworks in the Science of Complex Systems

Series on Complexity Science   ISSN: 1755-7453

Series Editor: Henrik Jeldtoft Jensen *(Imperial College London, UK)*

Published

Vol. 1: A Complexity Approach to Sustainability: Theory and Application
        by Angela Espinosa & Jon Walker

Vol. 2: Stochastic Dynamics of Complex Systems: From Glasses to Evolution
        by Paolo Sibani & Henrik Jeldtoft Jensen

Vol. 3: Hypernetworks in the Science of Complex Systems
        by Jeffrey Johnson

Series on Complexity Science – Vol. 3

Hypernetworks in the Science of Complex Systems

Jeffrey Johnson
The Open University, UK

Imperial College Press

Published by

Imperial College Press
57 Shelton Street
Covent Garden
London WC2H 9HE

Distributed by

World Scientific Publishing Co. Pte. Ltd.
5 Toh Tuck Link, Singapore 596224
USA office: 27 Warren Street, Suite 401-402, Hackensack, NJ 07601
UK office: 57 Shelton Street, Covent Garden, London WC2H 9HE

British Library Cataloguing-in-Publication Data
A catalogue record for this book is available from the British Library.

Series on Complexity Science — Vol. 3
HYPERNETWORKS IN THE SCIENCE OF COMPLEX SYSTEMS

Copyright © 2013 by Imperial College Press

All rights reserved. This book, or parts thereof, may not be reproduced in any form or by any means, electronic or mechanical, including photocopying, recording or any information storage and retrieval system now known or to be invented, without written permission from the Publisher.

For photocopying of material in this volume, please pay a copying fee through the Copyright Clearance Center, Inc., 222 Rosewood Drive, Danvers, MA 01923, USA. In this case permission to photocopy is not required from the publisher.

ISBN 978-1-86094-972-2

Typeset by Stallion Press
Email: enquiries@stallionpress.com

Printed in Singapore

To Ron Atkin

Preface

This book aims to persuade you that the theory of hypernetworks is necessary if not sufficient for the science of complex systems and its applications in the planning, design and management of social and technical systems. It is aimed at a wide audience interested in solving real world problems, especially policy makers and scientists. Although it makes extensive use of mathematical symbolism it assumes no prior mathematical knowledge. In the interest of accessibility, the mathematics is mostly descriptive, explicitly explained, and there are no difficult theorems.

All but the simplest systems have many heterogenous parts interacting in many complicated ways. Traditionally they are designed and managed in ordinary language, but this is not possible for complex systems of systems of systems. These require large heterogeneous data sets and computers to process these data on the basis of models embodied in their data structures and programs. The mathematical structures developed in this book are intended to enable scientists, designers and policy makers to work together to create small and large systems that work.

The inextricable entanglement of policy, science and design is a theme developed later in the book, but it underlies the driving rationale for the theory of hypernetworks – it must be applicable and useful for addressing practical problems.

Structures related to hypernetworks have long been known, including networks, hypergraphs, Galois structures, simplicial complexes, and so on. The theory developed here tries to unify all these structures in a coherent way. Hypernetworks bring something new through making the relational structure more explicit and providing new structures for analysing relational dynamics. Hypernetworks provide a new way of representing multilevel systems with the objective of integrating their micro-, meso- and macrolevel dynamics.

This book reports work in progress. No doubt it contains misconceptions and errors that must be corrected. It is hoped that readers will give their feedback and take the research forward with constructive criticism and new ideas. Hypernetworks in the Science of Complex Systems will be supported by the website: www.hypernetworks.info.

Foreword

New fundamental breakthroughs in our understanding of how the world functions can only happen if we have the appropriate language to describe and analyse the phenomena of interest. Complexity science is in a situation similar to the one encountered by physics around 1900 when the development of statistical mechanics was able to relate the phenomenological science of thermodynamics to the microscopic world of atoms and molecules. A few years later, the establishment of the formalism of quantum mechanics enabled scientists to begin to make sense out of experimental facts which were in direct contradiction to the established worldview of classical physics.

Today, complexity science is in many ways the frontier of our endeavour to comprehend the world surrounding us. Complexity tries to establish generalities across what have hitherto been considered disconnected subjects such as biological evolution, geophysics, neuroscience, sociology and economics. This is done by identifying shared trends such as intermittent and abrupt changes and probability distributions of event sizes that resemble power laws and certainly do not compare with the normal distribution. Many phenomenological similarities between different types of complex systems have been established, for example, the observation of similarities in the statistics describing the bursts of activity in the brain, in rain, in solar flare activity, mass extinction and economic crashes.

Theoretically we are eagerly trying to establish the conceptual and mathematical tools that will allow us to turn complexity into a quantitative science in which we are able to discuss the relation between systemic cooperative behaviour and the properties of the constituent parts. The purpose of the Series, on Complexity Science published by Imperial College Press is to contribute to this undertaking. The first book in the series, *A Complexity Approach to Sustainability: Theory and Application*, by Espinosa and Walker, develops and generalises ideas from cybernetics to be able to investigate sustainability – an issue that is of particular relevance if one accepts that complex systems typically undergo intermittent upheavals and transitions by their own accord. The second book in the series, *Stochastic Dynamics of Complex Systems: From Glasses to Evolution*, by Sibani and Jensen, discusses mathematical methods which can handle non-stationary slowly and episodically relaxing systems. Future planned books will address the dynamics of networks and brain dynamics investigated using a complexity science approach.

The present book, *Hypernetworks in the Science of Complex Systems*, by Jeffrey Johnson, is of a very broad scope and speaks to anyone interested in a methodology of sufficient generality that it may be able to handle aspects of complex systems such as openness and hierarchical structure. Although it is certainly concerned with networks, the approach and philosophy of the book is very different from most recently published books on the mathematics of network theory. Johnson's book is a unique combination of mathematics and systems theory with a wish to point towards sociology and political science. It is at the forefront of activities that seek to establish the relevant tool box of concepts and formalisms that allow us to see complex systems as networks of networks and to be able to make headway in our ability to analyse complex systems involving humans, for example, traffic, finance, climate change, political conflict etc. I believe the book will be of great help to whoever is participating or interested in this important research activity.

September 2013, London
Henrik Jeldtoft Jensen, Series Editor

Acknowledgments

Many people have made many contributions to this book. It is dedicated to Ron Atkin who was my teacher and doctoral supervisor at Essex University in the late nineteen sixties and early seventies, and the director of the Urban and Regional Research Projects I worked on until 1977. He was an amazingly creative and inspirational leader and many of the most important ideas developed here were initiated by him. I am very grateful to him for his many insights and for his friendship. Ron died on the second of August this year and I will miss him greatly.

Atkin's work was forty years ahead of its time and some people had the foresight to recognise and support it. They included Brian Gaines, Lionel March, Alan Wilson, Brian Griffiths, Mike Batty, Nigel Thrift, John Casti, and many others who championed the research, which at the time was considered very controversial. All have extended their support and generosity to me and I am very grateful for it.

In 1997 I moved from the Mathematics Department at Essex University to the Geography Department at Cambridge University to work with Graham Chapman and Peter Gould on the International Television Flows Project. I learned a lot from Graham as we developed the theory of Q-analysis in the context of collecting large data sets on European television programmes. He remains a close and highly stimulating colleague. Peter, who was based in the Geography Department at Pennsylvania State University, was a pioneer in quantitative geography and an early supporter of Q-analysis. When I visited him we had many what he called "blood on the wall" discussions on the multilevel coding scheme the project devised for television programmes. Unfortunately Peter died in 2000 but I often reflect on things he said to me. These included the visionary remark that "all human activity is spatially referenced" – how he would have loved the Big Data revolution.

In 1980 I moved from Cambridge to work in Lionel March's group in the Design Discipline of the Open University. Lionel was a great supporter of Atkin's ideas and I am eternally grateful for the opportunity he gave me to develop them since then. While in what became the Department of Design and Innovation I learned a lot about design, especially from Phil Steadman and Nigel Cross during many discussions and our doctoral student supervisions. I learned from Phil that, viewed the right way, research and academic life can be very funny and great fun.

During the nineteen eighties and nineties John Casti, Roland Thord, David Batten and Anders Karqvist organised annual meetings of the "Summer University of Southern Stockholm" around the theme of networks in action. This gave me an opportunity to develop many of the ideas in the book in the context of a critical audience. I was fortunate to meet many outstanding scientists at these meetings, including Chris Barrett who was then at Los Alamos National Laboratory and now heads the Virginia Bioinformatics Institute. The highly innovative work done by Chris and his colleagues has had a big impact on my understanding of the interplay between science, computation and policy.

Other early adopters of Atkin's ideas included Clive Downs, Max King and Senino Holtier. I am grateful for our collaborations in the early days.

In the late nineteen eighties George Rzevski joined the department, bringing many original ideas on complex systems and designing intelligent machines. At one departmental meeting someone suggested an interesting course of action. When asked if we were allowed to do it, George thumped the table and said "an entrepreneur does not ask permission!" I am grateful for this and many other insights. George's colleague Anthony Lucas-Smith joined the department in the nineties and he has been a tireless member of our Centre for Complexity Science. I am grateful to him for all the help he has given me.

In the nineteen nineties I worked closely with Jean-Claude Simon and his group in Paris on machine vision. He was another highly distinguished and generous colleague. Sadly he died in 2000 but I am grateful for his friendship the opportunity to have worked with him.

Early in 2000 Ralph Dum brought new ideas on complex systems to the European Commission and began the process of coordinating the nascent European complex systems research community. For more than a decade Ralph has been at the forefront of research in complex systems and its applications in policy and this has had a significant influence on my work. This continues with his Global Systems Science initiative which has had a big impact on my understanding of science in policy. In this context I have been very fortunate to work with Sander van der Leeuw, Carlo Jaeger, Steven Bishop, and Joan David Tàbara.

Paul Bourgine has had an enormous impact on this book. He suggested that I should write the paper on hypernetworks for the European Conference on Complex Systems which led directly to this book. With Mina Teicher, we created the Complex Systems Society in 2005 and we have all worked closely together since then. The zebra fish image analysis in Chapter 6 was a collaboration with Paul and Nadine Peyriéras. Paul is currently a driving force behind the creation of the international UNESCO UniTwin Complex Systems Digital Campus which is networking hundreds of institutions worldwide. Hypernetworks are being used in an experiment to design information systems able to support high quality personalised open education for the many thousands of students and researchers associated with the Digital Campus.

Acknowledgments

Of the many other people I have worked with in Europe, two have shown particular interest in and commitment to hypernetworks. Jorge Louçã invited me many times to Lisbon to discuss and present my ideas and we are beginning a new programme of research around his Observatorium of millions of newspaper articles collected over the last five years. Andrjez Nowack has also become very interested in hypernetworks and we are planning new programmes of research in multilevel social systems. Both Jorge and Andjrez have given me great encouragement and I am very grateful for it. Through Jorge I met João Fiedaro and participated in his Real Time Composition workshops. I am grateful for his help in preparing the example of Real Time Composition as evolving hypernetworks and permission to reproduce images from his website.

Various people have given me the opportunity to present the ideas in the book at residential schools including Carmen Costea, Robert MacKay, Eve Mitleton-Kelly and Tassos Bountis. I thank them for their support. Robert kindly read the manuscript and suggested many improvements, and I am very grateful to him. Of many other colleagues in the complex systems community I am grateful for the discussions I have had with David Lane on multilevel systems and ontology. Thanks are also due for the support we received from Irene Poli at our meetings in Venice thinking about new mathematics and statistical theory for complex systems science.

Closer to home, my OU colleagues Katerina Alexiou and Theodore Zamenopoulos have always taken an interest in my hypernetwork research and they have helped my build up the Centre for Complexity Science at the Open University. Jane Bromley is a tremendous help in everything I do, and I greatly appreciate her support. Claudia Eckert has helped me rethink my ideas on design. Bill Nuttall has given me useful feedback, as has Stephen Peake through our discussions on complexity and policy. Matthew Cook has given me a social science perspective on my work and this has been very helpful in the policy sections. Simon Buckingham Shum has given me many insights into education and data analytics. David Hales co-created the Realtime Composition example in Chapter 8, and shared his insights. David Rodrigues has been very active developing hypernetwork ideas and I am grateful for his enthusiasm and collaboration. Chris Earl was very supportive as the project came to a close. I am very grateful to them all.

Some of the examples in the book come from collaborations with former doctoral students. In particular, Pejman Iravani worked with me on the robot example, Joan Serras worked with me on the multilevel road traffic example, and Valerie Rose worked with me on multilevel machine vision. Currently Cristian Jimenez Romero is working with me on hypernetworks for scalable personalised education. They have all made a big contribution and I am very grateful.

The final stages of preparing this book were completed during the European TOPDRIM project (Topology Driven Methods for Complex Systems). I am very grateful to Mario Rasetti for his interest in hypernetworks and his personal support, and to Emanuela Merelli for giving me the opportunity to join the project.

Already TOPDRIM is giving new results in Q-analysis through our collaboration with Emanuela and Matteo Rucco. I am grateful to all my TOPDRIM colleagues for their input, and to the European Commission for the opportunity to develop and disseminate the ideas through the TOPDRIM project.

Through the NESS (Non-Equilibrium Social Science) and Étoile (Enhanced Technology for Open Intelligent Learning Environments) projects I have had the pleasure of discussing hypernetwork ideas with Paul Ormerod, Bridget Rosewell, Yi-Cheng Zhang, Andrjez Nowack, and Greg Fisher. My work has benefitted greatly from other research and coordination activities supported by the European Commission, and the many excellent scientists in the emerging field of complex systems science. This includes Dirk Helbing, Anna Carbone, Peter Richmond, Sylvie Occelli, Bruce Edmonds, Stephen Bishop, Julian Hunt, Jamie MacIntosh, Janet Smart, Paul Lukowicz, Andras Lorincz, Maxi Miguel, Alex Verspignani, Michel Morvan, Henry Wynn, Juergen Jost, Fatihcan Atay, Yasmin Merali, Jeremy Pitt, and many others.

Many people in the networks communities have worked with hypergraphs and at different times. I have had many informative discussions with Paul Bourgine's former students David Chavalarias and Camille Roth, and I also have benefited from the comments of Yin Seong Ho, Stephen Seidman, Douglas White, Guido Calderelli and many others.

Peter Checkland and Jim Scholes kindly allowed me to use Fig. 8.5 and John Wiley & Sons Ltd granted copyright permission.

I am grateful to Lance Sucharov of ICP for commissioning this book, the patience of Lizzie Bennet and Jackie Downs during the ups and downs of its production, to Matthew Judge who saw it through the editing stage to completion, and to Laurent Chaminade for its publication.

When writing acknowledgements pages like this it is inevitable that someone very important will be overlooked until it is too late. There are many special people that should be acknowledged as playing a major part in the research leading to this book. If you are one of them be assured that I will realise it almost immediately – please forgive my oversight.

I am particularly grateful to my family and friends for the support they have given me. My father-in-law Stuart Weir and my mother Eileen both took a keen interest in its progress but regrettably they did not live to add another incomprehensible tract to their collections. My daughters Lucy and Anna and son-in-law Ben have also taken a keen interest in progress and helped me to keep going. My wife Carol has been very indulgent, generous, supportive and patient – neither words nor mathematics can adequately express my gratitude to her and everyone else.

February 2013, Woburn Sands
Jeffrey Johnson

Contents

Preface vii

Foreword ix

Acknowledgments xi

1. Introduction 1
   - 1.1 The emergence of hypernetwork theory 1
   - 1.2 Complexity 6
   - 1.3 Complex systems science, design and policy 10
   - 1.4 Hypernetworks in the science of complex systems 11

2. Sets and Relations 13
   - 2.1 Elements of set theory 13
   - 2.2 Mappings and functions 18
   - 2.3 Relations 20

3. Hypergraphs and the Galois Lattice 31
   - 3.1 Hypergraphs 31
   - 3.2 The Hypergraphs of a Bipartite Network 33
   - 3.3 The Galois Connection 34
   - 3.4 Galois Pairs and Maximal Rectangles 37
   - 3.5 The Galois Lattice 38
   - 3.6 Weak and Strong Connectivity in Hypergraphs 39

4. Simplicial Complexes and Q-analysis 41
   - 4.1 From sets to simplices 41
   - 4.2 Simplices, polyhedra and their faces 46
   - 4.3 The intersection of simplices 47
   - 4.4 Representing social structure by connected simplices 48
   - 4.5 Simplicial families and simplicial complexes 49

4.6		Multidimensional connectivity in simplicial complexes	50
4.7		Q-analysis	52
4.8		Structure Vectors	56
4.9		The Difference Operator	57
4.10		Eccentricity	58
4.11		Q-graphs	60
4.12		Stars and hubs	62
4.13		Galois Families	65
4.14		Simplicial prisms	66
4.15		Galois Prisms	68
4.16		Descriptor Simplices and Antivertices	69
4.17		Networks and Simplicial Complexes	71
4.18		Examples	73
	4.18.1	Example: Sky and Water	73
	4.18.2	Example: Discriminating textured surfaces	75
	4.18.3	Example: Random Q-analysis	81
	4.18.4	Example: the Q-analysis of road intersections	84
	4.18.5	Example: The Wisdom of Crowds	92
	4.18.6	Example: Multidimensional Structure in Road Networks	95

5. Backcloth and traffic: dynamics constrained by topology   101

5.1	The space-time backcloth constrains physical and social traffic	102
5.2	Structuring the social backcloth to constrain the traffic	103
5.3	The backcloth allows and forbids but does not require	104
5.4	Traffic as patterns of numbers on the backcloth simplices	105
5.5	Measuring the traffic	106
5.6	Measuring the backcloth	108
5.7	Traffic-dependent backcloth dynamics	110
5.8	Matrices as backcloth supporting flow traffic	112
5.9	Matrices as backcloth supporting descriptor traffic	115
5.10	Q-transmission in simplicial complexes	119
5.11	Examples	122
5.12	Dimensions, probabilities and q-transmission	126
5.13	Supertraffic	128
5.14	The topology of the backcloth	129
5.15	Algebraic topology	130
5.16	From cohomology in physics to q-connectivity in social science	139
5.17	Shomotopy	143
5.18	From the q-graph to the q-complex	146
5.19	Designing the backcloth to carry the traffic	149

6. Hypernetworks   151

6.1	Hypersimplices and hypernetworks	152
6.2	Examples	153
6.3	The Fundamental Question of Hypernetworks	163
6.4	n-ary relations	168
6.5	Hyperfaces for defining intersections of hypersimplices	173
6.6	Intersecting Complicated Structures	176

7. Multilevel Systems — 177

7.1	Systems of Systems of Systems	177
7.2	The Intermediate Word Problem	179
7.3	Relational Hypersimplices and Aggregation	180
7.4	Structural aggregation in multilevel systems	181
7.5	AND aggregations and OR aggregations in multilevel systems	183
7.6	Taxonomic aggregation in multilevel systems	184
7.7	Emergence in Aggregation and Disaggregation Dynamics	191
7.8	Combinatorial Explosion in Downward Emergence	193
7.9	Defining hierarchical levels by disaggregation	194
7.10	Clustering and Hierarchical Set Definition	196
7.11	Multidimensional descriptor spaces	197
7.12	Multilevel Descriptor Simplices	197
7.13	Personal Constructs, Triadic Sorting and Clustering	198
7.14	Mereology	204
7.15	The intransitivity of meronymic relations	210
7.16	Hierarchical traffic aggregation	211
7.17	The Grand Challenge of multilevel systems	215
7.18	Example: Machine vision	216
7.19	Example: Robot Football	228
7.20	Multilevel Land-Use Transportation Systems	230
7.21	Example: hospital admission for assault	239
7.22	Example: the London riots of 2011	240
7.23	Example: street gangs in London	242

8. Time, Events, Prediction and Forecasting — 247

8.1	Prediction	247
8.2	Feedback, simulation and prediction	251
8.3	Time and Events	256
8.4	Mapping system time to clock time	259
8.5	Multilevel events	260
8.6	Rare and extreme events in multilevel systems	262
8.7	Tipping points, nudges, and multilevel cascades	263
8.8	Time horizons	264
8.9	Time, prediction and events in the science of complex systems	264

8.10	Example: Real Time Composition	265

9. Hypernetworks and Design 275

9.1	Design	276
9.2	Design and the Intermediate Word Problem	281
9.3	Example: designing an educational course module	286

10. Policy, Design, Planning and Science 297

10.1	The entanglement of policy, design, planning and science	297
10.2	Policy and Science	298
10.3	Policy and Design	304
10.4	Reasoning and logic in policy design	306
10.5	Narratives in policy science	310
10.6	Hypernetwork science for building policy narratives	310
10.7	Art in science and policy	312

11. Notes and reflections 315

11.1	Resumé	315
11.2	Networks and Hypernetworks	316
11.3	Design	317
11.4	Policy	318
11.5	Systems theory and complex systems science	318
11.6	Multilevel systems, modelling and computation	319
11.7	A research agenda for hypernetworks and their applications	319

Bibliography 321

Index 325

Chapter 1

Introduction

1.1 The emergence of hypernetwork theory

Network theory has revolutionised our understanding of social systems at every level, from individuals to organisations to nations. The simple idea that the interaction between two things can be represented by two vertices and an edge has turned out to be incredibly powerful. Strangely, the need for representation and study of the interactions between three or more things has been largely overlooked. But, as Figure 1.1 shows there are many examples of such relations.

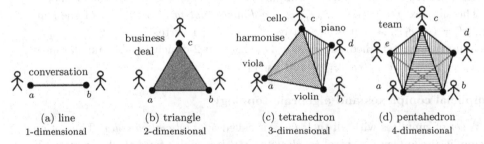

Fig. 1.1 Relationships can bind two or more things together

In the nineteen fifties C. H. Dowker published the paper *The homology groups of relations* [Dowker (1952)] which showed that relations between n things could be represented by multidimensional polyhedra with n vertices, such as those shown in Figure 1.1. This idea lay dormant for a quarter of a century until in the nineteen sixties R. H. Atkin introduced the revolutionary idea that social relations could be represented by polyhedra. For example, a business deal between three people can be represented by a triangle, written as $\langle a, b, c \rangle$, the relation of four people playing music together can be represented by a tetrahedron, $\langle a, b, c, d \rangle$, and the relationship between five people working together as a team can be represented by a pentahedron, $\langle a, b, c, d, e \rangle$. This idea is entirely compatible with network theory since, for example, a relationship between two people having a conversation can be represented by a polyhedron with two vertices, namely a line or an edge, $\langle a, b \rangle$.

These ideas first appeared in the article *A mathematical approach towards a social science*, published in the *Essex Review* in 1968 [Atkin et al (1968)]. The rest of this section will sketch out Atkin's ideas and show how they form the basis of hypernetwork theory. Everything is covered in more detail later in the book.

Polyhedra and simplices

In the early seventies Atkin and coworkers investigated the topological properties of relations in the context of town planning. The breakthrough came when Atkin suggested a new kind of connectivity based on the shared faces of social polyhedra.

To avoid confusion it is essential to understand that a p-dimensional polyhedron has $p+1$ vertices. The vertices of networks, $\langle v \rangle$, have *dimension* zero. Edges, $\langle v_0, v_1 \rangle$, have dimension one but two vertices. For higher dimensional polyhedra, a triangle $\langle v_0, v_1, v_2 \rangle$ has dimension two but three vertices. A tetrahedron $\langle v_0, v_1, v_2, v_3 \rangle$ has dimension three but four vertices. And so on. By labelling the first vertex v_0, the last vertex of a p-dimensional polyhedron can be labelled v_p, and this convention will be used as appropriate throughout this book. Thus the generality is that a p-dimensional polyhedron will be written as $\langle v_0, v_1, ..., v_p \rangle$.

Polyhedra are the geometric realisation of more abstract objects called *simplices*. Let V be a set of vertices. An abstract p-*simplex* is determined by a set of $p+1$ vertices, written as $\langle v_0, v_1, ..., v_p \rangle$. Simplices are often represented by the symbol σ.

The simplex $\sigma = \langle v_0, v_1, ..., v_q \rangle$ is a q-*dimensional face*, or q-*face*, of the simplex $\sigma' = \langle v_0, v_1, ..., v_p \rangle$ if every vertex of σ is also a vertex of σ'. For example, the 3-dimensional tetrahedron $\langle v_0, v_1, v_2, v_3 \rangle$ has four 2-dimensional triangular faces $\langle v_1, v_2, v_3 \rangle$, $\langle v_0, v_2, v_3 \rangle$, $\langle v_0, v_1, v_3 \rangle$ and $\langle v_0, v_1, v_2 \rangle$.

Simplicial complexes and algebraic topology

A set of simplices with all their faces is called a *simplicial complex*. The notion of simplicial complex is central to *algebraic topology* which uses algebraic methods to study the structure of topological spaces. Although the terminology may be unfamiliar, the ideas are intuitive and simple as illustrated by the disk with a hole in it in Figure 1.2(a). This is a simple topological space.

Topology is often described as rubber sheet geometry, with two spaces having the same topology if one can be stretched into the other without tearing. So the space in Figure 1.2(a) is topologically equivalent to that in Figure 1.2(b), and both are topologically equivalent to the space in Figure 1.2(c).

These three spaces are all equivalent to a disk with a hole in it, which is intuitively obvious but poses the question of how it could be demonstrated formally that any given topological space is equivalent to another or not. Algebraic topology answers many questions of this form by transforming the topological problem into an algebraic problem which, it turns out, is much easier to solve.

Introduction

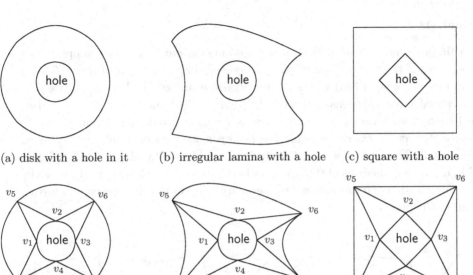

(a) disk with a hole in it   (b) irregular lamina with a hole   (c) square with a hole

(d) triangulation of (a)   (e) triangulation of (b)   (f) triangulation of (c)

Fig. 1.2   Representing topological properties by the connectivity of abstract algebraic simplices

Figure 1.2(f) shows eight 2-dimensional triangular simplices configured with a diamond shaped hole in the middle. It is said of be a *triangulation* or *triangular decomposition* of the space. This space is not just topologically equivalent to that in 1.2(c) but it is also geometrically equivalent. Thus the topological space can be represented by the algebraic objects $\langle v_1, v_2, v_5 \rangle$, $\langle v_2, v_5, v_6 \rangle$, $\langle v_2, v_3, v_6 \rangle$, $\langle v_3, v_6, v_7 \rangle$, $\langle v_3, v_4, v_7 \rangle$, $\langle v_4, v_7, v_8 \rangle$, $\langle v_1, v_4, v_8 \rangle$, $\langle v_1, v_5, v_8 \rangle$. In Chapter 5 an algorithm will be given that detects holes in configurations of simplices including triangulations such as this. It works irrespective of the way the space is triangulated.

Figure 1.2(e) shows a triangulation of the original disk with a hole. In this case the triangles have curved edges, but they are determined by their vertices as abstract algebraic simplices. These abstract simplices lose all the geometric information but retain the topological information of how the rubber-sheet triangles are connected to each other. The algorithm can be then be used to detect the hole.

Figure 1.2(e) shows a triangulation of the space in Fig. 1.2(b.) This has exactly the same abstract 2-simplices as the other two spaces. The algorithm shows that this space too has one 2-dimensional hole.

This example illustrates the idea that topological spaces can be transformed into simple algebraic spaces permitting algorithmic calculations that answer topological questions such as whether or not two spaces are topologically equivalent with the same number of n-dimensional holes.

Q-analysis

Two simplices are q-near if they share a q-dimensional face. Two simplices are q-connected if there is a chain of pairwise q-near simplices between them. This is illustrated in Figure 1.3(a) where the tetrahedra σ and σ' are 1-near because they share an edge, or 1-dimensional face. In Figure 1.3(b) the tetrahedra σ_1 and σ_4 are 1-connected, since σ_1 is 1-near σ_2, σ_2 is 1-near σ_3, and σ_3 is 1-near σ_4. A *Q-analysis determines* classes of *q-connected components*, sets of simplices that are all q-connected. The first paper on Q-analysis was published by [Atkin *et al* (1971)] in 1971. It gave a Q-analysis of the land uses in the town of Colchester. Q-connectivity is a generalisation of 0-dimensional connectivity in networks. It is covered in further detail in Chapter 4.

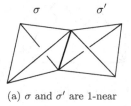
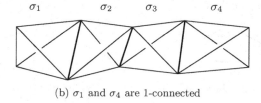

(a) σ and σ' are 1-near   (b) σ_1 and σ_4 are 1-connected

Fig. 1.3 q-connected polyhedra

Backcloth and traffic

The vertices and edges of networks often have numbers associated with them. For example in a social network the vertices may be associated with the amount of money a person has and the edges may be associated with how much money passes between pairs of people. In electrical networks the vertices have voltage associated with them and the edges have current. Although the network's voltages and currents may change, the network itself does not. Similarly in a road network the traffic flows may vary but the network infrastructure does not. The same holds for simplicial complexes when there are patterns of numbers across the vertices and the simplices. The numbers may change when the underlying simplicial complex does not.

Atkin suggested that the relatively unchanging network or simplicial complex structure be called a *backcloth* and that the numbers be called the *traffic* of activity on the backcloth. As an example, the airline network acts as a backcloth to the traffic of airline passengers. The term backcloth comes from the scenery painted on large canvas sheets used in theatres as a static backdrop behind the actors.

Atkin originally used simplicial complexes to characterise a wide variety of phenomena in physics by the *Cocycle Law* that the space-time backcloth supporting many physical phenomena has no holes [Atkin (2010)]. Atkin's conceptual leap "from cohomology in physics to q-connectivity in social science" was published in a landmark paper [Atkin (1972)].

Hypersimplices, hypernetworks, multilevel systems and structural events

In the nineteen eighties it became clear that for some purposes simplices were not rich enough to discriminate different things formed from the same set of vertices. This is illustrated in Figure 1.4 where the same set of graphical objects , $v_0, ..., v_p$ is assembled in different ways. Let R_1 and R_2 be the assembly relations. Then *relational simplices* can be formed by making the relations explicit in the representation as follows. For example, let $\sigma = \langle v_0, ..., v_p; R_1 \rangle$ be the house-like assembly (Fig. 1.4(a)) and and $\sigma' = \langle v_0, ..., v_p; R_2 \rangle$ be the other (Fig. 1.4(b)).

Then although σ and σ' have the same vertices, it is clear that $\sigma \neq \sigma'$ for $R_1 \neq R_2$. In this book a simplex augmented by its relation is called a *hypersimplex*. A set of hypersimplices is called a *hypernetwork*.

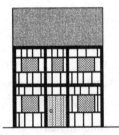
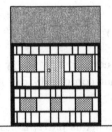

(a) Visual features assembled into a house-like configuration
$\sigma = \langle v_0, ..., v_p; R_1 \rangle$

(b) Visual features assembled into a fantasy configuration
$\sigma' = \langle v_0, ..., v_p; R_2 \rangle$

Fig. 1.4   Hypersimplices can discriminate different relations on the same set of parts

Hypersimplicies are structured sets of vertices, and in Chapter 6 it will be said that the hypersimplex exists at a *higher level of representation* than its vertices. This leads the way to a theory of multilevel systems in Chapter 7. The absence of a formalism able to aggregate and disaggregate multilevel traffic over multilevel backcloths is a major problem across the sciences. Hypernetworks are a step towards creating a formalism to represent the dynamics of multilevel systems.

A powerful idea suggested by Atkin is to define the formation of a p-simplex to be a *p-event* in *system time*, as opposed to what he called *clock time* [Atkin (1981)]. For example, a structural event occurs when the ingredients of a meal, the vertices, are assembled and processed to become the meal, a hypersimplex. This event can be associated with longer or shorter clock time intervals.

Preferential attachment suggests a mechanism for the formation of network events as new links are made [Barabási *et al* (1999); Barabási (2003)].

Figure 1.5 shows the event of assembling the blocks b_1, b_2, b_3 under the relation R into a cross configuration, $\langle b_1, b_2, b_3; R \rangle$. Before the event the blocks are not assembled and the cross does not exist. After the event the blocks are assembled into the cross, which now does exist. The formation of p-events marks system time, which is important for backcloth dynamics.

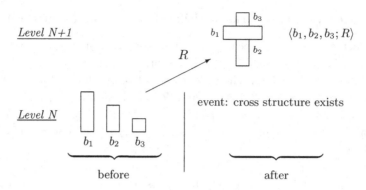

Fig. 1.5 Assembling vertex parts to form a hypersimplex whole as a multilevel structural event

1.2 Complexity

There is no agreement on how the word "complex" should be defined, but there is wide consensus that complexity can arise in systems that have one or more of the following properties:

- many heterogeneous parts, *e.g.* a city, a company, the climate, markets, riots
- complicated transition laws, *e.g.* economies, disease, gossip, behaviour change
- unexpected or unpredictable emergence, *e.g.* chemical systems, accidents
- sensitive dependence on initial conditions, *e.g.* weather systems, investments
- path-dependent dynamics, *e.g.* evolution, elections, personalities, famine, war
- network connectivities, *e.g.* gossip, epidemics, copying, company director networks
- multiple subsystem dependencies, *e.g.* cities, ecosystems, financial meltdown
- dynamics emerge from interactions of autonomous agents, *e.g.* traffic, markets
- self-organisation into new structures and behaviours, *e.g.* ghetto formation
- non-equilibrium and far-from equilibrium dynamics, *e.g.* share price changes
- discrete dynamics with combinatorial explosion, *e.g.* chess, policy, investments
- adaptation to changing environments, *e.g.* biology, business, opinions, agriculture
- co-evolving subsystems, *e.g.* land-use and transportation, virus software, friendship
- ill-defined boundaries, *e.g.* genetically modified crops, pollution, terrorism, health
- multilevel dynamics, *e.g.* companies, armies, governments, the Internet, revolutions
- feedback, *e.g.* profit from investment, opinion polls, critical blogs and tweets, copying
- confounding behaviour, *e.g.* disobedience, bucking the trend, breaking rules, anger
- unrepeatable experiments, *e.g.* electing a government, war, having children, policy
- misleading data, *e.g.* sensor errors, questionnaire flaws, disinformation, spin, lies
- globality – everything affects everything else *e.g.* biofuel & starvation, hegemony

Most systems exhibit many or all of these characteristics. Any one of them can make systems appear complex, but together they can make systems very difficult to understand, predict, design, control, and manage.

In his paper *From Complexity to Perplexity*, Horgan addresses the question whether science can achieve a unified theory of complex systems [Horgan (1995)]. He quotes the thirty one definitions of complexity given by Seth Lloyd and illustrates the diversity by a selection, including entropy, information, fractal dimension, effective complexity (degree of regularity rather than randomness), hierarchical complexity, grammatical complexity, thermodynamic depth, time computational complexity, spatial computational complexity (memory), and mutual information (between parts). Lloyd's list is ordered by three questions frequently asked to quantify the complexity of things: 1. How hard is it to describe? 2. How hard is it to create? and 3. What is its degree of organization? [Lloyd (2011)].

Edmonds gives over forty definitions of measures of complexity [Edmonds (1999)]. Many of these are technical relating to representation and computation, but the list includes more subjective measures such as the number of constructs an individual uses to describe a system.

Thus the contested term "complexity" may be a social construct rather than an absolute property, fitted within an interpretive epistemology that provides one of many possible ways to understand the world with all possessing some degree of validity. However there is much more agreement on the more detailed characteristics that can make a system appear complex. Furthermore, the existence of many interpretations suggests the possibility of knowledge and science at a metalevel above the particular interpretations, as discussed later in the context of policy.

Complex Systems Science

Conventional science is compartmentalised into domains such as physics, chemistry, biology, psychology, sociology, and so on. Some scientists specialise in a single domain, knowing every aspect of it, drilling down to great depth in research. In contrast to this in-depth *vertical* science, complex systems science involves *horizontal* concepts and integrative research across the domains, as illustrated in Figure 1.6. It is interdisciplinary and usually involves teams with complementary specialisms.

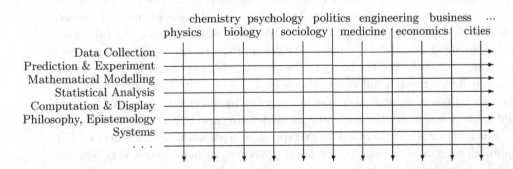

Fig. 1.6 Complex systems science works horizontally across the individual scientific domains

During the twentieth century it became apparent that many systems could not be investigated using the experimental or theoretical methods of the traditional physical sciences.

The realisation that many systems are sensitive to initial conditions has changed scientists attitudes to what it means to make a prediction. Theory alone tells us that there is a horizon beyond which prediction is not possible. None-the-less, to plan and manage systems it is necessary to be able to predict the consequences of policy interventions. The concept of prediction in social systems is different to that in physical systems. In the context of policy some predictions are intended to be self-fulfilling prophesies, *e.g.* "our policy was to build 1000 homes this year, we predicted that 1000 homes would be built this year, and we have built 1000 homes this year".

The science of complex systems attempts to provide methods of understanding the dynamics of systems where conventional methods fail. These methods apply across the domains, *e.g.* chaotic dynamics can be observed in biological systems, economic systems, chemical systems, road traffic systems, and many others. There are many systems in which the behaviour of the whole emerges from interactions between the parts, *e.g.* traders in markets, birds in flocks, people in cities, cars on roads, sportsmen in teams, and cells in bodies.

Confining scientific enquiry to one domain can give deep insights, but unexpected things can happen when a subsystem from one domain interacts with a subsystem from another. In 1956, W. Ross Ashby wrote

> Science stands today on something of a divide. For two centuries it has been exploring systems that are either intrinsically simple or that are capable of being analysed into simple components. The fact that such a dogma as 'vary the factors one at a time' could be accepted for a century, shows that scientists were largely concerned in investigating such systems as allowed this method; for this method is often fundamentally impossible in the complex systems. [Ashby (1956)]

Thus complex systems science is necessarily interdisciplinary, integrating knowledge from all domains including the humanities, social sciences, natural sciences and the sciences of the artificial. Complex systems science draws on all of these but adds something new. The computer revolution of the twentieth century enabled a new kind of science. For the first time in history it became possible to analyse the dynamic interactions of millions of things explicitly, *e.g.* it is possible to calculate the interactions of millions of drivers in a city and to observe the emergent tailbacks and traffic jams, where the simulated dynamics are close to those observed.

Much of our science is based on extrapolation from what has gone before. This approach to understanding the world and predicting its behaviour is very powerful and works well most of the time. Occasionally it fails to foresee rare but extreme events, and this is a challenge for complex systems science.

Introduction

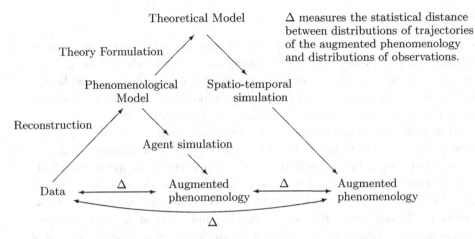

Fig. 1.7 The complex systems methodology for reconstructing models from data

Figure 1.7 gives an overview of the complex systems methodology for reconstructing models from data [Bourgine et al (2012)]:

> The scientific perspective of complex systems methodology is illustrated in [Figure 1.7]. As with all science it begins with data from which scientists reconstruct phenomenological models. For example, Kepler constructed a phenomenological model in which the planets sweep out equal areas in equal times which Newton formulated as a theory of planetary motion able to reproduce this phenomenology. In the case of the motion of two bodies, Newtons Laws produce equations that can be solved explicitly making it possible, for example, to predict precisely where a cannon ball will land. In the three-body case the equations cannot be integrated and the system is chaotic. Nonetheless the spatio-temporal behaviour of the system can be simulated by iterated computation providing an augmented phenomenology ([Figure 1.7], bottom right). The objective in this modelling is to produce an augmented phenomenology whose statistical difference from observation, Δ, is as small as possible (theoretically zero for a perfect model). In most cases simulations can at best sample the space of all system trajectories around given initial conditions and with an error, Δ, which measures the difference between the statistical distributions of the simulated trajectory and the statistical distributions of the data. ... As an example for a social system consider the people evacuating a building in an emergency. The motion of people in crowds is observed and a phenomenological model is created of the ways people move with respect to each other. Using this phenomenological model, an agent-based computer simulation can be used to create an augmented phenomenology for this system ([Figure 1.7], bottom-centre). A theoretical model of pedestrian flows can be proposed and permit spatio-temporal simulations to create another augmented phenomenology ([Figure 1.7], bottom-right).

Computer simulation

Computer simulation is a powerful new scientific method. There are many systems in which the meso- and macro-dynamics emerge from the discrete microlevel interactions of many autonomous agents. The resulting dynamics are too complicated to be captured by formulae, but computer simulation allows those dynamics to be played out at the microlevel to produce emergent dynamics at higher levels.

Despite having its own methodological problems, computer simulation is giving new insights into many kinds of system. Furthermore new sources of data are emerging about human beings, including the way individuals use mobile telephony and the Internet for much of their economic and social activities. Never in the history of humankind has so much been known about the microdynamics of whole populations, with emergent data sources eclipsing and augmenting traditional census, taxation and survey data. So called *Big Data* is revolutionising the science of socio-physical systems.

1.3 Complex systems science, design and policy

Complexity and Design

Designers are masters of complexity. They generate and predict the behaviour of new systems that did not previously exist. They work with clients who do not know precisely what they want. They know about components and the processes that can combine them to make new systems with emergent properties. They predict the effects of external forces such as regulation, fashion and costs. And they manage the process of creating and implementing new designs as working systems.

Chapter 9 of this book argues that hypernetworks have a special role to play in design as the creation of artificial systems. Design involves *predicting* how systems will behave before they exist, and this requires science. Just as important is the co-evolution in design between the *requirements* and the candidate designs generated to satisfy those requirements.

Complexity and Policy

It is argued in Chapter 10 that complex systems science has no meaning without applications, and that policy is the laboratory of complex systems science. Since scientists do not have the mandate or the money to do social experiments complex systems science must be developed within a policy context.

The logic of policy and the logic of science can be different. In principle science is neutral while policy is normative. In some case policy is driven by rhetoric where winning power is an essential prerequisite to instigating change. Thus the science-policy interface will always be difficult to manage, especially since the policy makers will always be the senior partners because they have both the moral right to make changes that affect people's lives, and they have the authority and resources necessary to implement those changes.

A further argument says that policy is *designing the future*. An important consequence of this is that policy makers must stay in the co-evolutionary loop between policy requirements and the generation and evaluation of potential systems to satisfice those requirements.

1.4 Hypernetworks in the science of complex systems

Unlike the physical world, social systems can change their organising principles and become completely different. For most people the world today is completely different to that of their parents and children. Social systems are subject to constant innovation that can make existing theories redundant and require new theories to support policy, organisation and management.

The socio-physical world is constantly being designed and redesigned at all levels. Such science as may have applied to the previous socio-physical world may no longer apply after design and policy events. In this case *new* science is required to understand the emerging new world.

This book suggests that hypernetworks provide an appropriate formalism for reconstructing phenomenological models from data as a step towards agent simulation and the development of theoretical models. This formalism is appropriate for the challenge of building science in an ever-changing world.

It will be argued that the science of complex socio-physical systems, policy and design are inextricably entangled with each other. Furthermore it is argued that they are inextricably entangled with hypernetworks, and the emergent whole is reflected in this book, as shown in Figure 1.8.

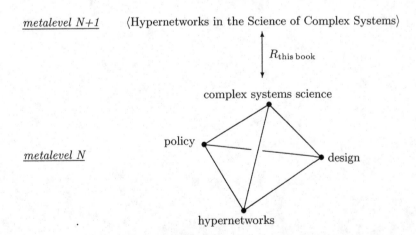

Fig. 1.8 ⟨complex systems science, design, policy, hypernetworks; $R_{\text{this book}}$⟩ → HITSOCS

Chapter 2

Sets and Relations

Sets and relationships between the elements of sets are fundamental to the study of systems. The theory of sets is foundational in mathematics and provides a formalism that can give great clarity when systems get complicated with many parts and many relationships between them.

Fortunately most of the sets in complex systems science are *finite* and this greatly simplifies the mathematical discussion. For example, the set of all people alive in the world at any point in time is large but not infinite. The same applies to the number of buildings, cars, and almost anything else.

The common number systems are a notable exception to this. The *integers* (whole numbers), *rationals* (fractions), and the *real numbers* (the rationals with irrational numbers such as $\sqrt{2}$ and π) are all *infinite*. Infinity can be very tricky as the Greeks discovered through paradoxes such as Achilles and the tortoise (the tortoise has a start in a race and by the time Achilles gets to where the tortoise was, the tortoise has moved on and still has a start, so Achilles never catches up with the tortoise). Fortunately, sets of observations are always finite and even when they involve numbers, those numbers are finite fractions (usually finite decimal fractions).

Thus although they are represented by numbers in infinite number sets, the amount of money circulating is finite, the number of emails is finite – in fact the universal set of all data existing in the world is finite.

This chapter will give some of the basic definitions of sets and relations. As noted above, numbers are special cases of sets, and mappings or functions are special cases of relations. The presentation is informal and its main purpose is to establish concepts and notation.

2.1 Elements of set theory

Sets

A *set* is a collection, possibly infinite, of distinct objects called the *members* or *elements* of the set. The set with no elements is called the *empty set* and is denoted by the symbol ∅.

Set membership

The symbol $\in$ is used to mean "is a member of" or "belongs to", so the notation $x \in X$ means that "x belongs to the set X". The symbol $\notin$ means "is not a member of", so $x \notin X$ means that x is not a member of X.

Set definition by intension and extension

A set is defined by *extension* when its elements are listed explicitly within brackets, e.g. {salt, pepper}. Sets defined by extension do not need to have any obvious relationship between their elements, e.g. {telephone, rose, laughter, music}.

A set is defined by *intension* when a rule is given for deciding its members, e.g. $\{x \,|\, x \text{ has property } p\}$. The elements of sets defined by intension are more obviously related since they all obey the defining rule, for example $M = \{x \,|\, x \text{ is a man}\}$ and $F = \{x \,|\, x \text{ is a woman}\}$.

Care is required when naming sets. For example, a scheme for coding television programmes included a set called "Sports Not Requiring Equipment" and gave the examples "boxing" and "wrestling". But "boxing" $\notin \{x \,|\, x \text{ is a sport not requiring equipment}\}$ because boxing requires shorts, a gum shield, a ring and other pieces of equipment. The problem here is that the name of the set is meaningful and implies an intensional definition that is not consistent with the given extension.

Operations on sets

Sets are often drawn as *Euler circles* in *Venn diagrams*, as shown in Fig. 2.1. The points inside the circle represent the elements of the set.

The set A is a *subset* of the set B if every member of A is also a member of B. This is written as $A \subseteq B$ and reads as "A is a subset of B". B is a *superset* of A if A is a subset of B. When B contains elements that A does not A is said to be a *proper subset* of B, written $A \subset B$, and B is a *proper superset* of A.

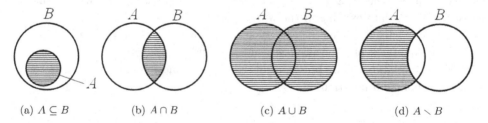

(a) $A \subseteq B$ (b) $A \cap B$ (c) $A \cup B$ (d) $A \setminus B$

Fig. 2.1 Representing sets and their properties by diagrams

The *intersection* of two sets A and B, written as $A \cap B$ is the set of elements belonging to them both, $A \cap B = \{x \,|\, x \in A \text{ and } x \in B\}$. When two sets have no elements in common, $A \cap B = \emptyset$, they are said to be *disjoint*.

The *union* of two sets A and B, written as $A \cup B$ is the set of elements that belong to A, to B, or both, $A \cup B = \{\, x \,|\, x \in A \text{ or } x \in B \,\}$.

The *difference* of two sets A and B, written as $A \smallsetminus B$ is the set of elements that belong to A but do not belong to B, $A \smallsetminus B = \{\, x \,|\, x \in A \text{ and } x \notin B \,\}$.

The *symmetric difference* of sets A and B is the set $A \triangle B \stackrel{\text{def}}{=} \{\, x \,|\, x \in A \text{ or } x \in B \text{ but } x \notin A \cap B \,\}$. It is the set of elements that belong to A or belong to B but do not belong to both. It follows that $A \triangle B = (A \smallsetminus B) \cup (B \smallsetminus A)$. This is also called *exclusive or*.

Classes

A *class* is a set whose members are other sets, *e.g* let M be the set of males in a population and F be the set of females. Then $\mathcal{C} = \{M, F\}$ is the class containing the set of males and the set of females. Note that $\mathcal{C}$ is not the set of all people — that is the set $M \cup F$.

Partitions

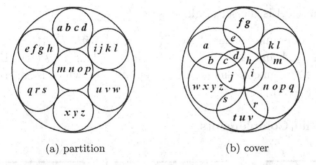

(a) partition   (b) cover

Fig. 2.2   Partitions and covers of the alphabet

Let $\mathcal{A} = \{A_i | i = 1, ..., n\}$ be a class of subsets of A, so $A_i \subseteq A$ for $i = 1, ..., n$. $\mathcal{A}$ is said to be a *partition* of A if every element of A is a member of exactly one of the subsets, *i.e.* $A = \bigcup_{i=1}^{n} A_i$ and $A_i \cap A_j = \emptyset$ for all $i \neq j$ for $i, j = 1, ..., n$.

For example $\{M, F\}$ is a partition of P, the set of people, because everyone in P is in one set or the other, but not both, *i.e.* $P = M \cup F$ and $M \cap F = \emptyset$. Figure 2.2(a) shows a partition of the alphabet.

Covers

Let C be the class of children. Let $P = M \cup F \cup C$. The class of sets $\{M, F, C\}$ is not a partition of P because $M \cap C \neq \emptyset$ and $F \cap C \neq \emptyset$. Instead it is an example of a "cover". Let $\mathcal{A} = \{A_i | i = 1, ..., n\}$ be a class of subsets of A. $\mathcal{A}$ is said to be a

16 Hypernetworks in the Science of Complex Systems

cover of A if every element of A is a member of at least one of the subsets A_i for some i in $\{1, 2, ..., n\}$. Figure 2.2(b) shows a cover of the alphabet.

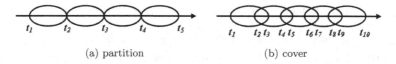

(a) partition (b) cover

Fig. 2.3 Time intervals as partitions and covers

Figure 2.3 shows two ways of converting a continuous scale into a discrete scale. Let the intervals be $[t_i, t_{i+1}) \stackrel{\text{def}}{=} \{t \mid t_i \leq t < t_{i+1}\}$. $[t_i, t_{i+1}) \cap [t_j, t_{j+1}) = \emptyset$ for $i \neq j$ in Fig. 2.3(a) and this is a partition. On the other hand in Fig. 2.3(b) $[t_i, t_{i+1}) \cap [t_j, t_{j+1}) \neq \emptyset$ for some $i \neq j$ and this is a cover. There is always this choice between a cover and partition when dividing continuous scales into discrete regions.

Power Sets

The *power set* of a set A is the set of all subsets of A.

The power set is sometimes denoted by $\mathcal{P}(A)$ or 2^A with $\mathcal{P}(A) = \{A' \mid$ for all $A' \subseteq A\}$. Since $\emptyset \subseteq A$ and $A \subseteq A$, both $\emptyset$ and A belong to $\mathcal{P}(A)$. If A has n elements, $\mathcal{P}(A)$ has 2^n elements (including the empty set), which is consistent with the notation 2^A for $\mathcal{P}(A)$.

Universal Sets and Complements

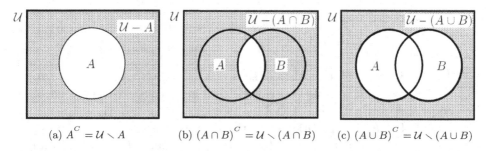

(a) $A^C = \mathcal{U} \smallsetminus A$ (b) $(A \cap B)^C = \mathcal{U} \smallsetminus (A \cap B)$ (c) $(A \cup B)^C = \mathcal{U} \smallsetminus (A \cup B)$

Fig. 2.4 Complements in universal sets

Observing any particular system involves local "universes" spanned by classes of sets and their elements. For example let $\mathcal{S} = \{S_i \mid i \in I_{\mathcal{S}}\}$ be a class of sets. The *universal set* for $\mathcal{S}$ is defined to be the union of all the sets S_i, i.e. the set $\mathcal{U}_{\mathcal{S}} \stackrel{\text{def}}{=} \bigcup_{S_i \in \mathcal{S}} S_i$.

Let $\mathcal{U}$ be a universal set and let A be a subset, $A \subseteq \mathcal{U}$. Figure 2.4(a) shows the *complement* of A in $\mathcal{U}$ as the set $A^C \stackrel{\text{def}}{=} \mathcal{U} \smallsetminus A$. Figure 2.4(b) shows $(A \cap B)^C$ and Fig. 2.4(c) shows $(A \cup B)^C$. Complements obey De Morgan's Laws:

$$(A \cap B)^C = A^C \cup B^C$$
$$(A \cup B)^C = A^C \cap B^C$$

Is the universal set universal, *i.e.* is there just one universal set that contains everything? If such a set existed it would contain itself, leading to Russell's Paradox. Generally one speaks of a *universe of discourse* which contains everything relevant to a particular application, and complements are defined with respect to the particular universe of discourse. There can be many universes of discourse and systems may have classes of universes, especially when they have many levels.

Set paradoxes are significantly avoided when dealing with finite sets, as is the case with data sets (though not necessarily the sets from which they are sampled).

Sets and Logic

Let P be a *proposition* concerning x which can be True or False. We write $P(x) =$ True when the proposition is true, and $P(x) =$ False when the proposition is false. Every set has a defining proposition, P_A, where $A = \{x \mid P_A(x) = \text{True}\}$. For extensionally defined sets, $P_A(x)$ is True when x belongs to the listed elements. For intensionally defined sets, P_A is the proposition that defines the elements.

In logic, the "wedge" symbol $\wedge$ is used to mean "and" while the "vee" symbol $\vee$ is used to mean "or". Given two propositions, p and q, $p \wedge q$ is true when p is true *and* q is true. $p \vee q$ is true when p is true *or* q is true or both p and q are true. The following identities hold:

$$A \cap B = \{x \mid P_A(x) \wedge P_B(x) \text{ is True}\}$$
$$A \cup B = \{x \mid P_A(x) \vee P_B(x) \text{ is True}\}$$

This illustrates how tightly set theory is bound up with logic and reasoning, and it is one reason for sets being so powerful for modelling systems.

Products of Sets

The *product* of two sets A and B, $A \times B$, is defined to be the set of ordered pairs (a, b) for all $a \in A$ and $b \in B$.

$$A \times B = \{(a,b) \mid a \in A \text{ and } b \in B\}$$

The product of a class of sets $A_1, ..., A_n$, $\prod_{i=1}^{n} A_i = A_1 \times A_2 \times ... \times A_n$, is defined to be set of all n-tuples $(a_1, ..., a_n)$ where $a_i \in A_i$ for $i = 1, ..., n$.

For illustration consider the set of all pairs of males and females, $M \times F = \{(x,y) \,|\, x \in M \text{ and } y \in F\}$. It has, for example, the subset $\{(x,y) \,|\, x \in M \text{ and } y \in F$ and x is married to $y\}$. Today this is only part of the story since in some countries there exist legally married couples (x,y) with $(x,y) \in M \times M$ or $(x,y) \in F \times F$.

In general, relationships can be represented by subsets of set products. For example, let P be a set of people. Then the set $G \subseteq P \times P = \{(x,y) \,|\, x$ plays golf with $y\}$ can represent the set of people who play golf together.

Let $\mathbb{R}$ be the set of real numbers. $\mathbb{R} \times \mathbb{R} = \{\,(x,y) \,|\, x \in \mathbb{R} \text{ and } y \in \mathbb{R}\}$ is the *Cartesian product* of $\mathbb{R}$ with itself, and is the set of all pairs of real numbers. More generally

$$\prod_{i=1}^{n} \mathbb{R}_i = \underbrace{\mathbb{R} \times \mathbb{R} \times ... \times \mathbb{R}}_{n \text{ times}}$$

is the set of n-tuples or *vectors* of the form $(x_1, x_2, ..., x_n)$ where $x_i \in \mathbb{R}$.

2.2 Mappings and functions

Although the terminology varies, the following definitions will be used. Let A and B be sets. A *mapping* is rule, f, that assigns an element $f(a) \in B$ to each $a \in A$. A is called the *domain* of the mapping and B is called its *codomain* or *range*. $f(a)$ is sometimes called the *image* of a under f. The notation $f : A \to B$ means that f is a mapping from A to B. A mapping to a set of numbers is often called a *function*.

Let $f : A \to B$ be a mapping from A to B. The *image* of A under f is defined as $f(A) = \{f(a)|$ for all $a \in A\}$. In general $f(A) \subset B$, as illustrated in Fig. 2.5(a) where, for example, $b_5 \notin f(A)$. In this case the mapping is said to be from A *into* B. When $f(A) = B$ the mapping f is said to be *onto*, as in Fig. 2.5(b).

A mapping is said to be *many-to-one* if two or more elements in the domain are mapped to the same element in the codomain. For example, in Fig. 2.5(a) $f(a_1) = f(a_2) = f(a_4) = b_2$.

A mapping $f : A \to B$ is said to be *one-to-one* if each member of A is mapped to a unique member of B. Thus Fig. 2.5(c) shows a one-to-one *into* mapping and Fig. 2.5(d) shows a one-to-one *onto* mapping. A 1-1 into mapping is called an *injection*, a many-1 onto mapping is called a *surjection*, and a 1-1 onto mapping is called a *bijection*.

Composition of mappings

The composition of the mappings $f : A \to B$ and $g : B \to C$ is the mapping $g \circ f : A \to C$ where $g \circ f(a) \stackrel{\text{def}}{=} g(f(a))$, as illustrated in Fig. 2.6. If f is one-to-one

Sets and Relations

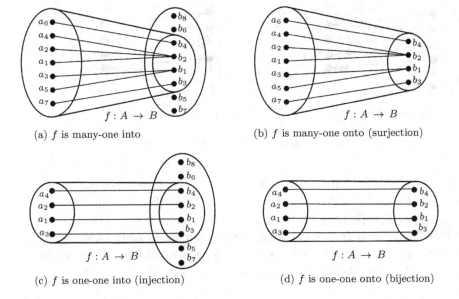

(a) f is many-one into

(b) f is many-one onto (surjection)

(c) f is one-one into (injection)

(d) f is one-one onto (bijection)

Fig. 2.5 Mappings

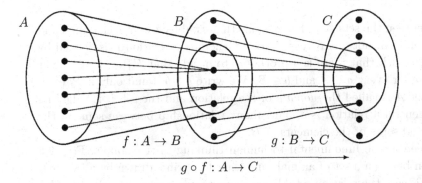

Fig. 2.6 Composition of mappings

and g is one-to-one, then $g \circ f$ is one-to-one. If either of f or g is many-to-one then $g \circ f$ is many-to-one. If both f and g are onto mappings then $g \circ f$ is an onto mapping.

Inverses

The *inverse* of a 1-1 mapping $f : A \to B$ is a mapping $g : f(A) \to A$ such that $g(f(a)) = a$ for all $a \in A$. g is usually written as f^{-1}. Thus $f^{-1} \circ f(a) = a$.

The *inverse* of a many-to-one mapping $f : A \to B$ is a mapping $f^{-1} : f(A) \to \mathcal{P}(A)$ such that $f^{-1}(b) = \{a \mid$ for all $a \in A$ with $f(a) = b\}$.

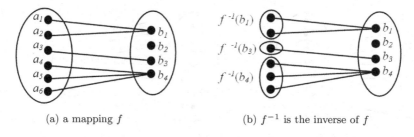

(a) a mapping f (b) f^{-1} is the inverse of f

Fig. 2.7 Mappings and inverses

Let $f^{-1}(B) = \{f^{-1}(b) \subseteq A|$ for all $b \in B\}$, with $f^{-1}(b) = \emptyset$ for b in $B \smallsetminus f(a)$. In Fig. 2.7(b) $f^{-1}(B) = \{\{a_1, a_2\}, \{a_3\}, \{a_4, a_5\}\}$. $f^{-1}(B)$ partitions A into disjoint subsets. This general property of the inverse mapping is more powerful than might appear at first sight, e.g. if f maps a set of people to their age in years, then $f^{-1}(n)$ is the set of people of age n, and the set of people is partitioned into sets of people of the same age. Everyone belongs to one of these sets, and no-one belongs to two or more sets.

2.3 Relations

A *relation* R between the sets A and B is a rule that establishes for each a in A and each b in B whether or not a is R-related to b. Generally a relation R is defined by a proposition P_R such that there is a procedure to decide that $P_R(a,b) =$ True or $P_R(a,b) =$ False for every $a \in A$ and $b \in B$. We write $a\,R\,b$ if and only if $P_R(a,b)$ = True. The set A is called the *domain* of the relation and the set B is called its *codomain*. In general a relation is a many-many relationship between some of the elements of A and some of the elements of B.

Relations seem to be fundamental to human thinking. For example, Fig. 2.8 shows a relation between a set of animals and their footprints drawn by a five year old child. In this case the relation establishes a one-to-one correspondence and this relation is also a mapping. In general mappings are special cases of relations.

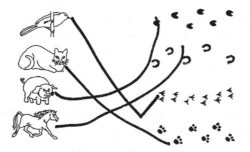

Fig. 2.8 A relation drawn by a five year old child

Sets and Relations

Incidence matrices and relations

Figure 2.9 gives a frequently cited relation between a set of eighteen women, W, and a set of fourteen activities, A. These data were collected by five ethnographers and published in the book *Deep South* in 1941 [Davies *et al* (1941)].

> In particular, they collected systematic data on the social activities of 18 women whom they observed over a nine month period. During that period, various subsets of these women had met in a series of 14 informal social events. The participation of the women was uncovered using "interviews, the records of participant observers, guest lists, and the newspapers" ... the data reflect joint activities like "a day's work behind the counter of a store, a meeting of a women's club, a church supper, a card party, a supper party, a meeting of the Parent-Teacher Association, etc. [Freeman (2003)]

The data are given as an *incidence matrix* where here an × in the ij^{th} cell denotes a relationship between the ith woman and the jth event. Let this relation be denoted R. Then, for example, 1. Mrs. Evelyn Jefferson R Event (1), 2. Mrs. Laura Mandeville R Event (1), and 4. Mrs Brenda Rogers R Event (1).

Names of Participants of Group I	Code Numbers and Dates of Social Events Reported in *Old City Herald*													
	(1) 6/27	(2) 3/2	(3) 4/12	(4) 9/26	(5) 2/25	(6) 5/19	(7) 3/15	(8) 9/16	(9) 4/8	(10) 6/10	(11) 2/23	(12) 4/7	(13) 11/21	(14) 8/3
1. Mrs. Evelyn Jefferson	×	×	×	×	×	×		×	×					
2. Miss Laura Mandeville	×	×	×		×	×	×	×						
3. Miss Theresa Anderson		×	×	×	×	×	×	×	×					
4. Miss Brenda Rogers	×		×	×	×	×	×	×						
5. Miss Charlotte McDowd			×	×	×		×							
6. Miss Frances Anderson			×		×	×		×						
7. Miss Eleanor Nye					×	×	×	×						
8. Miss Pearl Oglethorpe						×		×	×					
9. Miss Ruth DeSand					×		×	×	×					
10. Miss Verne Sanderson							×	×		×		×		
11. Miss Myra Liddell								×	×	×	×	×		
12. Miss Katherine Rogers								×	×	×		×	×	×
13. Mrs. Sylvia Avondale							×	×	×	×		×	×	×
14. Mrs. Nora Fayette						×	×		×	×	×	×	×	×
15. Mrs. Helen Lloyd							×	×		×	×	×		
16. Mrs. Dorothy Murchison								×	×					
17. Mrs. Olivia Carleton									×		×			
18. Mrs. Flora Price									×		×			

Fig. 2.9 A relation between women and events attended (Source: Davis *et al* 1941)

In this example the relation is shown by the presence or absence of a cross, ×. It is common to have incidence matrices in which the existence of a relation is shown by a 1 and its absence is shown by a 0. For example, let W be the set of women and E be the set of events. Then the incidence matrix M corresponding to Fig. 2.9 can be rewritten as shown in Fig. 4.25(a). The *transpose matrix* shown in Fig. 4.25(b) gives the relation between the social events and the women. The transpose matrix, M^T, is obtained by flipping round the matrix M.

22 Hypernetworks in the Science of Complex Systems

	Social Events																Women																
	1	2	3	4	5	6	7	8	9	10	11	12	13	14		1	2	3	4	5	6	7	8	9	10	11	12	13	14	15	16	17	18
w_1	1	1	1	1	1	1	0	1	1	0	0	0	0	0	e_1	1	1	0	1	0	0	0	0	0	0	0	0	0	0	0	0	0	0
w_2	1	1	1	0	1	1	1	1	0	0	0	0	0	0	e_2	1	1	1	0	0	0	0	0	0	0	0	0	0	0	0	0	0	0
w_3	0	1	1	1	1	1	1	1	1	0	0	0	0	0	e_3	1	1	1	1	1	1	0	0	0	0	0	0	0	0	0	0	0	0
w_4	1	0	1	1	1	1	1	1	0	0	0	0	0	0	e_4	1	0	1	1	1	0	0	0	0	0	0	0	0	0	0	0	0	0
w_5	0	0	1	1	1	0	1	0	0	0	0	0	0	0	e_5	1	1	1	1	1	1	1	0	1	0	0	0	0	0	0	0	0	0
w_6	0	0	1	0	1	1	0	1	0	0	0	0	0	0	e_6	1	1	1	1	0	1	1	1	0	0	0	0	1	0	0	0	0	0
w_7	0	0	0	0	1	1	1	1	0	0	0	0	0	0	e_7	0	1	1	1	1	0	1	0	1	1	0	0	1	1	1	0	0	0
w_8	0	0	0	0	0	1	0	1	1	0	0	0	0	0	e_8	1	1	1	1	0	1	1	1	1	1	1	1	1	0	1	1	0	0
w_9	0	0	0	0	1	0	1	1	1	0	0	0	0	0	e_9	1	0	1	0	0	0	0	1	1	1	1	1	1	1	0	1	1	1
w_{10}	0	0	0	0	0	0	1	1	1	0	0	1	0	0	e_{10}	0	0	0	0	0	0	0	0	0	1	1	1	1	1	1	0	0	0
w_{11}	0	0	0	0	0	0	0	1	1	1	0	1	0	0	e_{11}	0	0	0	0	0	0	0	0	0	0	0	0	0	1	1	0	1	1
w_{12}	0	0	0	0	0	0	0	1	1	1	0	1	1	1	e_{12}	0	0	0	0	0	0	0	0	0	1	1	1	1	1	1	0	0	0
w_{13}	0	0	0	0	0	0	1	1	1	1	0	1	1	1	e_{13}	0	0	0	0	0	0	0	0	0	0	0	1	1	1	0	0	0	0
w_{14}	0	0	0	0	0	1	1	0	1	1	1	1	1	1	e_{14}	0	0	0	0	0	0	0	0	0	0	0	1	1	1	0	0	0	0
w_{15}	0	0	0	0	0	0	1	1	0	1	1	1	0	0																			
w_{16}	0	0	0	0	0	0	0	1	1	0	0	0	0	0																			
w_{17}	0	0	0	0	0	0	0	0	1	0	1	0	0	0																			
w_{18}	0	0	0	0	0	0	0	0	1	0	1	0	0	0																			

(a) The incidence matrix M (b) The transpose incidence matrix M^T

Fig. 2.10 The incidence matrices of the Southern Women event relations

Bipartite relations

A relation between two sets A and B where $A \cap B = \emptyset$ is said to be *bipartite*, e.g. $W \cap A = \emptyset$ so the women-events relation is bipartite. Bipartite relations can be displayed by listing the two sets vertically and joining related pairs by lines, as shown in Fig. 2.11. At first sight bipartite relations may seem uninteresting, but it will be seen that they have many fundamental structures associated with them.

Relations and logic

Because they are defined by propositions, relations have their own logical structures. Let R_1 and R_2 be relations between the sets A and B is defined by the proposition P_1 and P_2. Let $(P_1 \wedge P_2)$ mean "P_1 and P_2" and $(P_1 \vee P_2)$ mean "P_1 or P_2". Then let the *conjunction* and *disjunction* of relations be defined as follows:

Conjunction: $a\,(R_1 \wedge R_2)\,b$ if and only if $(P_1 \wedge P_2)(a,b) = $ True.

Disjunction: $a\,(R_1 \vee R_2)\,b$ if and only if $(P_1 \vee P_2)(a,b) = $ True.

Figure 2.12 illustrates this. Note that the conventional way of representing relations as arrows does not give a natural way of representing disjunction (Fig. 2.12(d)).

The *negation* of the proposition P, not-P, is defined as $\neg P(a,b) = $ True if and only if $P(b,a) \neq $ True. Then the *negation* of R, not-R, denoted $\neg R$ can be defined as $a\,\neg R\,b$ if and only if $\neg P(a,b) = $ True.

Sets and Relations   23

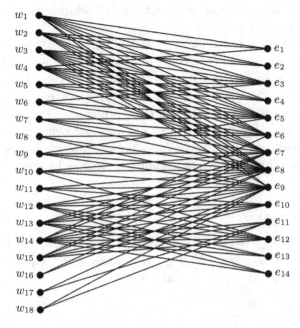

Fig. 2.11 The bipartite relation between the women and social events

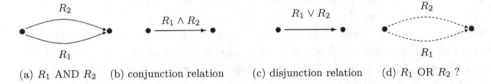

(a) R_1 AND R_2 (b) conjunction relation   (c) disjunction relation   (d) R_1 OR R_2 ?

Fig. 2.12 The conjunction and disjunction relations

A relation R with defining proposition P will be said to be *Boolean* when $P(a,b)$ = True implies and is implied by $\neg P(a,b)$ = False.

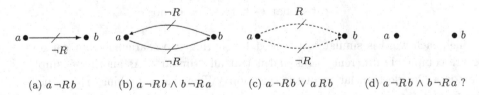

(a) $a\neg R\,b$ (b) $a\neg R\,b \wedge b\neg R\,a$ (c) $a\neg R\,b \vee a R\,b$ (d) $a\neg R\,b \wedge b\neg R\,a$?

Fig. 2.13 Making "not-related" explicit

Figure 2.13(a) illustrates the negated relation $\neg R$ represented by an arrow with a line though it. In the case that R is a relation from a set to itself, suppose a is not R-related to b and b is not R-related to a. Then $a\neg R\,b$ and $b\neg R\,a$ as shown in

Fig. 2.13(b). When the defining proposition for R is Boolean, either $a\,R\,b$ or $a\,\neg R b$ as shown in Fig. 2.13(c).

What is the difference between Fig. 2.13(b) and Fig. 2.13(d)? In both cases a is not R-related to b and b is not R-related to a. The distinction is potentially important for application in which the state of relationships is explicitly represented in the database. Figure 2.13(a) makes the non-relatedness *explicit* while Fig. 2.13(d) allows the non-relatedness to be *implicit*: if there is no line in the drawing between a and b then implicitly a is not R-related to b. This is analogous to saying that "if $a\,R\,b$ is not in the database then a is not R-related to b", which may be not be true – not knowing if two things are related is not the same as knowing that they are not related.

Composition of Relations

The composition of two relations, R_{AB} between sets A and B and R_{BC} between sets B and C, is the relation $R_{AC} \stackrel{\text{def}}{=} R_{BC} \circ R_{AB}$ where $a\,R_{BC} \circ R_{AB}\,c$ if and only if there exists $b \in B$ such that $a\,R_{AB}\,b$ and $b\,R_{BC}\,c$.

Equivalence and Partitions

When are two things the same? For example if I write $2 = 2$, is the two on the left *the same as* the two on the right? Are you *the same* person as you were one second ago or yesterday? Is every dollar "the same" as every other?

When are two things similar? For example, is the music of Beethoven "similar" to that of Mozart or Haydn? When is something sufficiently similar to something else as to be considered to be the same. For example, a car hire company may offer you a different model to the one you ordered but so similar that it is in the same performance or comfort class. Thus this symbol 2 is certainly *similar* to this symbol 2, and for some purposes it may *equivalent*, but it is not exactly the *same* symbol.

Similarity does not define classes in a simple way. For example, consider the sequence

$$\text{cat} \leftrightarrow \text{hat} \leftrightarrow \text{hot} \leftrightarrow \text{hog} \leftrightarrow \text{dog}$$

in which each word is similar to the next by sharing two characters, but cat and dog are completely different words and not at all "similar". As another example, below is a Google translation of the question "When are two things the same?" from English into French and back again:

$$\text{When are two things the same?} \xrightarrow{\text{English to French}} \text{Quand deux choses sont les mêmes?}$$

$$\text{Quand deux choses sont les mêmes?} \xrightarrow{\text{French to English}} \text{When two things are the same?}$$

Sets and Relations

The result is not the same as the original, even though they are all supposed to be equal. It can get even more complicated:

When are two things the same? $\stackrel{\text{English to French}}{\longrightarrow}$ Quand deux choses sont les mêmes?

Quand deux choses sont les mêmes? $\stackrel{\text{French to German}}{\longrightarrow}$ Wenn zwei Sachen sind die gleichen?

Wenn zwei Sachen sind die gleichen? $\stackrel{\text{German to English}}{\longrightarrow}$ If two things are the same?

Here the result of the translation is almost metaphysical, if it makes sense at all, and to English speakers carries a different meaning. It illustrates the possibility of words being put together in apparently meaningful but baffling ways, as in the student graffito "why is a mouse when it spins?".

By its imprecision on equality and similarity, natural language glosses over many subtle distinctions and allows apparently disjoint sets to "leak" into others creating ambiguity and shifting meaning. Without great care one does not know what one is talking about. The following mathematical account of "equivalence" replaces the troublesome concept of "equals" with something much more subtle and precise. At the same time it also gives precision to the concept of "similarity".

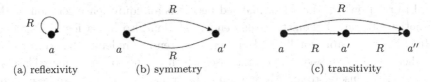

Fig. 2.14 Properties of relations

Let R be a relation on the set A. Figure 2.14 illustrates the definitions:

(a) R is *reflexive* if and only if $a\,R\,a$ for all a A.

(b) R is *symmetric* if and only if $a\,R\,a'$ implies $a'\,R\,a$ for all a and a' in A.

(c) R is *transitive* iff $a\,R\,a'$ and $a'\,R\,a''$ imply $a\,R\,a''$ for all a, a' and a'' in A.

These three simple properties can be combined in ways that give great precision to the concept of equality and equivalence:

A reflexive, symmetric, transitive relation R on A is an *equivalence relation*.

An equivalence relation R on set A partitions A into a class of subsets of A called *equivalence classes*. Every element in an equivalence class is R-related to every other element in that class. No element in an equivalence class is R-related to an element in a different equivalence class. Thus two elements can be "equivalent" with respect to R without being equal or being the same.

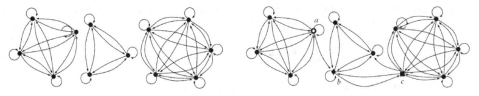

(a) equivalence relations partition the vertices   (b) the relation does not partition the vertices

Fig. 2.15 Equivalence relations partition sets into equivalence classes

This definition is illustrated in Fig. 2.15(a) where the elements are partitioned into three equivalence classes by the relation, and there are no links between the equivalence classes (they are *disconnected*). The relation in Fig. 2.15(b) does not partition the vertices because it is not reflexive on c, it is not symmetric between a and b and, for example, it is not transitive between a and c.

Examples of equivalence relations include the gender relation (everyone is either male or female), the same-age relation, the same-height relation, and so on. Many relations are treated as if they were equivalence relations, even though the relationship may be ambiguous for a small number of cases, *e.g.* in most countries the "married-to relation" on the set of married people is an equivalence relation, with the exception of a few bigamists. The gender relation also has a few ambiguous cases due to physical abnormalities. In practice we tend to redefine the set A to be $A \smallsetminus A'$ where A' is the subset of aberrant cases which belong to two or more classes.

It is sometimes useful to force a relation to be transitive. The *transitive closure* of a relation R on the set A is the relation R_t with $a\,R_t\,a''$ if there exists a' with $a\,R\,a'$ and $a'\,R\,a''$. This rule is applied until no more pairs become related. The transitive closure of a reflexive symmetric relation is an equivalence relation.

Example

Although the foregoing definitions may seem rather abstract, and despite their simplicity, they provide some powerful methods of disambiguation. For example, the "part of" relation between a whole and its parts has troubled philosophers for thousands of years. The subset relation is transitive, $A \subseteq B$ and $B \subseteq C$ implies $A \subseteq C$, which suggests the "part of" relation should be transitive. However, this can be problematic. Figure 2.16(a) provides an amusing illustration in which:

> Simpson's finger is part of Simpson,
> Simpson is part of the Philosophy Department, whence
> Simpson's finger is part of the Philosophy Department.

This sounds odd, but the problem is easily resolved. Figure 2.16(b) shows Simpson's finger being R_1 aggregated into Simpson, and Simpson being R_2 aggregated into the Philosophy Department. Of course R_1 and R_2 are completely different "part of" relations, and they are both different to $R_2 \circ R_1$ which aggregates Simpson's finger into the Philosophy Department. The use of the notation $R_{\text{part of}}$ for all of R_1, R_2 and $R_2 \circ R_1$ is misleading and causes unnecessary confusion.

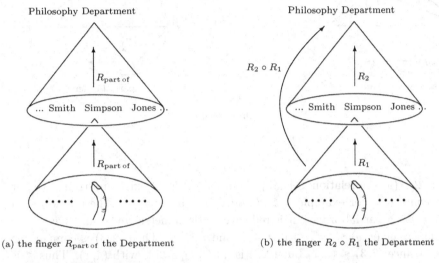

(a) the finger $R_{\text{part of}}$ the Department   (b) the finger $R_2 \circ R_1$ the Department

Fig. 2.16 Simpson's finger is part of the Philosophy Department

Order Relations

Let R be a relation on a set A.

R is *asymmetric* if $a\,R\,a'$ implies $a'\,\neg R\,a$.

R is *antisymmetric* if $a\,R\,a'$ and $a'\,R\,a$ implies $a = a'$.

For example, the relation "father of" is asymmetric while the relation "brother of" is not (it is symmetric). The relation $\leq$ on a set of numbers is antisymmetric since $x \leq y$ and $y \leq x$ implies $x = y$.

The elements a and a' are *comparable* under R if $a\,R\,a'$ or $a'\,R\,a$. If a and a' are not comparable they are said to be *incomparable*. R obeys the *trichotomy law* for A if $a\,R\,a'$, $a'\,R\,a$ or $a = a'$ for all a and a' in A.

A reflexive and transitive relation is a *quasi order*.

A reflexive, transitive and antisymmetric relation is a *partial order*.

A reflexive, transitive and antisymmetric relation with all pairs of elements comparable is a *total order*.

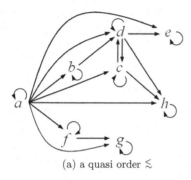 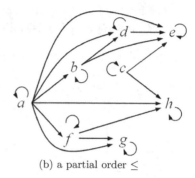

(a) a quasi order $\lesssim$ (b) a partial order $\leq$

Fig. 2.17   Order relations

In Figure 2.17(a) the relation $\lesssim$ is a quasi order – it is reflexive and transitive but not antisymmetric (for example, $c \lesssim d$ and $d \lesssim c$, but $c \neq d$). The relation $\leq$ in Fig. 2.17(b) is a partial order – it is reflexive, antisymmetric and transitive.
For example, let $(x, y) \leq (x', y')$ if $x \leq x'$ and $y \leq y'$. Then $(2, 3)$ is comparable with $(5, 7)$ since $(2, 3) \leq (5, 7)$ but $(2, 3)$ is not comparable with $(4, 1)$. Thus $\leq$ is not a total order. It is reflexive and antisymmetric. Since it is also transitive, it is a partial order.

A total order on a set A has the property that the elements can be set out on a line with a to the left of a' if $a \leq a'$. For this reason total orders are sometimes called *linear orders*. The number systems $\mathbb{N}$, $\mathbb{Q}$ and $\mathbb{R}$ are all totally ordered.

An order can be induced on any set by mapping it into a number system. For example, IQ tests map people to their intelligence quotient, and the set of people is ordered by their IQ score. This is sometimes taken to mean that the set of people can be ordered and compared according to their intelligence. Less contentious is the mapping that takes people to their age, and the population is ordered by the "older than or equal age as".

Order relations can play an important role in complex systems. In particular they play an important role in discriminating the levels in multilevel systems. Much of this follows from the fact that any class of subsets of a set is partially ordered, as illustrated in Fig. 2.18. In this diagram composite lines are omitted, for example there is no line between $\{a, b, c, d, e, f, g\}$ and $\{a, b, c\}$, even though $\{a, b, c\} \subseteq \{a, b, c, d, e\} \subseteq \{a, b, c, d, e, f, g\}$.

In Fig. 2.18 the sets are arranged in *levels*. The levels are defined as follows. All the set that have no subsets in the class are placed at the lowest level, *Level 1*. At the next level, *Level 2* put the sets that have no intermediate subsets between them and the *Level 1* sets, e.g. $\{a, b, c\}$. At the next level, *Level 3* put the sets that have no intermediate sets between them and the *Level 2* sets, e.g. $\{a, b, c, d, e\}$. This leaves $\{a, b, c, d, e, f, g\}$ which is placed at *Level 4*.

Sets and Relations

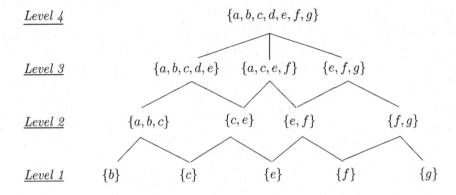

Fig. 2.18 A partially ordered class of subsets under $\subseteq$

The set $\{a, b, c, d, e, f, g\}$ is a *supremum* in this ordered class of sets, and appears at the top in the diagram. If the empty set were added, at what might be defined to be *Level 0* it would be an *infimum*, since it is a subset of all the other sets. In this case the infimum would be at the lowest level in the diagram.

Suppose the sets $\{a, b, c, d, e, f\}$ and $\{a, c, d, e, f, g\}$ entered this system. How would they fit in the diagram between *Levels 3* and *4* ? One possibility is to define a new intermediate *Level $3\,{}^1\!/_2$*. Another is to renumber all the levels and to create a new top *Level 5'* as shown in Fig. 2.19. It is common for new inter-level objects to evolve in complex systems, and this example suggests that the level of any object is *relative* to the others rather than being absolute.

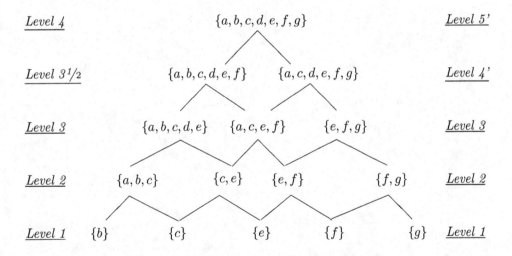

Fig. 2.19 The evolution of new structures

Chapter 3

Hypergraphs and the Galois Lattice

3.1 Hypergraphs

A *hypergraph* is a set of *vertices*, V, and a set of subsets of V, E called *hypergraph edges*. In general the members of E can have more than two elements.

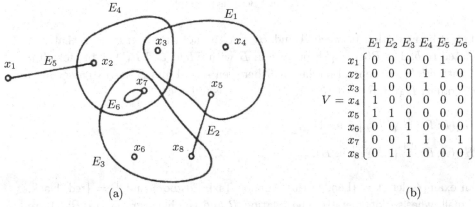

Fig. 3.1 The Berge hypergraph

The hypergraph shown in Figure 3.1 is taken from the book *Hypergraphs* by Claude Berge [Berge (1989)]. The vertices are $V = \{x_1, x_2, x_3, x_4, x_5, x_6, x_7, x_8\}$ and there are six edges $E = \{E_1, E_2, E_3, E_4, E_5, E_6\}$. The relation, R, between the edges and vertices is given in Figure 3.1(b). The edges are $E_1 = \{x_3, x_4, x_5\}$, $E_2 = \{x_5, x_8\}$, $E_3 = \{x_6, x_7, x_8\}$, $E_4 = \{x_2, x_3, x_7\}$, $E_5 = \{x_1, x_2\}$ and $E_6 = \{x_7\}$. The set $H_E \stackrel{\text{def}}{=} \{E_1, E_2, E_3, E_4, E_5, E_6\}$ will be called a *hypergraph*.

Interestingly, in Figure 3.1(a)) the conventional dot and line representation of graphs for the loop $E_6 = \{x_7, x_7\}$ and lines $E_2 = \{x_5, x_8\}$ and $E_5 = \{x_1, x_2\}$ is mixed with the Euler circle method of representing sets used for E_1, E_3, and E_4. It would be more consistent to draw the singleton and two-element sets as Euler circles as shown in Figure 3.2(a), and this is how hypergraphs will be drawn here.

31

Every hypergraph has a *dual* hypergraph as illustrated in Figure 3.2(b). Here the "edges" are sets of edges associated with the vertices, e.g. x_2 is associated with the dual edge $\{E_4, E_5\}$.

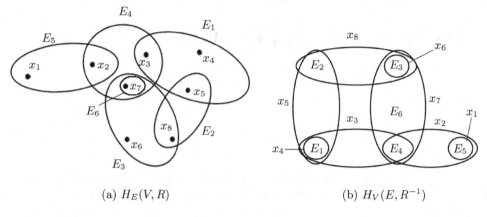

(a) $H_E(V, R)$

(b) $H_V(E, R^{-1})$

Fig. 3.2 The Dual Berge hypergraphs

Let R be a relation between A and B. Let the notation $a\,R\,b$ mean that a is R-related to b. Let $R(a) = \{b \,|\, \text{for all } b \text{ in } B \text{ with } a\,R\,b\}$ and $R(b) = \{a \,|\, \text{for all } a \text{ in } A \text{ with } a\,R\,b\}$. In general a relation R between sets A and B has two associated hypergraphs, $H_A(B; R)$ and $H_B(A; R)$, defined as follows:

$$H_A(B; R) \stackrel{\text{def}}{=} \{R(a) \,|\, a \text{ in } A),$$

$$H_B(A; R) \stackrel{\text{def}}{=} \{R(b) \,|\, b \text{ in } B).$$

As an example let $A = \{$London Bus, London Taxi, Postbox$\}$ and $B = \{$red, black, big, small, wheels, slot, metal$\}$. The relation R and the hypergraph $H_A(B; R)$ are shown in Figure 3.3.

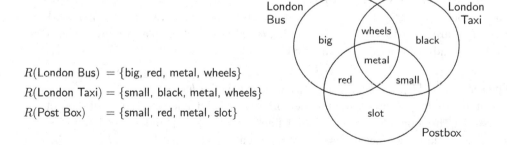

Fig. 3.3 The Hypergraph $H_A(B; R)$

3.2 The Hypergraphs of a Bipartite Network

A relation R between sets A and B has an associated network with vertices $A \cup B$ and edges (a,b) where $a\,R\,b$. Bipartite relations with $A \cap B = \emptyset$ have a much richer connectivity structure than might appear at first sight. As before let $R(a) = \{\,b\,|\,b \in B$ with $a\,R\,b\}$, e.g. $R(a) = \{b_1, b_2, b_3, b_4\}$ in Figure 3.4(a). Then

$$H_A(B;R) = \{R(a_1), R(a_2), R(a_3)\}$$
$$= \{\,\{b_1, b_2, b_3\}, \{b_2, b_3, b_4, b_5, b_6\}, \{b_5, b_6, b_7, b_8\}\,\}$$

$$H_B(A;R) = \{R(b_1), R(b_2), R(b_3), R(b_4), R(b_5), R(b_6), R(b_7), R(b_8)\}$$
$$= \{\,\{a_1\}, \{a_1, a_2\}, \{a_2\}, \{a_2, a_3\}, \{a_3\},\,\}$$

are the hypergraph edges of R in Figure 3.4(b). The sets $R(a)$ and $R(b)$ are called *hyperedges*. The number of vertices in a hyperedge $R(a)$ is called its *extent*, written $|R(a)|$.

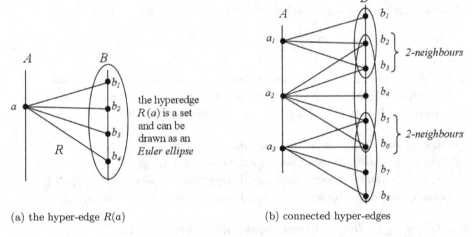

(a) the hyper-edge $R(a)$ (b) connected hyper-edges

Fig. 3.4 Connected hyperedges in a bipartite network

Let the hyperedge $R(a)$ be a *neighbour* of the hyperedge $R(a')$ if their intersection is non-empty, $R(a) \cap R(a') \neq \emptyset$. $R(a)$ is an h-*neighbour* of $R(a')$ if $|R(a) \cap R(a')| \geq h$.

Hyperedges $R(a)$ and $R(a')$ are said to be h-*connected* under R if there exists a sequence $a_1, a_2, ..., a_\ell$ with $a = a_1$, $a' = a_\ell$, with $R(a_i)$ being an h-neighbour of $R(a_{i+1})$ for $i = 1, ..., \ell - 1$.

Figure 3.4(b) illustrates this. $R(a_1)$ is a 2-neighbour of $R(a_2)$ and $R(a_2)$ is a 2-neighbour of $R(a_3)$. Thus a_1 and a_3 are h-connected for $h = 2$. More generally Figure 3.5 shows a *chain* of h-connected hyperedges where h is the smallest value of $|R(a_i) \cap R(a_{i+1})|$.

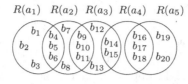

Fig. 3.5 a_1 is 2-connected to a_5

Being h-neighbours is an important property in networks. Generally $R(a) \cap R(a')$ provides structure for a_1 to interact with a_2 and the set of all pairwise intersections can play an important role in the dynamics of systems. To establish notation let $R(\{a, a'\}) \stackrel{def}{=} R(a) \cap R(a')$. Although pairwise intersections are clearly important, why stop there? For example, why not consider $R(\{a, a', a''\}) \stackrel{def}{=} R(a) \cap R(a') \cap R(a'')$?

3.3 The Galois Connection

Let R be a relation between A and B, and let A' be a subset of A, $A' \subseteq A$. Then

$$R(A') \stackrel{def}{=} \bigcap_{a \in A'} R(a).$$

This definition allows the intersection of the $R(a)$ to be formed from any subset of A. For example, Figure 3.6 shows a relation between a set of animals, A, and a set of their features, F. Let $A' = \{$ mouse, hare, deer, camel $\}$, where

$R(\text{mouse}) = \{$ tiny, brown, quadruped, vegetarian $\}$,
$R(\text{hare}) \quad = \{$small, brown, quadruped, vegetarian $\}$,
$R(\text{deer}) \quad = \{$large, brown, quadruped, vegetarian, hooves, antlers$\}$, and
$R(\text{camel}) = \{$large, brown, quadruped, vegetarian, hooves, hump$\}$.

Then $R(A') = \bigcap_{a \in A'} R(a) = \{$ brown, quadruped, vegetarian $\}$.

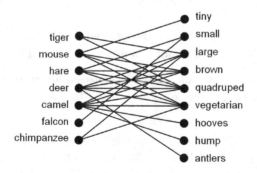

Fig. 3.6 An animal – characteristic bipartite network

Of course for many subsets A' of A the intersection $\bigcap_{a \in A'} R(a)$ will be empty. For example, in Figure 3.7 $R(\{\text{tiger, mouse, chimpanzee}\}) = \{\text{large, quadruped}\} \cap \{\text{tiny, brown, quadruped, vegetarian}\} \cap \{\text{small, vegetarian}\} = \emptyset$.

Although R is a relation between sets A and B, by an abuse of notation the same symbol is used to define a mapping from the power set of A (set of all subsets) to the power set of B, $R : \mathcal{P}(A) \to \mathcal{P}(B)$ with $R : A' \to R(A')$ for all A' in $\mathcal{P}(A)$. In general R is many-one, e.g. $R(\{\text{tiger, mouse}\}) = \{\text{quadruped}\} = R(\{\text{tiger, hare}\})$.

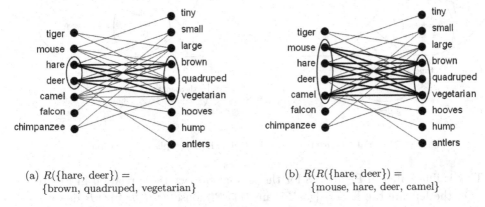

(a) $R(\{\text{hare, deer}\}) =$
{brown, quadruped, vegetarian}

(b) $R(R(\{\text{hare, deer}\}) =$
{mouse, hare, deer, camel}

Fig. 3.7 {hare, deer} $\subset R^2(\{\text{hare, deer}\}) = \{\text{mouse, hare, deer, camel}\}$

By another abuse of notation let R also represent a mapping from the power set of B to the power set of A, $R : \mathcal{P}(B) \to \mathcal{P}(A)$ with $R : B' \to R(B')$. In the literature if R is a relation between A and B, the relation "going the other way" from B to A is sometimes written as R^{-1}, so that $a R b$ if and only if $b R^{-1} a$. However, our abuse of notation makes the development simpler. In particular, the symbol R^2 can be used for the double application of R, $\mathcal{P}(A) \xrightarrow{R} \mathcal{P}(B) \xrightarrow{R} \mathcal{P}(A)$ to give $R^2 : \mathcal{P}(A) \to \mathcal{P}(A)$.

As illustrated in Figure 3.7 $A' \subseteq R^2(A')$ for all $A' \subseteq A$ (assuming that A has no isolated vertices). $A' \subseteq A$ is defined to be *maximal* under R if $A' = R^2(A')$, and $B' \subseteq B$ is *maximal* under R if $B' = R^2(B')$. Then

If A' is a maximal subset of A then $R(A')$ is a maximal subset of B.
If B' is a maximal subset of B then $R(B')$ is a maximal subset of A.

To see this, let $R(A') = B'$. If A' is maximal then $A' = R^2(A') = R(R(A')) = R(B')$. Then $R(A') = R(R(B'))$ so $B' = R^2(B')$ and B' is maximal. A similar argument shows $R(B')$ is maximal.

The hypergraph $\mathcal{H}_A(B; R) \stackrel{\text{def}}{=} \{R(A') \mid \text{for all maximal } A' \subseteq A\}$ will be called the *Galois* hypergraph of $H_A(B; R)$. The hypergraph $\mathcal{H}_B(A, R) \stackrel{\text{def}}{=} \{R(B') \mid \text{for all maximal } B' \subseteq B\}$ will be called the *Galois* hypergraph of $H_B(A; R)$.

The mappings $R : \mathcal{H}_A(B;R) \to \mathcal{H}_B(A;R)$ and $R : \mathcal{H}_A(B;R) \to \mathcal{H}_B(A;R)$ are one-to one. Together they form what is called a *Galois Connection*.

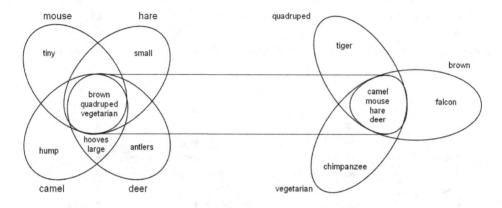

Fig. 3.8   Maximal sets are paired in the Galois Hypergraphs of a relation

This is illustrated in Figure 3.8 for the animal-characteristic relation in Figure 3.7. On the left are the hyperedges $R(\text{camel}) \cap R(\text{mouse}) \cap R(\text{hare}) \cap R(\text{deer}) = \{\text{brown, quadruped, vegetarian}\}$ in $\mathcal{H}_A(B,R)$. On the right right are the hyperedges $R(\text{brown}) \cap R(\text{quadruped}) \cap R(\text{vegetarian}) = \{\text{camel, mouse, hare, deer}\}$ in $\mathcal{H}_A(B,R)$. The Galois connection establishes the *Galois pair* relationship $A' \leftrightarrow B'$ where $R(A') = B'$ and $R(B') = A'$, for example

$$\{\text{brown, quadruped, vegetarian}\} \leftrightarrow \{\text{camel, mouse, hare, deer}\}$$

The Galois connection is considered by many to be a particularly beautiful structure. Among many elegant properties it has the following:

For all maximal A' and A'' either $R(A') \cap R(A'') = \emptyset$ or $A' \cup A''$ and $A' \cap A''$ are maximal. For all maximal B' and B'' either $R(B') \cap R(B'') = \emptyset$ or $B' \cup B''$ and $B' \cap B''$ are maximal. Furthermore

$$R(A' \cup A'') = R(A') \cap R(A''), \quad R(A' \cap A'') = R(A') \cup R(A'').$$

$$R(B' \cup B'') = R(B') \cap R(B''), \quad R(B' \cap B'') = R(B') \cup R(B'').$$

3.4 Galois Pairs and Maximal Rectangles

Fig. 3.9 Arches related to the blocks used to construct them

Figure 3.9 shows a set of arches, $A = \{a_1, a_2, a_3, a_4, a_5, a_6, a_7\}$ with each arch made from a subset of the blocks $B = \{b_1, b_2, b_3, b_4, b_5, b_6, b_7, b_8, b_9, b_{10}, b_{11}, b_{12}\}$. Let a be R-related to b if it contains block b. This bipartite relation can be represented by an incidence matrix as shown in Figure 3.10. The entry in the ith row and the jth column of the matrix is one if a_i is related to b_j, and it zero otherwise.

In a Galois pair $A' \leftrightarrow B'$ every a in A' is R-related to every b in B'. Therefore the rows and columns of the matrix can be rearranged so that all the a_i in A' are contiguous and all the b_j in B' are contiguous, with the corresponding rectangle of entries in the matrix all ones. For example, let $A' = \{a_1, a_2, a_3\}$ and $B' = \{b_3, b_4\}$. Then as shown in Figure 3.10 the corresponding rectangle is filled with ones because each of a_1, a_2 and a_3 is related to b_3 and b_4.

The rectangle corresponding to $A' = \{a_1, a_2, a_3\} \leftrightarrow B' = \{b_3, b_4\}$ is *maximal*. Two other maximal rectangles are shown in Figure 3.10 corresponding to the Galois pairs $\{a_3, a_4\} \leftrightarrow \{b_4, b_5\}$ and $\{a_5, a_6, a_7\} \leftrightarrow \{b_7, b_8, b_9\}$. The maximal rectangles $A' \leftrightarrow B'$ where A' has just one element or B' has just one element are not shown.

	b_1	b_2	b_3	b_4	b_5	b_6	b_7	b_8	b_9	b_{10}	b_{11}	b_{12}
a_1	1	0	1	1	0	0	0	0	0	0	0	0
a_2	0	1	1	1	0	0	0	0	0	0	0	0
a_3	0	0	1	1	1	0	0	0	0	0	0	0
a_4	0	0	0	1	1	1	1	0	0	0	0	0
a_5	0	0	0	0	0	0	1	1	1	1	0	0
a_6	0	0	0	0	0	0	1	1	1	0	1	0
a_7	0	0	0	0	0	0	1	1	1	0	0	1

Fig. 3.10 Maximal rectangles in the arch-block structure

3.5 The Galois Lattice

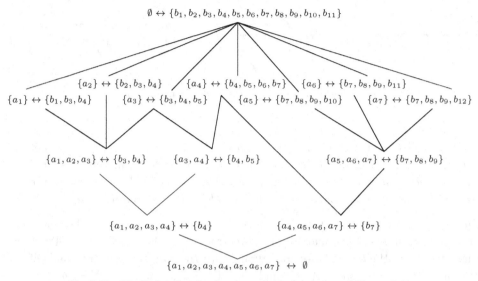

Fig. 3.11 The Galois Lattice for the arch-block relation of Figure 3.10

The Galois pairs form a partially ordered set induced by set ordering. Let $A' \leftrightarrow B'$ and $A'' \leftrightarrow B''$ be Galois pairs. Then $A' \subset A''$ if and only if $B' \supset B''$. Thus the Galois pairs can be arranged as a lattice, also called a *Hasse diagram* or a *construct lattice*. The Galois lattice for the arch-block structure is shown in Figure 3.11. Figure 3.12 gives another example. In this case the *supremum* of the lattice is the Galois pair $\{1, 2, 3, 4, 5\} \leftrightarrow \emptyset$ and the *infimum* is $\emptyset \leftrightarrow \{a, b, c, d, e, f, g, h, i\}$.

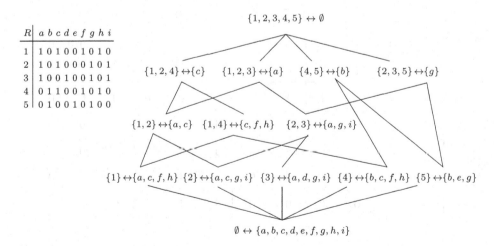

Fig. 3.12 A relation R and its Galois Lattice

3.6 Weak and Strong Connectivity in Hypergraphs

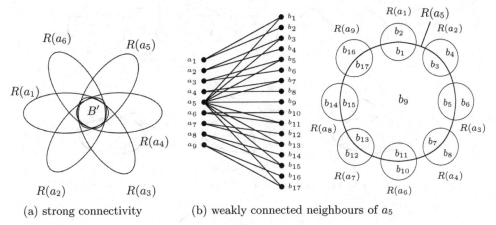

(a) strong connectivity (b) weakly connected neighbours of a_5

Fig. 3.13 Strong and weak connectivity

Galois pairs are sites of connectivity and potential interaction in hypergraphs. For the pair $A' \leftrightarrow B'$ with $A' = \{a_1, a_2, a_3, a_4, a_5, a_6\}$ shown in Figure 3.13(a) the larger the set B' the more highly connected are the elements of A'. Let the *hub* of a set of hypergraph edges be their intersection, $\mathrm{hub}(A_1, A_2, ..., A_n) \stackrel{\mathrm{def}}{=} \cap_{i=1}^{n} A_i$.

Let the *neighbourhood* of a in A be the set $\mathcal{N}_A(a) \stackrel{\mathrm{def}}{=} \{a' \mid R(a) \cap R(a') \neq \emptyset\}$. Figure 3.13(b) shows an extreme case in which none of the members of the neighbourhood $\mathcal{N}_A(a_5)$ intersects any of the others, apart from $R(a_5)$, so that $\mathrm{hub}(\mathcal{N}_A(a)) = \emptyset$. $\mathcal{N}_A(a_5)$ is said to have *weak* connectivity.

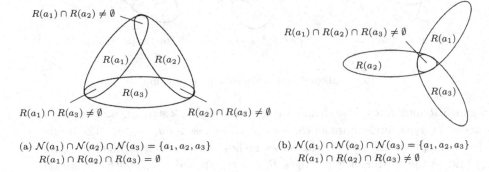

(a) $\mathcal{N}(a_1) \cap \mathcal{N}(a_2) \cap \mathcal{N}(a_3) = \{a_1, a_2, a_3\}$
$R(a_1) \cap R(a_2) \cap R(a_3) = \emptyset$

(b) $\mathcal{N}(a_1) \cap \mathcal{N}(a_2) \cap \mathcal{N}(a_3) = \{a_1, a_2, a_3\}$
$R(a_1) \cap R(a_2) \cap R(a_3) \neq \emptyset$

Fig. 3.14 Strongly and weakly connected neighbourhoods

Figure 3.14 illustrates a fundamental difference in the way hypergraph hyperedges can be configured. In Figure 3.14(a) the hyperedges intersect each other pairwise, but their hub is empty. In this case the configuration is not as highly connected as the configuration in Figure 3.14(b). When the hub of a neighbourhood is non-empty, the neighbourhood will be said to be *strongly connected*.

These ideas are related to "shomotopy" discussed in Chapter 5.

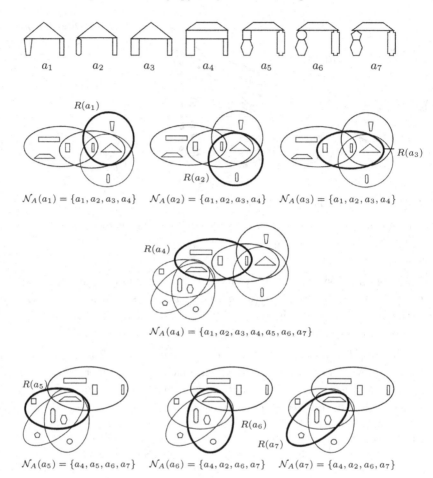

Fig. 3.15   Neighbourhoods in the arch-block structure

Figure 3.15 illustrates these definitions for the arch-block structures. The neighbourhoods for a_1, a_2 and a_3 are all the same, as are those for a_5, a_6, a_7. The neighbourhood of a_4 contains all the other other arches. As can be seen, the hyperedge $R(a_4)$ bridges the cluster of hyperedges $R(a_1)$, $R(a_2)$, and $R(a_3)$ with the cluster of hyperedges $R(a_5)$, and $R(a_6)$ and $R(a_7)$. Thus $R(a_4)$ connects the hypergraph.

Chapter 4

Simplicial Complexes and Q-analysis

4.1 From sets to simplices

Although hypergraphs provide a method of representing relationships between more than two things they are not rich enough to make some basic distinctions, *e.g.* in Fig. 4.1 the arches a_1 and a_2 are represented by the same set of blocks, $\{x_1, x_2, x_3\}$, but they are different structures.

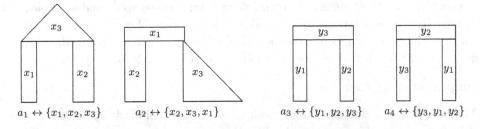

Fig. 4.1 Limitations on the representational power of hypergraph edges

Suppose the rule for forming the arch were "(i) take a set of three blocks, (ii) take an element from the set and put it on the left; (iii) take another element from the set and put it on the right; (iv) take another element from the set and put it on top of the others". Selecting elements from a set is similar to pulling their elements out of a black bag with your eyes closed. As far as the set is concerned, all the elements are *equivalent*, and the order in which they appear is not relevant. When all the elements are the same there is no problem, *e.g.* a_3 and a_4 are the same.

However, suppose one wanted to build the arch a_1 and not a_2. Then the elements have to be selected in the right order. Let the construction be modified as "(i) order the elements as x_1, x_2, and x_3. (ii) take element x_1 and put it on the left; (iii) take x_2 and put it on the right; (iv) take x_3 and put on top of x_1 and x_2". This gives the arch a_1 as desired. It is associated with an *ordered* set of vertices, which can be written as $\langle x_1, x_2, x_3 \rangle$. This is different to $\langle x_2, x_3, x_1 \rangle$ which represents a_2.

41

Simplices

Let V be a set whose element are called *vertices*. Any subset of V, $\{v_0, v_1, ..., v_p\}$ determines an object called an *abstract p-simplex*, written $\sigma = \langle v_0, v_1, ..., v_p \rangle$. A p-simplex can be represented by a p-dimensional *polyhedron* in $(p+k)$-dimensional space, where $k \geq 0$.

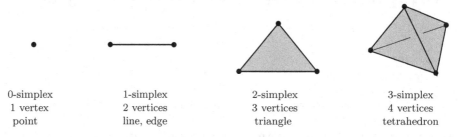

0-simplex	1-simplex	2-simplex	3-simplex
1 vertex	2 vertices	3 vertices	4 vertices
point	line, edge	triangle	tetrahedron

Fig. 4.2 An n-dimensional simplex has $n+1$ vertices

Although they can be viewed as abstract objects determined by their vertices, simplices have a *geometric representation* as polyhedra in multidimensional space, *e.g.* a simplex with three vertices is a triangle in 2-dimensional space and a simplex with four vertices is a tetrahedron in 3-dimensional space. Let the notation $|\sigma|$ mean the number of vertices of a simplex σ. The *dimension* of simplex σ, $dim(\sigma)$, is defined as the number of vertices of σ minus one (Figure 4.2), $dim(\sigma) = |\sigma| - 1$.

Examples of simplicies

In Molière's play *The Bourgeois Gentleman*, Monsieur Jourdain is delighted to learn that he has been speaking prose all his life. Hopefully some readers will be equally delighted to realise that they have been using simplices all their lives.

Family meals as simplices

Consider a family with Mum, Dad, Daughter and Son eating Dinner together. This can be represented by the simplex $\sigma = \langle$Mum, Dad, Daughter, Son, Dinner$\rangle$. This is a different structure to each person eating on their own, $\langle$Mum, Dinner$\rangle$, $\langle$Dad, Dinner$\rangle$, $\langle$Daughter, Dinner$\rangle$, and $\langle$Son, Dinner$\rangle$. The simplex σ can support all kinds of social activity. For example, the daughter might tell a funny story about something that happened at school; or Mum might tell everyone she is thinking of getting a new job; or the family might discuss their preferences for a forthcoming holiday with a favoured suggestion emerging that no-one had previously considered. This emergent suggestion depends on the simplex as a whole – if anyone had been missing it might not have appeared, and possibly dinner was the ideal or only time to have such a discussion. Most readers will have experienced the need to get *all* the right people together in the right environment to get new ideas and make decisions.

Phone calls as simplices

(a) three pairwise phone calls (b) a three-way phone call

Fig. 4.3 Three pairwise phone calls ≠ one three-way phone call

Figure 4.3 illustrates how simplices with higher dimensions can support different kinds of interactions. In Fig. 4.3(a) the unsuspecting father gets a phone call from his daughter via the simplex ⟨Daughter, Dad⟩. He hears her say: "I am in a shop and Mum said you would pay for my new dress", to which he replies: "OK, it will be a pleasure". Shortly afterwards the ⟨Daughter, Mum⟩ simplex is activated, and Mum gets the message that: "I just called Dad and he says he will be happy to pay for my new dress". Following this Dad gets a call via the simplex ⟨Mum, Dad⟩ and is admonished as: "Are you crazy! That girl spends far too much on clothes already without you encouraging her. Why didn't you ask me first?" Of course there's no way out for Dad, who is left to reflect that if only there had been a three-way phone call as shown in Fig. 4.3(b) then none of this would have happened.

Social dynamics on simplices

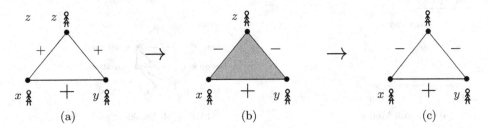

Fig. 4.4 Social dynamics depend on structure

Figure 4.4(a) shows three girls, x, y and z. Girls x and y are close friends and initially they each have a mildly positive relationship with z. Fig. 4.4(b) represents an occasion on which all three girls are playing together but x and y start ganging up on z and being unpleasant to her. Thus the relationships ⟨z, x⟩ and ⟨z, y⟩ become negative. After this z may have enduring negative feelings with each of x and y as shown in Fig. 4.4.

Business meetings as simplices

Fig. 4.5 Simplices constrain what can be said during business meetings

Business conversations are constrained by simplices. For example, in Fig. 4.5(a) ⟨Jill, Tom⟩ represents the marketing team from one company and ⟨Bill, Sue⟩ is the purchasing team of another. Suppose Jill phones Bill to tell him about a new product and they arrange a meeting represented by the tetrahedron in Fig. 4.5(b). During the meeting each side must be very careful. For example, if the sellers reveal that they have a surplus of the product, the buyers could force a better deal for themselves. On the other hand, if the buyers reveal that they have a problem that the new product will resolve, the sellers may hold back on a reduced price offer they were planning. The dynamics of the meeting depend on the various faces of the tetrahedron, e.g. if the buyers say the product is too expensive, Jill may request a private conversation with Tom to discuss a possible reduction in the price on the isolated structure ⟨Jill, Tom⟩. If things are getting tricky for the buyers they may excuse themselves and use the ⟨Bill, Sue⟩ structure to discuss their position. This illustrates that making appropriate use of the simplices can be very important.

The Full Monty

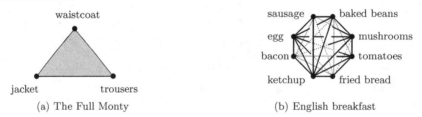

(a) The Full Monty   (b) English breakfast

Fig. 4.6 Examples of simplices

The phrase "The Full Monty" has come to mean "complete" or everything that could be expected. It is said to come from the Montague Burton tailoring chain which hired three-piece suits to men getting married ⟨jacket, trousers, waistcoat⟩ (Fig. 4.6(a)). Another, less likely, explanation is that it comes from the full English breakfasts General Montgomery recommended for his troops ⟨egg, bacon, sausage, fried bread, baked beans, mushrooms, tomatoes, ketchup⟩ ((Fig. 4.6(b)).

Vertex parts and polyhedral wholes

In his book on Gestalt psychology [Katz (1951)] rejects the equation

Vanilla Ice Cream = Cold + Sweet + Vanilla Aroma + Softness + Yellow

which suggests that each attribute can be sensed separately and put together in a linear way. In our terms, Vanilla Ice Cream is a polyhedron with five vertices bound together by an *indivisible* 5-ary relation. This can be written as

Vanilla Ice Cream = ⟨Cold, Sweet, Vanilla, Softness, Yellow⟩

≠ ⟨ Cold ⟩ + ⟨Sweet⟩ + ⟨Vanilla⟩ + ⟨ Softness⟩ + ⟨Yellow⟩

with the "Gestalt" construct of *Vanilla Ice Cream* represented by a polyhedron with five vertices. Figure 4.7 illustrates the distinction between an unrelated set of vertices and the "Gestalt" polyhedron. It also illustrates the difference between a polyhedron with five vertices embedded in a 4-dimensional space and a network-theoretic *clique* embedded in 2-dimensional space in which every vertex is connected to every other by a 1-dimensional link. The clique is the worst representation, since ice-cream is experienced as a whole, not as combinations of pairs of senses.

Vanilla Ice Cream = ⟨Cold, Sweet, Vanilla, Softness, Yellow⟩   *Polyhedron ≠*

≠ {⟨Cold ⟩, ⟨Sweet⟩, ⟨Vanilla⟩, ⟨Soft⟩, ⟨Yellow⟩}   *Set of Vertices*

≠ {⟨Cold, Sweet ⟩, ⟨Cold, Vanilla⟩, ⟨Cold, Soft⟩,   *≠Set of Lines*
⟨Cold, Yellow⟩, ⟨Sweet, Vanilla⟩, ⟨Sweet, Soft⟩,
⟨Sweet, Yellow⟩, ⟨Vanilla, Soft⟩, ⟨Soft, Yellow⟩,
⟨Vanilla, Yellow⟩}

The polyhedron ⟨Cold, Sweet, Vanilla, Soft, Yellow⟩ here expresses the concept of *whole* which is clearly more than the sum of its parts:

⟨Cold, Sweet, Vanilla, Soft, Yellow⟩ ≠ ⟨Cold⟩+⟨Sweet⟩+⟨Vanilla⟩+⟨Soft⟩+⟨Yellow⟩

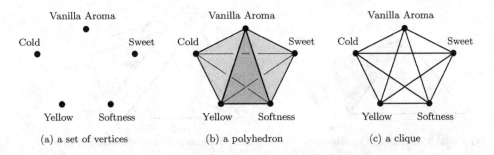

Fig. 4.7   Set of vertices ≠ polyhedron ≠ clique

46   Hypernetworks in the Science of Complex Systems

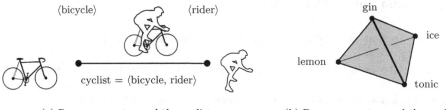

(a) Remove a vertex and the cyclist simplex ceases to exist

(b) Remove a vertex and the perfect gin and tonic ceases to exist

Fig. 4.8   Remove a vertex and the simplex ceases to exist

The essential feature of a polyhedron is that it ceases to exist if any of the vertices are removed. For example, consider a cyclist represented as the combination ⟨rider, bicycle⟩. Remove either the man or the bicycle and what is left ceases to be a cyclist. Removing a vertex is like sticking a pin in a balloon, causing the structure to collapse and whatever is left is not the whole simplex. Remove any vertex from ⟨gin, tonic, ice, lemon⟩ and it ceases to be the perfect gin and tonic. Generalising edges to polyhedra allows a distinction to be made between the *parts* of things represented by vertices, and *wholes* represented by polyhedra.

4.2   Simplices, polyhedra and their faces

Connectivity is a one of the most powerful concepts for analysing complex systems as illustrated by the widespread use of networks. The vertices of networks are 0-dimensional simplices, $\langle v \rangle$ and the edges are 1-dimensional simplices, $\langle v, v' \rangle$. Two edges are "connected" if they share a vertex, and paths can be defined as chains of connected edges.

Simplices allow a natural multidimensional generalisation of this well-established concept of connectivity. For example, Figure 4.9 shows the four faces of a tetrahedron (3-simplex). This common use of the term "face" generalises. The 2-dimensional faces of a 3-dimensional tetrahedron are 2-dimensional triangles, the

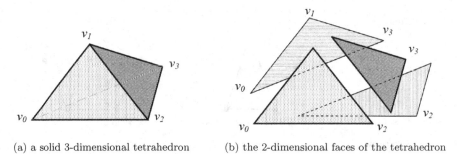

(a) a solid 3-dimensional tetrahedron

(b) the 2-dimensional faces of the tetrahedron

Fig. 4.9   The 2-dimensional triangular faces of a 3-dimensional tetrahedron

Simplicial Complexes and Q-analysis   47

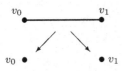

(a) vertices are 0-dimensional faces
of 1-dimensional edges

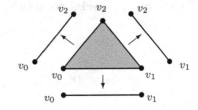

(a) edges are 1-dimensional faces
of 2-dimensional triangles

Fig. 4.10   The faces of edges and triangles

1-dimensional faces of a 2-dimensional triangle are its 1-dimensional edges, and the the 0-dimensional faces of a 1-dimensional edge are its 0-dimensional vertices (Fig. 4.10).

The simplex $\sigma = \langle v'_0, v'_1, ..., v'_q \rangle$ is defined to be a q-*dimensional face* of the simplex $\sigma' = \langle v_0, v_1, ..., v_p \rangle$ if $\{v'_0, v'_1, ..., v'_q\} \subseteq \{v_0, v_1, ..., v_p\}$. This is written as $\sigma \lesssim \sigma'$. For example, $\sigma = \langle v_0, v_2, v_3 \rangle$ is a 2-dimensional triangular face of the 3-dimensional tetrahedron $\sigma' = \langle v_0, v_1, v_2, v_3 \rangle$.

4.3   The intersection of simplices

In networks, links and arrows are connected by vertices. For multidimensional polyhedra, connectivities can have higher dimension than than the zero-dimensions of a vertex. Two simplices are q-*near* if they share a q-dimensional face. The *intersection* of two simplices σ and σ' is defined to be their highest dimensional shared face, σ''. We write $\sigma \cap \sigma' = \sigma''$.

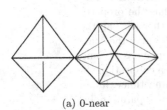

(a) 0-near

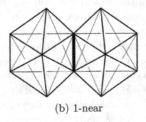

(b) 1-near

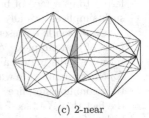

(c) 2-near

Fig. 4.11 q-near simplices

In Fig. 4.11(a) the simplices share a vertex, which is a 0-dimensional face so they are 0-near. In Fig. 4.11(b) the simplices share an edge, which is a 1-dimensional face so they are 1-near. In Fig. 4.11(c) the simplices share a triangle, which is a 2-dimensional face so they are 2-near.

4.4 Representing social structure by connected simplices

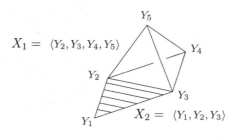

Fig. 4.12 Atkin's method of representing social structure by simplices

The idea of representing social structure by connected polyhedra is due to R. H. Atkin, and was first published in the article *A Mathematical Approach Towards Social Science* in the Essex University Review:

> To examine the idea of connectivity in more detail consider, for example, a collection of people and the sociological roles which they are said to be playing. Let the role-set be denoted by Y and let it contain a finite number of roles Y_1, Y_2, ... ; similarly let there be a finite number of persons X_1, X_2, ... in the collection of people X. An individual person X_1 plays, say, roles Y_1, Y_2, Y_3 and a second person X_2 plays the roles Y_2, Y_3, Y_4, Y_5. We now define an *abstract p-simplex* to be a subset of Y containing $(p+1)$ roles provided that there is at least one individual who plays all these roles. Thus the 2-simplex $\langle Y_1, Y_2, Y_3 \rangle$ exists since X_1 plays the three roles represented therein, so also do the 3-simplex $\langle Y_2 Y_3 Y_4 Y_5 \rangle$, the 0-simplex $\langle Y_5 \rangle$, and many others. The two simplices $\langle Y_1, Y_2, Y_3 \rangle$ and $\langle Y_2, Y_3, Y_4, Y_5 \rangle$ are clearly joined by the 1-simplex $\langle Y_2, Y_3 \rangle$ - which is referred to as a face of both the 2- and 3-simplices. The collection of all such simplices actually forms a complex $K(Y)$ which has the property that a person is represented by one of its simplices together with all the faces of that simplex. We may note that two people who play the same roles are indistinguishable in this model. ... Thus our persons X_1 and X_2 can communicate with each other because they have a connection via their common simplices $\langle Y_2, Y_3 \rangle$, $\langle Y_2 \rangle$ and $\langle Y_3 \rangle$. This connection exhibits the fact that X_1 "sees" X_2 via the common faces of their separate polyhedra. On the other hand X_2 has many faces (15 in all, if we include the whole tetrahedron) any one of which might serve as a connecting face between himself and someone else. [Atkin *et al* (1968)]

Atkin developed the theory of *Q-analysis* based on these ideas [Atkin (1974, 1977, 1981, 2010)]. Originally interested in mathematical physics, he developed a theory showing that much of physics can be captured by what he called the "Law of the Trivial Cocycle". Atkin began by a fundamental evaluation of the nature of space, making a distinction between *real* space as Euclidean 3-space with the Pythagorean metric and *actual space* for what we can observe. For example, $\sqrt{2}$ belongs to real space but does not belong to actual space. An important idea in Atkin's work is that the structure of the observation space constrains the nature of the dynamics of the system. In particular he showed how the Cocycle Law bridges continuum theory and discrete theory for a wide range of physical phenomena. The idea of transferring these ideas to social systems is expressed in his paper "*From cohomology in Physics to q-connectivity in Social Science*". The theory of hypernetworks developed in this book is an extension of Atkin's theory and builds on the many powerful and highly original ideas he developed in the nineteen sixties, seventies and eighties.

4.5 Simplicial families and simplicial complexes

A set of simplices with all their faces forms a *simplicial complex*, i.e. a set of simplices K is defined to be a simplicial complex if $\sigma \in K$ implies $\sigma' \in K$ for all $\sigma' \lesssim \sigma$.

For some applications the requirement of having all the faces of all the simplices is too restrictive, and what will be called "simplicial families" do not have this restriction, *i.e.* any set of simplices forms a *simplicial family*. Every simplicial family determines a simplicial complex, namely the simplices of the family with all their faces.

Simplicial systems and bipartite relations

Let K be a simplicial family with simplices A and vertices B. Then a bipartite relation can be defined between A and B with $a\,R\,b$ if b is a vertex of a.

Alternatively, every bipartite relation $A \xleftrightarrow{R} B$ defines two simplicial families. For each a in A let $\sigma(a)$ be the simplex with vertex set $\{b\,|\,a\,R\,b\}$ and for each b in B let $\sigma(b)$ be the simplex with vertex set $\{a\,|\,a\,R\,b\}$.

The *conjugate families* of $A \xleftrightarrow{R} B$ are

$$F_A(B, R) = \{\sigma(a)\,|\, \text{for all } a \in A\} \text{ and}$$
$$F_B(A, R) = \{\sigma(b)\,|\, \text{for all } b \in B\}.$$

Let $K_A(B, R) = \{\sigma\,|\,\sigma \lesssim \sigma(a) \text{ for all } a \in A\}$ be the simplices in $F_A(B, R) = \{\sigma(a)\,|\,\text{for all } a \in A\}$ together with all their faces, and let $K_B(A, R) = \{\sigma\,|\,\sigma \lesssim \sigma(b) \text{ for all } b \in B\}$ be the simplices in $F_B(A, R) = \{\sigma(b)\,|\,\text{for all } b \in B\}$ with all their faces.

The *conjugate simplicial complexes* of $A \xleftrightarrow{R} B$ are

$$K_A(B, R) = \{\sigma\,|\,\sigma \lesssim \sigma(a) \text{ for any } a \in A\} \text{ and}$$
$$K_B(A, R) = \{\sigma\,|\,\sigma \lesssim \sigma(b) \text{ for any } b \in B\}.$$

4.6 Multidimensional connectivity in simplicial complexes

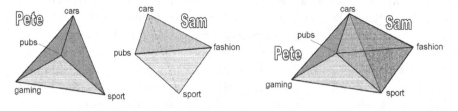

(a) Pete and Sam as tetrahedra   (b) Pete and Sam share a triangular face

Fig. 4.13 People connected through their interests

Figure 4.13(a) shows two simplices representing the interests of two friends. Pete's tetrahedron (3-simplex) is $\sigma(\text{Pete}) = \langle \text{gaming, pubs, sport, cars} \rangle$ while Sam's simplex is $\sigma(\text{Sam}) = \langle \text{pubs, sport, cars, fashion} \rangle$. These friends share the triangular face $\langle \text{pubs, cars, sport} \rangle$ and are 2-near.

Imagine these friends in a pub. Pete tells Sam about his successful poker game last night. Sam no doubt would listen politely, before telling Pete about a new style of shoes he'd seen in a magazine. Not interested in fashion, Pete might mention the car driven by his favourite soccer star, sparking Sam's interest in both cars and sport and lead to a more intense discussion.

In Fig. 4.14 Sue has the simplex $\langle \text{fashion, history, painting, literature} \rangle$. She shares just the vertex $\langle \text{fashion} \rangle$ with Sam, but has more in common with Jane, being 1-near through the face $\langle \text{history, literature} \rangle$.

Table 4.1 shows the relation between the people and their interests as an incidence matrix. The vertices are listed across the top and the polyhedra down the side. The entry 1 means "is a vertex of" and 0 means "is not a vertex of".

The set of connected simplices in Figure 4.14 is a structure that supports different kinds of interaction. Whereas Pete and Sam can enjoy conversations in pubs about fast cars and their favourite team, Sue and Jane are more likely to have conversations combining history and literature such as the accuracy of Shakespeare's

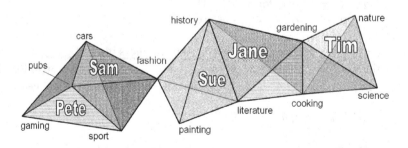

Fig. 4.14 A simplicial family of people and their interests, $F_{\text{People}}(\text{Interests})$

Simplicial Complexes and Q-analysis

	gaming	pubs	cars	sport	fashion	painting	history	literature	gardening	Cooking	nature	science
Pete	1	1	1	1	0	0	0	0	0	0	0	0
Sam	0	1	1	1	1	0	0	0	0	0	0	0
Sue	0	0	0	0	1	1	1	1	0	0	0	0
Jane	0	0	0	0	0	0	1	1	1	1	0	0
Tim	0	0	0	0	0	0	0	0	1	1	1	1

Table 4.1 The incidence matrix of the relation between the people and their interests

historical plays. In contrast Jane's conversations with Tim are likely to combine gardening with cooking, possibly discussing the seasonable implications of herbs and vegetables for the dishes they like to make.

In this micro-society, Pete and Sam are the closest sharing three interests. They form a relatively disconnected substructure from the rest, and they can be imagined chatting easily at a party. Tim is also rather peripheral, being connected only to Jane. In comparison, Sue and Jane are the most integrated, each being connected to two other people. They seem to be the most central people in this system.

The simplices σ and σ' are q-connected in a simplicial family F if there is a chain of simplices $\sigma_1, \sigma_2, ..., \sigma_\ell$ with $\sigma = \sigma_1$, $\sigma' = \sigma_\ell$, and σ_i being at least q-near σ_{i+1} for $i = 1, ..., \ell - 1$. The chain $\sigma_1, \sigma_2, ..., \sigma_\ell$ is called a *chain of connection* between σ and σ'. The simplices σ and σ' are said to be q-connected. By this definition, if σ and σ' are q-connected then they are p-connected for all $p \leq q$.

Simplicial families and complexes extend the idea of connectivity in networks to higher dimensions. For example, Sue is 1-near Jane, Jane is 1-near Tim, so Sue is 1-connected to Tim through Jane. This shows that two simplices can be be connected, even though they have no vertices in common, $\sigma(\text{Sue}) \cap \sigma(\text{Tim}) = \emptyset$.

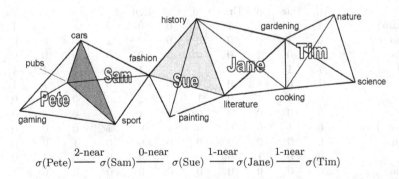

Fig. 4.15 A 0-dimensional chain of connection between Pete and Tim

4.7 Q-analysis

In general, being q-connected is an equivalence relation on a set of simplices and partitions them into q-connected components. A listing of the components for each dimensional q-value is called a *Q-analysis*, e.g the Q-analysis for $F_{\text{People}}(\text{Interests})$ in Fig. 4.15 is

q = 3: $\{\sigma(\text{Pete})\}$, $\{\sigma(\text{Sam})\}$, $\{\sigma(\text{Sue})\}$, $\{\sigma(\text{Jane})\}$, $\{\sigma(\text{Tim})\}$

q = 2: $\{\sigma(\text{Pete}), \sigma(\text{Sam})\}$, $\{\sigma(\text{Sue})\}$, $\{\sigma(\text{Jane})\}$, $\{\sigma(\text{Tim})\}$

q = 1: $\{\sigma(\text{Pete}), \sigma(\text{Sam})\}$, $\{\sigma(\text{Sue}), \sigma(\text{Jane}), \sigma(\text{Tim})\}$

q = 0: $\{\sigma(\text{Pete}), \sigma(\text{Sam}), \sigma(\text{Sue}), \sigma(\text{Jane}), \sigma(\text{Tim})\}$

For a small system, Q-analysis can be presented as a *skyscraper diagram* as shown in Fig. 4.16, as suggested in [Atkin (1977)].

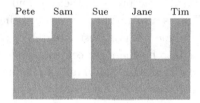

Fig. 4.16 A Q-analysis skyscraper diagram

As another example, the table below taken from [de Nooy *et al* (2005)] shows the major trade network in South America. This relation is asymmetric and there are two Q-analyses. The first analyses the connectivity between the exporting countries. This structure is dominated by Brazil which exports to all the countries, so that every other country is a face of Brazil.

	Arg	Bar	Bol	Bra	Chi	Col	Ecu	T–T	Par	Per	Uru	Ven
Argentina	1	0	1	1	1	0	0	0	1	0	1	0
Barbados	0	1	0	0	0	0	0	1	0	0	0	0
Bolivia	0	0	1	0	0	0	0	0	0	0	0	0
Brazil	1	1	1	1	1	1	1	1	1	1	1	1
Chile	1	0	1	0	1	0	0	0	1	1	1	0
Colombia	0	0	1	0	1	1	1	0	0	1	0	1
Ecuador	0	0	0	0	0	1	1	0	0	1	0	0
Trinidad–Tobago	0	1	0	0	0	0	0	1	0	0	0	0
Paraguay	0	0	0	0	0	0	0	0	1	0	0	0
Peru	0	0	1	0	0	0	0	0	0	1	0	0
Uruguay	0	0	0	0	0	0	0	0	0	0	1	0
Venezuela	0	0	1	0	1	1	1	1	1	1	0	1

Table 4.2 Major trading partners in South America (Source:[de Nooy *et al* (2005)]).

Simplicial Complexes and Q-analysis

q = **11** to **8**   {Brazil}
q = **7** to **6**   {Venezuela, Brazil}
q = **5** to **3**   {Venezuela, Brazil, Chile, Colombia, Argentina}
q = **2**   {Venezuela, Brazil, Colombia, Chile, Ecuador, Argentina}
q = **1**   {Venezuela, Brazil, Colombia, Chile, Ecuador, Peru, Argentina, Bolivia, Trinidad–Tobago}
q = **0**   {Venezuela, Brazil, Chile, Colombia, Ecuador, Peru, Barbados, Trinidad–Tobago, Argentina, Bolivia, Paraguay, Uruguay}

The skyscraper diagram for $K_{\text{Exporters}}(\text{Importers})$ is shown in Fig. 4.17(a). The structure is dominated at $q=11$ by Brazil, Venezuela at $q=7$, and Chile, Colombia, and Argentina at $q=5$.

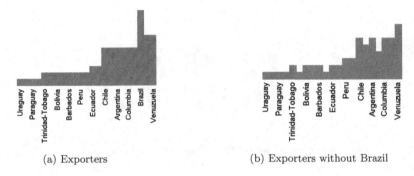

(a) Exporters   (b) Exporters without Brazil

Fig. 4.17 Skyscraper diagrams for the Q-analysis of trade within South America

When a Q-analysis is dominated by one or more simplices, as is the case for Brazil in Fig. 4.17(a), it can be interesting to analyse the structure with that simplex removed. When Brazil is removed the new Q-analysis becomes:

q = **7** to **6**   {Venezuela}
q = **5**   {Venezuela, Colombia}, {Chile}, {Argentinia}
q = **4**   {Venezuela, Colombia}, {Chile, Argentina}
q = **3**   {Venezuela, Chile, Colombia, Argentina}
q = **2**   {Venezuela, Chile, Colombia, Argentina, Ecuador}
q = **1**   {Venezuela, Chile, Colombia, Ecuador, Peru Argentina}, {Bolivia}, {Trinidad–Tobago},
q = **0**   {Venezuela, Chile, Colombia, Ecuador, Peru, Barbados, Trinidad–Tobago, Argentina, Bolivia, Paraguay, Uruguay}

with skyscraper diagram shown in Fig. 4.17(b). This Q-analysis shows that, besides Brazil, Venezuela, Chile, Argentina and Colombia being the dominant exporters, their markets are different combinations of countries. Colombia and Venezuela share the 5-dimensional face ⟨Bolivia, Chile, Colombia, Ecuador, Peru, Venezuela⟩ while Argentina and Chile share the 4-dimensional face ⟨Argentina, Bolivia, Chile, Paraguay, Uruguay⟩. Colombia is the simplex ⟨Bolivia, Chile, Colombia, Ecuador, Peru, Venezuela⟩ which is a 5-dimensional face of Venezuela.

54   *Hypernetworks in the Science of Complex Systems*

The following Q-analysis shows the structure between the importing countries:

q = 6   {Bolivia}
q = 5   {Bolivia}, {Peru}
q = 4   {Bolivia, Chile, Peru}, {Paraguay}
q = 3   {Bolivia Chile Paraguay, Peru, Colombia, Ecuador Argentina}
        {Trinidad–Tobago}, {Uruguay}
q = 2   {Bolivia, Chile, Colombia, Ecuador, Paraguay, Peru, Venezuela, Argentina, Uruguay}, {Trinidad–Tobago, Barbados}
q = 1   {Bolivia, Chile, Colombia, Ecuador, Trinidad–Tobago, Paraguay, Peru, Venezuela, Barbados, Argentina, Uruguay, Brazil}

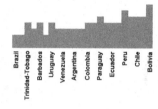

Fig. 4.18   Skyscraper diagram for the Q-analysis of importing countries

This Q-analysis shows that Bolivia ($q = 6$) and Peru ($q = 5$) are the main importing countries within South America. Their relatively high connectivity ($q = 4$) shows that they have a core of the same countries exporting to them. Inspection of Table 4.2 shows this to be ⟨Brazil, Chile, Colombia, Peru, Venezuela⟩. At $q = 4$ Chile is a face of Bolivia, ⟨Argentinia, Brazil, Chile, Colombia, Venezuela⟩, so that Bolivia, Peru and Chile form a 4-connected component.

As will be seen, Q-analysis has interesting clustering properties. Depending on the application, when two simplices are q-near for relatively high values of q it means they are relatively similar with respect to the vertex set. On the other hand, it is possible for two simplices to be q-connected and share no vertices at all. Usually q-connected components contain clusters of simplices that are all similar but not precisely in the same way, and the internal structure of q-connected components can be very interesting.

For example, the intersection of the importing simplices in the 4-component {Bolivia, Chile, Peru} is their 3-face ⟨Brazil, Chile, Colombia, Venezuela⟩, whose vertices are the countries as exporters. Thus each of Brazil, Chile, Colombia, and Venezuela exports to each of Bolivia, Chile, and Peru. Jumping ahead of the story a little, this corresponds to a *Galois pair* of simplices

⟨Brazil, Chile, Colombia, Venezuela⟩   ↔   ⟨Bolivia, Chile, Peru⟩

	Arg	Bar	Bol	Per	Chi	Col	Ecu	T–T	Par	Bra	Uru	Ven
Argentina	1	0	1	0	1	0	0	0	1	1	1	0
Barbados	0	1	0	0	0	0	0	1	0	0	0	0
Bolivia	0	0	1	0	0	0	0	0	0	0	0	0
Brazil	1	1	1	1	1	1	1	1	1	1	1	1
Chile	1	0	1	1	1	0	0	0	1	0	1	0
Colombia	0	0	1	1	1	1	1	0	0	0	0	1
Venezuela	0	0	1	1	1	1	1	1	1	0	0	1
Trinidad–Tobago	0	1	0	0	0	0	0	1	0	0	0	0
Paraguay	0	0	0	0	0	0	0	0	1	0	0	0
Peru	0	0	1	1	0	0	0	0	0	0	0	0
Uruguay	0	0	0	0	0	0	0	0	0	0	1	0
Ecuador	0	0	0	1	0	1	1	0	0	0	0	0

Table 4.3 The maximal rectangle for ⟨Brazil, Chile, Colombia, Venezuela⟩ ↔ ⟨Bolivia, Peru, Chile⟩

Table 4.3 is obtained by swapping the Ecuador row of Table 4.1 with the Venezuela row, and swapping the Brazil column with Peru column. As can be seen, a maximal rectangle emerges corresponding to the Galois pair ⟨Brazil, Chile, Colombia, Venezuela⟩ ↔ ⟨Bolivia, Peru, Chile⟩. This relationship between q-connected components and Galois pairs is interesting because it gives a way to approach the combinatorial explosion when computing the Galois connection.

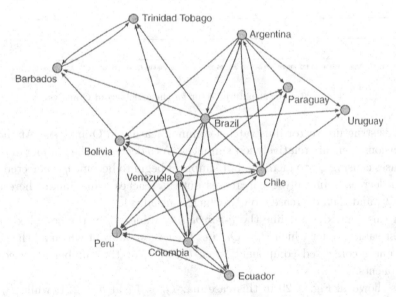

Fig. 4.19 The South American trade data viewed as a network

4.8 Structure Vectors

In a Q-analysis things cluster together through their shared vertices, and the pattern of components gives an insight into the connectivity of a simplicial family. The *structure vector* of a Q-analysis is a list of the number of components, Q_q at each dimension q. For example, the structure vector for the South American importers structure is $(\overset{0}{1}, \overset{1}{1}, \overset{2}{2}, \overset{3}{3}, \overset{4}{2}, \overset{5}{2}, \overset{6}{1})$, where here the dimension appears above the number of components.

For big data sets listing the number of components is impractical and it can be more useful to display the structure vectors as a graph. For example, the Observatorium project at the University of Lisbon is storing online newspapers from various countries. The web pages they are archiving have a lot of subtle structure, and there are many hundreds of thousands of them going back a year or more. As an experiment we analysed the *Australian* online newspaper articles over a period of three days. These 104 web pages used 8816 words, and there were 81,825 occurrences of these words in the 104 articles.

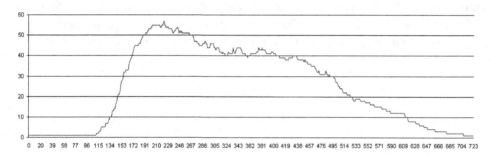

Fig. 4.20 The structure vector for the article-word Q-analysis

This structure vector illustrates a common feature in Q-analysis. At the higher dimensions there are relatively few simplices. As q decreases the number of simplices increases causing Q_q to increase, but simplices begin to become q-connected causing Q_q to decrease. Initially Q_q increases until it reaches a maximum, here denoted max-Q_q, and then decreases to Q_0, which is usually 1.

In this context we define the *q-percolation* value, P_q of the complex to be the highest value of q for which $P_q = Q_0$, *i.e.* the largest value at which all the simplices form one q-connected component when $Q_0 = 1$, or the number of disconnected components.

As shown in Fig. 4.20 in this case max-$Q_q = 57$ at $q = 221$, while $P_q = 110$. Thus the 104 articles form a maximum of 57 components at $q = 221$ and these all become connected at $q = 110$. Thus the percolation from maximum to minimum number of components occurs relatively rapidly between $q = 221$ and $q = 110$, which is about one sixth of the dimension range.

4.9 The Difference Operator

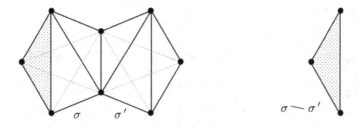

Fig. 4.21 The difference operator

It is useful to be able to consider the "difference" of two simplices with the vertices of one removed from the other. For example, Fig. 4.22(a) shows two 1-connected simplices σ and σ'. Figure 4.22(b) shows σ without the vertices its shares with σ'.

The *difference* between the simplices σ and σ', σ minus σ', is written $\sigma \frown \sigma'$, and defined to be the simplex with

$$\langle x \rangle \lesssim \sigma \frown \sigma' \text{ if and only if } \langle x \rangle \lesssim \sigma \text{ and } \langle x \rangle \not\lesssim \sigma'.$$

It follows that $\sigma \frown \sigma' = \sigma \frown (\sigma \cap \sigma')$, so the difference between σ and σ' is the same as σ with the shared face removed.

The difference operator is non-commutative, $\sigma \frown \sigma' \neq \sigma' \frown \sigma$. Also it can be shown that $(\sigma \frown \sigma') \cap (\sigma' \frown \sigma) = \emptyset$ for all σ and σ'.

The difference operator is not associative, and it is necessary to use brackets to show the order of applications, *i.e.* in general

$$(\sigma \frown \sigma') \frown \sigma'' \neq \sigma \frown (\sigma' \frown \sigma'')$$

This is illustrated in Fig. 4.22 where $\sigma = \langle x_1, x_2, x_3, x_4, x_5 \rangle$, $\sigma' = \langle x_3, x_4, x_8 \rangle$ and $\sigma'' = \langle x_4, x_5, x_6, x_7 \rangle$. Figure 4.22(a) shows σ, σ' and σ'', Fig. 4.22(b) shows $\sigma \frown \sigma' = \langle x_1, x_2, x_3 \rangle$ and σ''. On subtracting σ'' this becomes $\langle x_1, x_2 \rangle$ as shown in Fig. 4.22(c). In contrast, Fig. 4.22(d) also shows σ, σ' and σ'', Fig. 4.22(e) shows σ with $\sigma'' \frown \sigma' = \langle x_3, x_8 \rangle$. Subtracting $\langle x_3, x_8 \rangle$ from σ gives $\langle x_1, x_2, x_3, x_4 \rangle$ as shown in Fig. 4.22(f). Thus $(\sigma \frown \sigma') \frown \sigma'' \neq \sigma \frown (\sigma' \frown \sigma')$.

Although difference is not associative, the order in which simplices are subtracted does not matter,

$$(\sigma \frown \sigma') \frown \sigma'' = (\sigma \frown \sigma'') \frown \sigma', \text{ for all } \sigma, \sigma' \text{ and } \sigma''.$$

58   Hypernetworks in the Science of Complex Systems

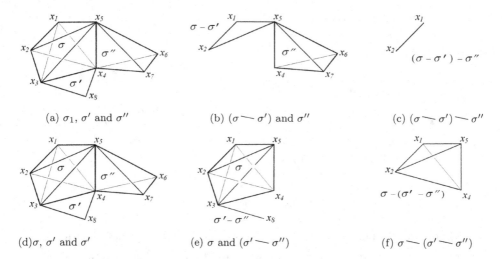

Fig. 4.22 The difference operator is not associative

The difference operator behaves well with intersections of simplices:

$$\sigma'' \cap (\sigma \frown \sigma') = (\sigma'' \cap \sigma) \frown (\sigma'' \cap \sigma') = (\sigma \cap \sigma'') \frown (\sigma' \cap \sigma'') = (\sigma \frown \sigma') \cap \sigma''$$

for all σ, σ' and σ''.

4.10 Eccentricity

Some simplices are highly connected to other simplices while some simplices are relatively disconnected. Those simplices that do not share many of their vertices with other simplices are relatively eccentric. This is not always clear from the Q-analysis. For example, Fig. 4.23 shows descriptive simplices for six wine glasses.

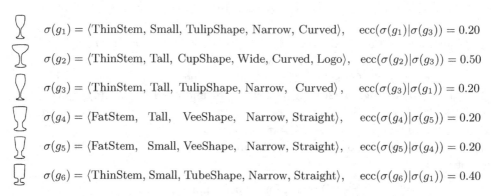

$\sigma(g_1) = \langle \text{ThinStem, Small, TulipShape, Narrow, Curved} \rangle,\quad \text{ecc}(\sigma(g_1)|\sigma(g_3)) = 0.20$

$\sigma(g_2) = \langle \text{ThinStem, Tall, CupShape, Wide, Curved, Logo} \rangle,\quad \text{ecc}(\sigma(g_2)|\sigma(g_3)) = 0.50$

$\sigma(g_3) = \langle \text{ThinStem, Tall, TulipShape, Narrow, Curved} \rangle,\quad \text{ecc}(\sigma(g_3)|\sigma(g_1)) = 0.20$

$\sigma(g_4) = \langle \text{FatStem, Tall, VeeShape, Narrow, Straight} \rangle,\quad \text{ecc}(\sigma(g_4)|\sigma(g_5)) = 0.20$

$\sigma(g_5) = \langle \text{FatStem, Small, VeeShape, Narrow, Straight} \rangle,\quad \text{ecc}(\sigma(g_5)|\sigma(g_4)) = 0.20$

$\sigma(g_6) = \langle \text{ThinStem, Small, TubeShape, Narrow, Straight} \rangle,\quad \text{ecc}(\sigma(g_6)|\sigma(g_1)) = 0.40$

Fig. 4.23 A set of wine glasses, their descriptive simplices, and their eccentricities

Simplicial Complexes and Q-analysis

Let $F = \{\sigma(g_1), \sigma(g_2), \sigma(g_3), \sigma(g_4), \sigma(g_5), \sigma(g_6)\}$. The Q-analysis is:

$Q = 5$: $\{\sigma(g_2)\}$
$Q = 4$: $\{\sigma(g_1)\}$ $\{\sigma(g_2)\}$ $\{\sigma(g_3)\}$ $\{\sigma(g_4)\}$ $\{\sigma(g_5)\}$ $\{\sigma(g_6)\}$
$Q = 3$: $\{\sigma(g_1), \sigma(g_3)\}$ $\{\sigma(g_2)\}$ $\{\sigma(g_4), \sigma(g_5)\}$ $\{\sigma(g_6)\}$
$Q = 2$: $\{\sigma(g_1), \sigma(g_2), \sigma(g_3), \sigma(g_4), \sigma(g_5)\}, \sigma(g_6)\}$

Let the eccentricity of a simplex with respect to another be:

$$\text{ecc}(\sigma|\sigma') \stackrel{\text{def}}{=} \frac{|\sigma \frown \sigma'|}{|\sigma|} = \frac{\text{number of } \sigma \text{ vertices not shared with } \sigma'}{\text{number of vertices of } \sigma}$$

Let the eccentricity of a simplex with respect to a family of simplices F be

$$\text{ecc}(\sigma|F) \stackrel{\text{def}}{=} \min\{\text{ecc}(\sigma|\sigma')| \ \sigma' \text{ belongs to } F\}$$

Finally, let the eccentricity of a simplicial family F with respect to family F' be:

$$\text{ecc}(F|F') \stackrel{\text{def}}{=} \min\{\text{ecc}(\sigma|\sigma')| \ \sigma' \text{ in F'}\}$$

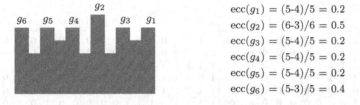

$\text{ecc}(g_1) = (5\text{-}4)/5 = 0.2$
$\text{ecc}(g_2) = (6\text{-}3)/6 = 0.5$
$\text{ecc}(g_3) = (5\text{-}4)/5 = 0.2$
$\text{ecc}(g_4) = (5\text{-}4)/5 = 0.2$
$\text{ecc}(g_5) = (5\text{-}4)/5 = 0.2$
$\text{ecc}(g_6) = (5\text{-}3)/5 = 0.4$

Fig. 4.24 The skyscraper diagram and eccentricitities for the glasses Q-analysis

The Q-analysis suggests that $\sigma(g_2)$ and $\sigma(g_6)$ are less integrated in F than the other simplices. As Figs. 4.23 and 4.24 show, these have the highest eccentricities (0.5 and 0.4 compared to 0.2 for the other simplices.

Neither connectivity nor eccentricity are absolute concepts. Adding a vertex or a simplex to a simplicial family can change either. For example, let F be a simplicial family with vertex set $\{v_0, v_1, ..., v_n\}$. Adding the simplex $\langle v_0, v_1, ..., v_n \rangle$ "swamps" all the other simplices so that they all become faces of this new simplex and all have eccentricity zero.

Similarly, adding vertices can change the structure, *e.g.* adding the vertex $\langle$Sherry_Glass$\rangle$ increases the dimensions of $\sigma(g_1)$, $\sigma(g_5)$, and $\sigma(g_6)$ and changes their connectivity and eccentricities. This illustrates that connectivity is sensitive to the vertices used to represent the system, and using an inappropriate vocabulary to describe a system can cause distortion.

4.11 Q-graphs

Recall from Chapter 2 the relation between eighteen women and fourteen events that they attended. The Q-analysis of $K_{\text{Women}}(\text{Events})$ is given below:

q=7 $\{w_1\}$ $\{w_3\}$ $\{w_{14}\}$
q=6 $\{w_1, w_3\}$ $\{w_2, w_4\}$ $\{w_{13}\}$ $\{w_{14}\}$
q=5 $\{w_1, w_2, w_4, w_3\}$ $\{w_{12}, w_{13}, w_{14}\}$
q=4 $\{w_1, w_2, w_4, w_3\}$ $\{w_{12}, w_{13}, w_{14}\}$ $\{w_{15}\}$
q=3 $\{w_1, w_2, w_4, w_3, w_6, w_{11}, w_9\}$ $\{w_{10}, w_{13}, w_{14}, w_{15}, w_{11}, w_{12}\}$
q=2 $\{w_1, w_2, w_4, w_3, w_5, w_6, w_7, w_9, w_8, w_{14}, w_{10}, w_{13}, w_{15}, w_{11}, w_{12}\}$
q=1, 0 $\{w_1, w_2, w_4, w_3, w_5, w_6, w_7, w_9, w_8, w_{14}, w_{10}, w_{11}, w_{12}, w_{13}, w_{16}, w_{15}, w_{17}, w_{18}\}$

Let W be the set of women and E be the set of events. Then the incidence matrix M of Fig. 2.9 which gives the relation between the women and the events can be written as shown in Fig. 4.25(a). The *transpose matrix* gives the relation between the social events and the women. It is denoted as M^T, and is obtained by flipping the matrix M around, as shown in Fig. 4.25(b).

Social Events

	1	2	3	4	5	6	7	8	9	10	11	12	13	14
w_1	1	1	1	1	1	1	0	1	1	0	0	0	0	0
w_2	1	1	1	0	1	1	1	1	0	0	0	0	0	0
w_3	0	1	1	1	1	1	1	1	1	0	0	0	0	0
w_4	1	0	1	1	1	1	1	1	0	0	0	0	0	0
w_5	0	0	1	1	1	0	1	0	0	0	0	0	0	0
w_6	0	0	1	0	1	1	0	1	0	0	0	0	0	0
w_7	0	0	0	0	1	1	1	1	0	0	0	0	0	0
w_8	0	0	0	0	0	1	0	1	1	0	0	0	0	0
w_9	0	0	0	0	1	0	1	1	1	0	0	0	0	0
w_{10}	0	0	0	0	0	1	1	1	0	0	1	0	0	0
w_{11}	0	0	0	0	0	0	1	1	1	0	1	0	0	0
w_{12}	0	0	0	0	0	0	0	1	1	1	0	1	1	1
w_{13}	0	0	0	0	0	0	1	1	1	1	0	1	1	1
w_{14}	0	0	0	0	0	1	1	0	1	1	1	1	1	1
w_{15}	0	0	0	0	0	0	1	1	0	1	1	1	0	0
w_{16}	0	0	0	0	0	0	0	1	1	0	0	0	0	0
w_{17}	0	0	0	0	0	0	0	1	0	1	0	0	0	0
w_{18}	0	0	0	0	0	0	0	1	0	1	0	0	0	0

Women

	1	2	3	4	5	6	7	8	9	10	11	12	13	14	15	16	17	18
e_1	1	1	0	1	0	0	0	0	0	0	0	0	0	0	0	0	0	0
e_2	1	1	1	0	0	0	0	0	0	0	0	0	0	0	0	0	0	0
e_3	1	1	1	1	1	1	0	0	0	0	0	0	0	0	0	0	0	0
e_4	1	0	1	1	1	0	0	0	0	0	0	0	0	0	0	0	0	0
e_5	1	1	1	1	1	1	1	0	0	0	0	0	0	0	0	0	0	0
e_6	1	1	1	1	0	1	1	1	0	0	0	0	1	0	0	0	0	0
e_7	0	1	1	1	1	0	1	0	1	1	0	0	1	1	1	0	0	0
e_8	1	1	1	1	0	1	1	1	1	1	1	1	1	0	1	1	0	0
e_9	1	0	1	0	0	0	0	1	1	1	1	1	1	1	0	1	1	1
e_{10}	0	0	0	0	0	0	0	0	0	1	1	1	1	1	0	0	0	0
e_{11}	0	0	0	0	0	0	0	0	0	0	0	0	0	1	1	0	1	1
e_{12}	0	0	0	0	0	0	0	0	0	1	1	1	1	1	1	0	0	0
e_{13}	0	0	0	0	0	0	0	0	0	0	1	1	1	1	0	0	0	0
e_{14}	0	0	0	0	0	0	0	0	0	0	1	1	1	1	0	0	0	0

(a) The incidence matrix M (b) The transpose incidence matrix M^T

Fig. 4.25   The incidence matrices of the Southern Women event relations

The Skyscraper diagram for the Q-analysis of $K_{\text{Women}}(\text{Events})$ is given in Fig. 4.26. This shows two groups of the women clustering around w_{14}, w_{13} and w_{12} on the left, and w_4, w_2, w_3 and w_1 on the right. Such diagrams give a good overview of the q-connectivity of simplicial complexes, but of course there are many subtle connectivities that they cannot capture.

Another way to display the connectivity between the simplices of a complex is to use the *q-graph* of the complex. The vertices of the q-graph represent the simplices, and there are edges between pairs of q-near simplices. To do this it is first necessary to compute all the values of q-nearness for all pairs of simplices. This can be achieved by forming the matrix product $M \times M^T$ and subtracting 1 from each entry.

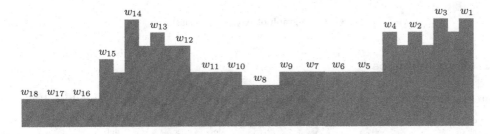

Fig. 4.26 Skyscraper diagram for the Q-analysis of $K_{\text{Women}}(\text{Events})$

	w_1	w_2	w_3	w_4	w_5	w_6	w_7	w_8	w_9	w_{10}	w_{11}	w_{12}	w_{13}	w_{14}	w_{15}	w_{16}	w_{17}	w_{18}
w_1	8																	
w_2	7	7																
w_3	7	6	8															
w_4	6	6	6	7														
w_5	3	3	4	4	4													
w_6	4	4	4	4	2	4												
w_7	3	4	4	4	2	3	4											
w_8	3	2	3	2	0	2	2	3										
w_9	3	3	4	3	2	2	3	2	4									
w_{10}	2	2	3	2	1	1	2	2	3	4								
w_{11}	2	1	2	1	0	1	1	2	2	3	4							
w_{12}	2	1	2	1	0	1	1	2	2	3	4	6						
w_{13}	2	2	3	2	1	1	2	2	3	4	4	6	7					
w_{14}	2	2	2	2	1	1	1	2	2	3	3	5	6	8				
w_{15}	1	2	2	2	1	1	2	1	2	3	3	3	4	4	5			
w_{16}	2	1	2	1	0	1	1	2	2	2	2	2	2	1	1	1		
w_{17}	1	0	1	0	0	0	0	1	1	1	1	1	1	2	1	1	2	
w_{18}	1	0	1	0	0	0	0	1	1	1	1	1	1	2	1	1	2	2

Fig. 4.27 The lower triangular matrix of MM^T shows the number of events shared by the women

Figure 4.28 shows the 3-graph of $K_{\text{Women}}(\text{Events})$. It has vertices those simplicies with dimension 3 or more (*i.e.* four or more vertices).

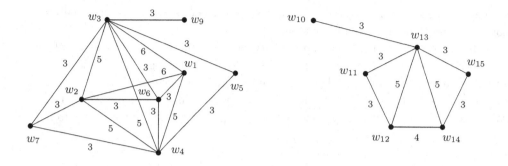

Fig. 4.28 The 3-graph of $K_{\text{Women}}(\text{Events})$

4.12 Stars and hubs

Figures 4.29(a) and (b) show two simplicial families, $F_1 = \{\sigma_{5,1}, \sigma_{5,2}, \sigma_{5,3}\}$ and $F_2 = \{\sigma_{5,4}, \sigma_{5,5}, \sigma_{5,6}\}$, each made of three 5-dimensional simplices. The simplices of F_1 are all pairwise 2-near sharing the triangles $\sigma_{2,1} \stackrel{\text{def}}{=} \sigma_{5,1} \cap \sigma_{5,2}$, $\sigma_{2,2} \stackrel{\text{def}}{=} \sigma_{5,2} \cap \sigma_{5,3}$, and $\sigma_{2,3} \stackrel{\text{def}}{=} \sigma_{5,3} \cap \sigma_{5,1}$.

The simplices of F_2 are also pairwise 2-near sharing the triangles $\sigma_{2,4} \stackrel{\text{def}}{=} \sigma_{5,4} \cap \sigma_{5,5}$, $\sigma'_{2,4} \stackrel{\text{def}}{=} \sigma_{5,5} \cap \sigma_{5,6}$, and $\sigma''_{2,4'} \stackrel{\text{def}}{=} \sigma_{5,6} \cap \sigma_{5,4}$. However, they are also three-wise 2-near since $\sigma_{2,4} = \sigma'_{2,4} = \sigma''_{2,4}$.

Let two q-graphs G and G' be *equivalent* if there is a bijection ϕ between their vertices such that $\langle v_1, v_2 \rangle \in G$ if and only if $\langle \phi(v_1), \phi(v_2) \rangle \in G'$.

As shown in Fig. 4.29(c) and (d) the q-graphs of F_1 and F_2 are equivalent (*e.g.* let $\phi(\sigma_1) = \sigma_4$, $\phi(\sigma_2) = \sigma_5$, and let $\phi(\sigma_3) = \sigma_6$. Also Fig. 4.29(e) and (f) show that the Q-analysis of F_1 is the same as that of F_2. However, these simplicial families have different topologies because the simplices of F_1 form a configuration with a "hole" while those of $F2$ are all connected by the same triangular face, $\sigma_{2,4}$. This common face acts as a *hub* of the star-like configuration.

Figure 4.30(a) shows six 3-simplices as tetrahedra sharing a common triangular face. Figure 4.30(a) shows these simplices brought together into what will be called a *star-hub* configuration. Let F be a simplicial family. *The hub of F* is defined as

$$hub(F) \stackrel{\text{def}}{=} \cap_{\sigma \in F} \sigma$$

When the hub is non-empty, $hub(F) \neq \emptyset$, F is said to be the *star of hub(F)*. More generally, given a face $\langle v_0, ..., v_p \rangle$ of any simplex in F, its *star* is defined as

$$star \langle v_0, ..., v_p \rangle \stackrel{\text{def}}{=} \{\sigma \in F \mid \langle v_0, ..., v_p \rangle \lesssim \sigma\}$$

These definitions allow F to be any simplicial family. Suppose F is a subfamily of a simplicial family $\mathcal{F}$ and that $hub(F) = \langle v_0, ..., v_p \rangle$. Then it is possible that there

Simplicial Complexes and Q-analysis

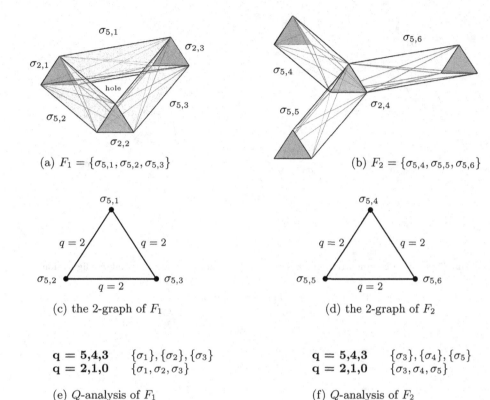

(a) $F_1 = \{\sigma_{5,1}, \sigma_{5,2}, \sigma_{5,3}\}$

(b) $F_2 = \{\sigma_{5,4}, \sigma_{5,5}, \sigma_{5,6}\}$

(c) the 2-graph of F_1

(d) the 2-graph of F_2

| **q = 5,4,3** | $\{\sigma_1\}, \{\sigma_2\}, \{\sigma_3\}$ | **q = 5,4,3** | $\{\sigma_3\}, \{\sigma_4\}, \{\sigma_5\}$ |
| **q = 2,1,0** | $\{\sigma_1, \sigma_2, \sigma_3\}$ | **q = 2,1,0** | $\{\sigma_3, \sigma_4, \sigma_5\}$ |

(e) Q-analysis of F_1

(f) Q-analysis of F_2

Fig. 4.29 Q-graphs cannot discriminate different topologies

exists a simplex σ in $\mathcal{F}$ with $\langle v_0, ..., v_p \rangle$ as a face, but σ does not belong to F. Thus in general

$$F \subseteq \text{star}(\text{hub}(F))$$

and for some F

$$F \subset \text{star}(\text{hub}(F)).$$

For example, in Fig. 4.30, let $\mathcal{F} = \{\sigma_1, \sigma_2, \sigma_3, \sigma_4, \sigma_5, \sigma_6\}$ and let $F = \{\sigma_1, \sigma_2, \sigma_3\}$. Then $\text{hub}(F)$ is the shaded triangle, but $\text{star}((\text{hub}(F))$ also includes the simplices σ_1, σ_2, and σ_3 so that

$$F \subset \{\sigma_1, \sigma_2, \sigma_3, \sigma_4, \sigma_5, \sigma_6\} = \text{star}((\text{hub }(F))$$

When $\text{star}(\text{hub}(F)) = F$ the family F will be called a *maximal star*.

Let $\langle v_0, ..., v_p \rangle$ be any face of a simplex in family F. Then, by definition, $\text{star}(\langle v_0, ..., v_p \rangle) = \{\sigma \,|\, \langle v_0, ..., v_p \rangle \lesssim \sigma\}$. It is possible that $\text{hub}(\{\sigma \,|\, \langle v_0, ..., v_p \rangle \lesssim \sigma\})$ is "larger" than $\langle v_0, ..., v_p \rangle$.

64   Hypernetworks in the Science of Complex Systems

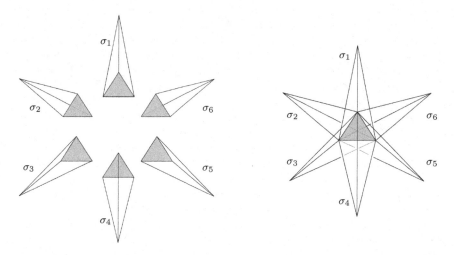

(a) six 3-simplices with a common triangular face   (b) the simplices in a star-hub configuration

Fig. 4.30   A star-hub configuration

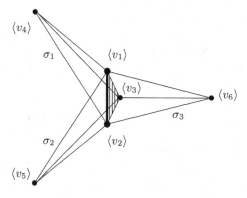

Fig. 4.31 $\langle v_1, v_2 \rangle \lesssim hub(star\langle v_1, v_2 \rangle) = \langle v_1, v_2, v_3 \rangle$

For example, in Figure 4.31 the star of the 1-simplex $\langle v_1, v_2 \rangle$ is $\{\sigma_1, \sigma_2, \sigma_3\}$. But the intersection of these simplices is the triangular face $\langle v_1, v_2, v_3 \rangle$. Thus in general

$$\langle v_0, v_1, ..., v_p \rangle \subset hub(star\langle v_0, v_1, ..., v_p \rangle)$$

When $\langle v_0, v_1, ..., v_p \rangle = hub(star\langle v_0, v_1, ..., v_p \rangle)$, the simplex $\langle v_0, v_1, ..., v_p \rangle$ is said to be a *maximal hub*.

4.13 Galois Families

Let F be a family of simplices, $\{\sigma(a_1), \sigma(a_2), ..., \sigma(a_m)\}$ with vertices $B = \{b_1, b_2, ..., b_n\}$. Let $A = \{a_1, a_2, ..., a_m\}$. Then there is a bipartite relation R between A and B defined as follows

$$a\,R\,b \text{ if } \langle b \rangle \lesssim \sigma(a) \text{ for } a \text{ in } A \text{ and } b \text{ in } B.$$

By an abuse of notation we will write $b\,R\,a$ if $a\,R\,b$. Let A' be any subset of A. Then

$$R(A') \stackrel{def}{=} \cap_{a \in A'} \sigma(a) \stackrel{def}{=} \sigma(A') \stackrel{def}{=} hub(A'),$$

and

$$R^2(A') \stackrel{def}{=} star(hub(A')), \text{ where } A' \subseteq R^2(A').$$

Similarly

$$R(B') \stackrel{def}{=} \{\sigma(a) \mid B' \lesssim \sigma(a)\} \stackrel{def}{=} star(B').$$

and

$$R^2(b') \stackrel{def}{=} hub(star(B')) \text{ where } B' \subseteq R^2(B').$$

The following hold:

For all $A' \subseteq A$, $R^2(A')$ is a maxmal star.

For all $B' \lesssim \sigma(a)$ for any a in A, $R^2(B')$ is a maximal hub.

The maximal stars $R^2(A')$ and maximal hubs $R^2(B')$ are in 1-1 correspondence.

The 1-1 correspondence is $R^2(A') \leftrightarrow R(A')$ or, equivalently, $R(B') \leftrightarrow R^2(B')$ is a *Galois connection* and $R^2(A') \leftrightarrow R(A')$ and $R(B') \leftrightarrow R(B')$ are called *star-hub Galois pairs*. The animal-characteristics relation in Fig. 4.32 has the Galois pair $\langle$brown, vegetarian, quadruped$\rangle \longleftrightarrow \langle$deer, hare, mouse, camel$\rangle$.

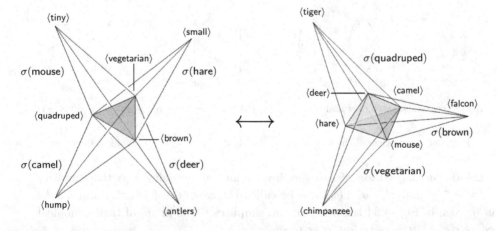

Fig. 4.32 The Galois pair $\langle$brown, vegetarian, quadruped$\rangle \longleftrightarrow \langle$deer, hare, mouse, camel$\rangle$.

4.14 Simplicial prisms

The *prism* between σ and σ', written $\sigma \diamond \sigma'$, is the simplex with the property that $\langle x \rangle \lesssim \sigma \diamond \sigma'$ if and only if $\langle x \rangle \lesssim \sigma$ or $\langle x \rangle \lesssim \sigma'$ or both.

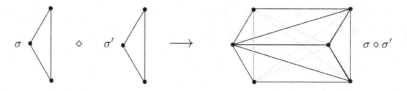

Fig. 4.33 The prism operator

Unless stated otherwise it is assumed that σ, σ' and $\sigma \diamond \sigma'$ all have the same standard vertex ordering (in this book it is assumed that any finite set of vertices is ordered by a *standard ordering*). Then the prism operator has the following properties:

Idempotency: $\quad \sigma \diamond \sigma = \sigma$
Commutativity: $\quad \sigma \diamond \sigma' = \sigma' \diamond \sigma$
Associativity: $\quad (\sigma \diamond \sigma') \diamond \sigma'' = (\sigma \diamond (\sigma' \diamond \sigma''))$

The prism operator also has the property

$$(\sigma \frown \sigma') \diamond (\sigma \cap \sigma') = \sigma$$

When simplices σ and σ' intersect they create what are here called their *eccentric faces*, $\sigma \frown \sigma'$ and $\sigma' \frown \sigma$, as illustrated in Fig. 4.34.

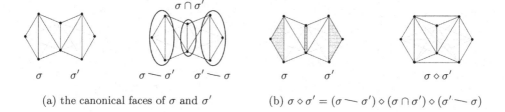

(a) the canonical faces of σ and σ' $\qquad$ (b) $\sigma \diamond \sigma' = (\sigma \frown \sigma') \diamond (\sigma \cap \sigma') \diamond (\sigma' \frown \sigma)$

Fig. 4.34 The prism of the simplices σ and σ' is the prism of their canonical faces

Let the *decomposition* of two simplices σ and σ' be defined as the simplices $\sigma - \sigma'$, $\sigma \cap \sigma'$ and $\sigma' - \sigma$. Let these be called the *canonical faces* of σ and σ'. As can be seen in Fig. 4.34 the prism of two simplices is the prism of their canonical faces, $\sigma \diamond \sigma' = (\sigma \frown \sigma') \diamond (\sigma \cap \sigma') \diamond (\sigma' \frown \sigma)$.

The prism operator works well with intersection and difference, obeying the distributive laws:

(i) $(\sigma \diamond \sigma') \cap \sigma'' = (\sigma \cap \sigma'') \diamond (\sigma' \cap \sigma'')$
(ii) $(\sigma \cap \sigma') \diamond \sigma'' = (\sigma \diamond \sigma'') \cap (\sigma' \diamond \sigma'')$
(iii) $(\sigma \diamond \sigma') \frown \sigma'' = (\sigma \frown \sigma'') \diamond (\sigma' \frown \sigma'')$
(iv) $\sigma \frown (\sigma' \diamond \sigma'') = (\sigma \frown \sigma') \diamond (\sigma \frown \sigma'')$

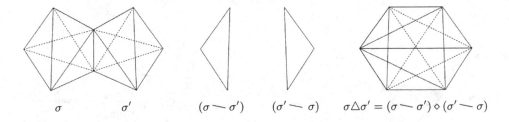

Fig. 4.35 The difference prism, $\sigma \triangle \sigma'$

The *difference prism*, $\sigma \triangle \sigma'$, of two simplices σ and σ' is defined to be the prism between their eccentric faces, $\sigma \triangle \sigma' = (\sigma \frown \sigma') \diamond (\sigma' \frown \sigma)$.

The difference prism has the property that: $\sigma \triangle \sigma' = \sigma \diamond \sigma'$ for $\sigma \cap \sigma' = \emptyset$.

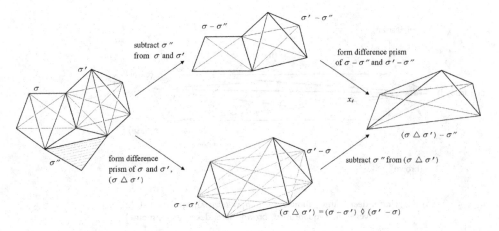

Fig. 4.36 Difference is distributive over the difference prism operator

Assuming standard vertex ordering, it also has the properties:

Nilpotency: $\sigma \triangle \sigma = \emptyset$
Commutativity: $\sigma \triangle \sigma' = \sigma' \triangle \sigma$
Associativity: $(\sigma \triangle \sigma') \triangle \sigma'' = \sigma \triangle (\sigma' \triangle \sigma'')$

68 Hypernetworks in the Science of Complex Systems

Distributivity: $(\sigma \triangle \sigma') - \sigma'' = (\sigma - \sigma'') \triangle (\sigma' - \sigma'')$
Distributivity: $\sigma - (\sigma' \triangle \sigma'') = (\sigma - \sigma'') \triangle (\sigma - \sigma'')$
Distributivity: $(\sigma \triangle \sigma') \cap \sigma'' = (\sigma \cap \sigma'') \triangle (\sigma' \cap \sigma'')$
Distributivity: $\sigma \cap (\sigma' \triangle \sigma'') = (\sigma \cap \sigma') \triangle (\sigma \cap \sigma'')$

4.15 Galois Prisms

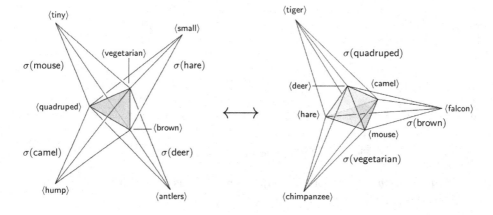

(a) Dual star-hub pairs

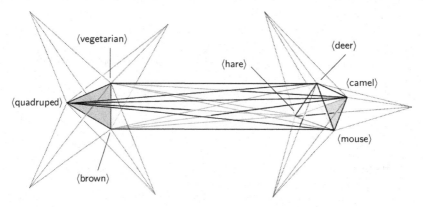

(b) $hub(\{$hare, deer, camel, mouse$\}) \diamond hub\{$vegetarian, quadruped, brown$\}$
= ⟨vegetarian, quadruped, brown, hare, deer, camel, mouse⟩

Fig. 4.37 The Galois prism formed from the hubs of dual stars

The *Galois prism* of a Galois pair $\sigma \leftrightarrow \sigma'$ is defined to be their prism, $\sigma \diamond \sigma'$.

Let R be a relation between sets A and B. Let $A' = \{a'_0, a'_1, ...\}$ and $B' = \{b'_0, b'_1, ...\}$, and let $\langle a'_0, a'_1, ...\rangle \leftrightarrow \langle b'_0, b'_1, ...\rangle$ be a Galois pair. Then the notation

$\langle a'_0, a'_1, ...; b'_0, b'_1, ...\rangle$ will be used for the Galois prism. We will write

$$\langle a'_0, a'_1, ...\rangle \diamond \langle b'_0, b'_1, ...\rangle \stackrel{\text{def}}{=} \langle a'_0, a'_1, ...; b'_0, b'_1, ...\rangle \stackrel{\text{def}}{=} \langle A'; B'\rangle.$$

The expression on the right of this equality is an abuse of notation since a strict substitution of A' and B' would give $\langle A'; B'\rangle = \langle \{a'_0, a'_1, ...\}; \{b'_0, b'_1, ...\}\rangle$ rather than $\langle a'_0, a'_1, ...; b'_0, b'_1, ...\rangle$. It would be possible to embellish the set with a symbol to show that the curly brackets are to be suppressed. For example given $X = \{x_1, x_2, ..., x_n\}$ one could define $\widetilde{X} = x_1, x_2, ..., x_n$ and write the Galois prism as $\langle \widetilde{A'}; \widetilde{B'}\rangle$. However, this makes the notation complicated and the abuse of language, which will be called the *bracket suppression convention*, is here preferred.

Figure 4.37(a) shows the star-hub pairs associated with the Galois pair

⟨hare, deer, camel, mouse⟩ ↔ ⟨vegetarian, quadruped, brown, hare, deer, camel, mouse⟩.

Figure 4.37(b) shows the Galois prism

⟨ hare, deer, camel, mouse ; vegetarian, quadruped, brown, hare, deer, camel, mouse⟩.

4.16 Descriptor Simplices and Antivertices

Let B be a set of birds with $B = \{$robin, ostrich, penguin, magpie, hawk, owl $\}$ and D be a set of *descriptors* with

$d_1 = $ has feathers	$d_2 = $ lays eggs	$d_3 = $ has long legs
$d_4 = $ is brown	$d_5 = $ flies well	$d_6 = $ is black & white
$d_7 = $ swims well	$d_8 = $ has talons	$d_9 = $ has good night vision

The bipartite relation between the birds and the descriptors is given in the table

	d_1	d_2	d_3	d_4	d_5	d_6	d_7	d_8	d_9
magpie	1	1	0	0	1	1	0	0	0
penguin	1	1	0	0	0	1	1	0	0
ostrich	1	1	1	0	0	1	0	0	0
stork	1	1	1	0	1	0	0	0	0
sparrow	1	1	0	1	1	0	0	0	0
falcon	1	1	0	1	1	0	0	1	0
tawny owl	1	1	0	1	1	0	0	1	1

Table 4.4 The relation between the animals and descriptors

Each bird has a *descriptor simplex*, *e.g.* $\sigma(\text{magpie}) = \langle d_1, d_2, d_5, d_6\rangle$. All of these birds share the face $\langle d_1, d_2\rangle$. The first three of these birds are also black and white, so they share the face $\langle d_1, d_2, d_6\rangle$ and they are 2-near. However, they all have discriminating features not possessed by the others: the magpie flies well, the penguin swims well and the ostrich has long legs. They all have dimension 3 and since they are 2-near they have eccentricity $(3 - 2)/(3 + 1) = 0.25$. The stork is 2-near the ostrich with the shared face $\langle d_1, d_2, d_3 \rangle$, and it also has eccentricity 0.25.

For the other birds $\sigma(\text{sparrow}) \lesssim \sigma(\text{falcon}) \lesssim \sigma(\text{tawny owl})$, so the sparrow and falcon have zero eccentricity. Suppose these descriptors were being used to classify the birds. Then falcons and owls would be recognised as sparrows since they have the face $\langle d_1, d_2, d_4, d_5 \rangle$, and owls would also be recognised as falcons due to the shared face $\langle d_1, d_2, d_4, d_5, d_8 \rangle$. To be a satisfactory classifier, the descriptor simplex for a class must be eccentric.

The problem here is caused by information being lost where there are zeros in the matrix. For example, the zero in the d_8 columns means that the sparrow *does not* have talons. The implicit information in the matrix can be made explicit by defining a set of *antivertices*, $\tilde{D} = \{\sim d_1, ..., \sim d_8\}$ with b_i related to $\sim d_j$ if and only if b_i is not related to d_j.

	d_1	d_2	d_3	d_4	d_5	d_6	d_7	d_8	d_9	$\sim d_1$	$\sim d_2$	$\sim d_3$	$\sim d_4$	$\sim d_5$	$\sim d_6$	$\sim d_7$	$\sim d_8$	$\sim d_9$
magpie	1	1	0	0	1	1	0	0	0	0	0	1	1	0	0	1	1	1
penguin	1	1	0	0	0	1	1	0	0	0	0	1	1	1	0	0	1	1
ostrich	1	1	1	0	0	1	0	0	0	0	0	0	1	1	0	1	1	1
stork	1	1	1	0	1	0	0	0	0	0	0	0	1	0	1	1	1	1
sparrow	1	1	0	1	1	0	0	0	0	0	0	1	0	0	1	1	1	1
falcon	1	1	0	1	1	0	1	0	0	0	0	1	0	0	1	1	0	1
tawny owl	1	1	0	1	1	0	0	1	1	0	0	1	0	0	1	1	0	0

Figure 4.5 The relation between the animals and the augmented vertex set

$D \cup \tilde{D}$ will be called the *augmented vertex set*. The Q-analysis of the relation between B and $D \cup \tilde{D}$ is:

$q = 8$
{magpie}, {penguin}, {ostrich}, {stork}, {sparrow}, {falcon}, {tawny owl},

$q = 7$
{magpie}, {penguin}, {ostrich}, {stork}, {sparrow, falcon, tawny owl},

$q = 6$ to 0
{magpie, penguin, ostrich, stork, sparrow, falcon, tawny owl},

Now the sparrow, falcon and tawny owl are eccentric (ecc = $(8\text{-}7)/(7+1) = 0.125$).

Another problem with the original vertex set is that "has feathers" and "lays eggs" are descriptors for all the birds, and that they overwhelm the connectivity of the conjugate structure, with everything else as a face. The matrix on the left below shows the relation between the descriptors and the augmented set of birds, and the Q-analysis is shown on its right.

Simplicial Complexes and Q-analysis

	d_1	d_2	d_3	d_4	d_5	d_6	d_7	d_8	d_9
magpie	1	1	0	0	1	1	0	0	0
penguin	1	1	0	0	0	1	1	0	0
ostrich	1	1	1	0	0	1	0	0	0
stork	1	1	1	0	1	0	0	0	0
sparrow	1	1	0	1	1	0	0	0	0
falcon	1	1	0	1	1	0	0	1	0
tawny owl	1	1	0	1	1	0	0	1	1
~magpie	0	0	1	1	0	0	1	1	1
~penguin	0	0	1	1	1	0	0	1	1
~ostrich	0	0	0	1	1	0	1	1	1
~stork	0	0	0	1	0	1	1	1	1
~sparrow	0	0	1	0	0	1	1	1	1
~falcon	0	0	1	0	0	1	1	0	1
~tawny owl	0	0	1	0	0	1	1	0	0

Q-analysis of $F_B(D)$

q = 6
$\{d_1, d_2\}\ \{d_3\}\ \{d_4\}\ \{d_5\}\ \{d_6\}\ \{d_7\}\ \{d_8\}\ \{d_9\}$

q = 5
$\{d_1, d_2\}\ \{d_4, d_8, d_9\}, \{d_5\}\{d_6\}, \{d_7\}\ \{d_3\}$

q = 4
$\{d_1, d_2\}\ \{d_4, d_8, d_9, d_5, d_6, d_7\}\ \{d_3\}$

q = 3 to 0
$\{d_1, d_2, d_4, d_8, d_9, d_5, d_6, d_7, d_3\}$

Table 4.6 The relation between the augmented set of animals and descriptors

With the augmentation by the antivertices, the descriptors d_1 and d_2 no longer dominate the structure, e.g. there is 5-component $\{d_4, d_8, d_9\}$. This has the associated Galois pair $\langle d_4, d_8, d_9 \rangle \leftrightarrow \langle$tawny owl, ~magpie, ~penguin, ~ostrich, ~stork$\rangle$, where d_4 = brown, d_8 =has talons and d_9 = has good night vision.

Let M be the incidence matrix of the relation between B and D and $\tilde{M}$ be the matrix with $\tilde{m}_{i,j} = 1$ if $m_{i,j} = 0$ and $\tilde{m}_{i,j} = 0$ if $m_{i,j} = 1$. Then the augmented matrices are shown in Fig. 4.38.

Fig. 4.38 The augmented matrices for the bird–descriptor relation

4.17 Networks and Simplicial Complexes

Incidence matrices can be used to produce both networks and graphs. Figure 4.39 illustrates the relationship between network and simplicial representations:

Figure 4.39(a) shows part of an incidence matrix. x_1 is related to y_1, y_2 and y_3. x_2 is related to y_1, y_2 and y_3.

Figure 4.39(b) shows the network edges for x_1, while Fig. 4.39(c) shows the pair $\langle x_1 \rangle \leftrightarrow \langle y_1, y_2, y_3 \rangle$.

Figure 4.39(d) shows the prism $\langle x_1 \rangle \diamond \langle y_1, y_2, y_3 \rangle$. The x_1 network edges are edges on this prism.

72   Hypernetworks in the Science of Complex Systems

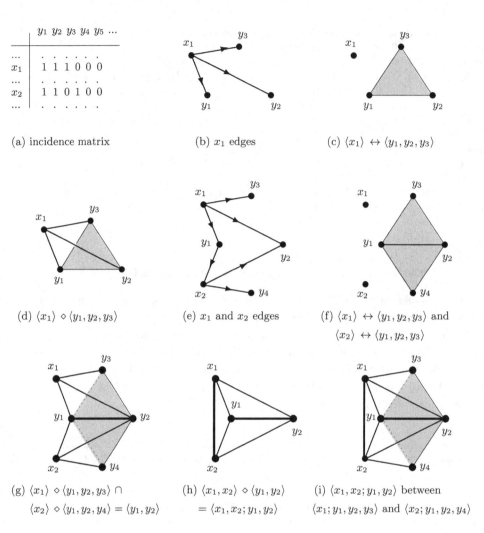

Fig. 4.39   Network edges, simplices and prisms of bipartite relations

Figure 4.39(e) shows the network edges for x_1 and x_2. Figure 4.39(f) shows the pairs $\langle x_1 \rangle \leftrightarrow \langle y_1, y_2, y_3 \rangle$ and $\langle x_2 \rangle \leftrightarrow \langle y_1, y_2, y_4 \rangle$. Note x_1 and x_2 both have edges with y_1 and y_2. In contrast $\sigma(x_1)$ and $\sigma(x_2)$ are 1-near, sharing the edge $\langle y_1, y_2 \rangle$.

Figure 4.39(g) shows the prisms $\langle x_1 \rangle \diamond \langle y_1, y_2, y_3 \rangle$ and $\langle x_2 \rangle \diamond \langle y_1, y_2, y_4 \rangle$ with the network structure on its 1-dimensional faces. The intersection of the prisms is $\langle y_1, y_2 \rangle$.

Figure 4.39(h) shows the Galois prism of $\langle x_1, x_2 \rangle$ and $\langle y_1, y_2 \rangle$ and Fig. 4.39(i) shows how this prism sits in the structure of Fig. 4.39(g).

4.18 Examples

4.18.1 Example: Sky and Water

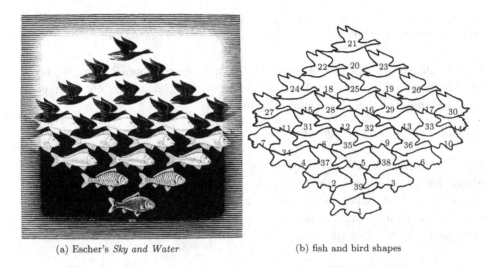

(a) Escher's *Sky and Water*   (b) fish and bird shapes

Fig. 4.40 The shapes and features abstracted from Escher's *Sky and Water*

Figure 4.40(a) shows Escher's picture *Sky and Water* in which the birds at the top of the picture seems to change into fish at the bottom. Figure 4.40(b) shows the various shapes that appear in the picture. The table below shows a relation between the shapes and a set of twelve descriptors.

Shapes:	1	2	3	4	5	6	8	9	10	11	12	13	7	21	22	23	24	25	26	28	29	27	31	32	33	30	34	35	36	37	38	14	15	16	17	18	19	20	39
scales	1	1	1	1	1	1	0	0	0	0	0	0	0	0	0	0	0	0	0	0	0	0	0	0	0	0	0	0	0	0	0	0	0	0	0	0	0	0	0
mouth	1	1	1	1	1	1	1	1	1	0	0	0	1	0	0	0	0	0	0	0	0	0	0	0	0	0	0	0	0	0	0	0	0	0	0	0	0	0	0
gills	1	1	1	1	1	1	1	1	1	0	0	0	1	0	0	0	0	0	0	0	0	0	0	0	0	0	0	0	0	0	0	0	0	0	0	0	0	0	0
fish-tail	1	1	1	1	1	1	1	1	1	1	1	1	0	0	0	0	0	0	0	0	0	0	0	0	0	0	0	0	0	0	0	0	1	0	0	0	0	0	0
fins	1	1	1	1	1	1	1	1	1	1	1	1	1	0	0	0	0	0	0	0	0	0	0	0	0	0	0	0	0	0	0	0	0	0	0	0	0	0	0
fish-shape	1	1	1	1	1	1	1	1	1	1	1	1	0	0	0	0	0	0	0	0	0	0	0	0	0	0	0	0	0	0	0	0	0	0	1	1	1	1	0	0
eye	1	1	1	1	1	1	1	1	1	1	1	1	1	1	1	1	1	1	1	1	1	1	1	1	1	1	0	0	0	0	0	0	0	0	0	0	0	0	0
duck-shape	0	0	0	0	0	0	0	0	0	0	0	0	0	1	1	1	1	1	1	1	1	1	1	1	1	0	1	1	1	1	1	0	0	0	0	0	0	0	0
two-wings	0	0	0	0	0	0	0	0	0	0	0	0	0	1	1	1	1	1	1	1	1	1	1	1	1	1	0	0	0	0	0	0	0	0	0	0	0	0	0
feathers	0	0	0	0	0	0	0	0	0	0	0	0	0	1	1	1	1	1	1	1	1	1	1	1	1	1	0	0	0	0	0	0	0	0	0	0	0	0	0
beak	0	0	0	0	0	0	0	0	0	0	0	0	0	1	1	1	1	1	1	1	1	1	1	1	1	1	0	0	0	0	0	0	0	0	0	0	0	0	0
legs	0	0	0	0	0	0	0	0	0	0	0	0	0	1	1	1	1	1	1	1	1	1	1	1	1	0	0	0	0	0	0	0	0	0	0	0	0	0	0

Table 4.7 The relation between descriptors and shapes for Escher's *Sky and Water*

Inspection of the incidence matrix in Table 4.7 reveals a number of maximal rectangles corresponding to star-hub Galois pairs, including:

⟨1, 2, 3, 4, 5, 6⟩ ⟷ ⟨scales, mouth, gills, fish-tails, fins, fish-shape, eye⟩
⟨1, 2, 3, 4, 5, 6, 8, 9, 10, 11, 12, 13⟩ ⟷ ⟨fish-tails, fins, fish-shape, eye⟩

⟨21, 22, 23, 24, 25, 26, 28, 29⟩ ⟷ ⟨eye, duck-shape, two-wings, feathers, beak, legs⟩
⟨21, 22, 23, 24, 25, 26, 28, 29, 27, 31, 32, 33⟩ ⟷ ⟨eye, duck-shape, two-wings⟩

Of course there are many more Galois pairs than this, *e.g.*

⟨1, 2, 3, 4, 5, 6, 8, 9, 10⟩ ⟷ ⟨mouth, gills, fish-tails, fins, fish-shape, eye⟩
⟨21, 22, 23, 24, 25, 26, 28, 29, 27⟩ ⟷ ⟨eye, duck-shape, two-wings, feathers, beak⟩

Some of the columns of the incidence matrix have been swapped to make the maximal rectangles more obvious. Even so there are other Galois pairs not forming maximal rectangles in this version, for example

⟨1, 2, 3, 4, 5, 6, 8, 8, 10, 7⟩ ⟷ ⟨mouth, gills, fins, eye⟩

Figure 4.41(a) shows the skyscraper diagram for the Q-analysis of the shape-descriptor family. As can be seen, the shapes fall into two major components corresponding to bird shapes and the fish shapes. Figure 4.41(b) shows the conjugate Q-analysis with σ(eye) having the largest dimension ($q = 24$) followed by σ(duck-shape) and σ(fish-shape) at $q = 16$.

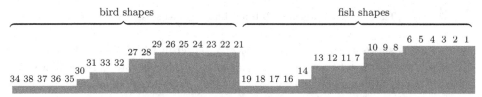

(a) Skyscraper diagram for the Q-analysis of F_{Shapes}(Descriptors).

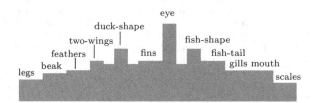

(b) Skyscraper diagram for the Q-analysis of $F_{\text{Descriptors}}$(Scales).

Fig. 4.41 The skyscraper diagrams for the shapes – features Q-analyses

Removing the "eye" descriptor creates two disconnected subfamilies, one with fish shapes and the other with duck shapes. Thus the transition from ducks at the top of Escher's picture to the fish at the bottom does not involve morphing from one shape to the other. Instead the picture is tiled by shapes, half of which get more duck-like towards the top and half of which get more fish-like towards the bottom.

4.18.2 Example: Discriminating textured surfaces

Figure 4.42 shows digital images of five different surfaces, paving slabs, large stones, concrete, small stones, and tarmac. There are four samples of each and the objective is to classify the images on the basis of a set of measurements. A first approach to analysing these images involves a statistical analysis of the pixel greyscale values, where black pixels have a greyscale of 0 and white pixels have a greyscale of 255. The distributions of the pixels is shown in Fig. 4.43.

This greyscale analysis shows that greyscale frequency of the paving slab pixels peak around 141 to 149 while that for concrete peaks around 136 to 170. The other surfaces have much greater non-normal greyscale spreads of across the whole 0–255 range. The small and large stone images having peaks at the highest greyscale because some of the stones are white. The standard deviations of the distributions

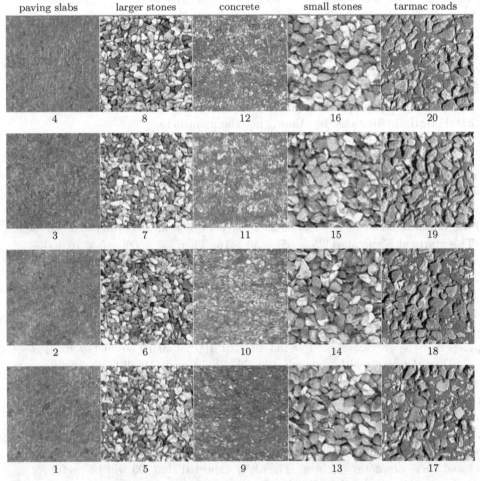

Fig. 4.42 Samples of textured road and pedestrian surfaces

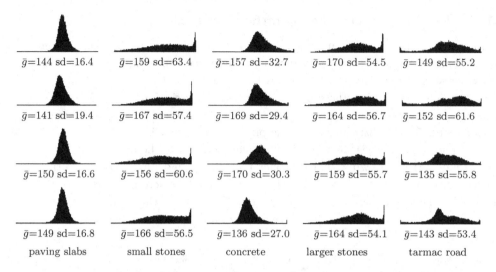

Fig. 4.43   Pixel greyscale distributions for the surface in Figure 4.42

are in the range 16–20 for the paving slabs and 27–33 for the concrete surfaces, while for the others the standard deviations are much higher. Thus, in principle the paving slabs and concrete can be discriminated from the others on the basis of their statistical distributions, but that is not the point here.

Ten measurements were made for each image. The first two are the mean greyscale and its standard deviation. The next two measurements concern "runs" of similar pixels. A *run* of pixels is a set of contiguous pixels along a row or a column. Descriptor D3 counts the number of vertical runs of pixels which pairwise have a greyscale difference less than 16. For example, if a column of pixels in an image had the values 25 55 60 59 70 88 90 80 33 20 ... D3 would identify the run 55 60 59 70. This starts at 55 because $|25 - 55| = 30 > 16$. Then $|55 - 60| = 5 < 16$, $|60 - 59| = 1 < 16$, $|59 - 70| = 11 < 16$. The run ends here because $|70 - 88| = 18 > 16$. This is followed by the run 88 90 80, and another run starts at 20 ...

Figure 4.44 shows the vertical runs for the twenty surfaces. The ends of the runs are shown as black pixels. In high contrast regions it can happen that adjacent pixels are black because they form single pixel runs. For example, suppose a vertical row of pixels had the greyscales ... 150 153 157 151 100 70 40 10 50 80 110 153 157 151 ... Then the pixels with greyscale 100, 70, 10, 50, and 110 all have differences greater than 16 with both neighbours, and they each form a single pixel run and are displayed in black. This accounts for the black areas seen especially in the second, fourth and fifth columns of surfaces. In the first column the surfaces have relatively low contrast and the runs of similar pixels are relatively long so there are fewer of them. By comparison, in the second columns the runs of similar pixels are relatively short and there are many of them. Thus it is expected that D3 will be useful for discriminating the paving slabs from the surface made up of large stones.

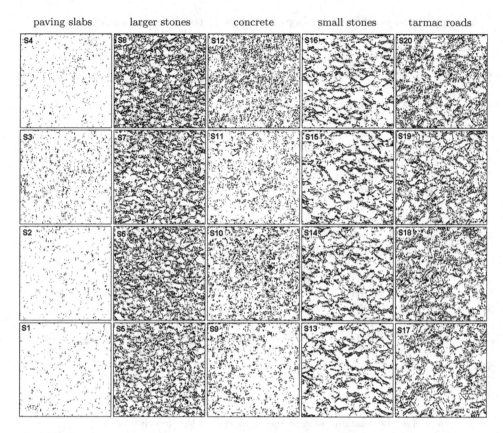

Fig. 4.44 Low contrast vertical runs greater than one pixel in length shown in white

Descriptor D4 is similar, but uses a less demanding similarity value of 20. The other descriptors count other things that are easy to find in the images. There is no guarantee that any of these descriptors is useful for clustering the surfaces, and the question addressed here is whether clusters can be found when some of the descriptors may provide little useful information or even act as noise. The descriptors D1 to D10 are defined below:

Descriptor 1: mean greyscale
Descriptor 2: standard deviation
Descriptor 3: number of long vertical runs with greyscale differences less than 16
Descriptor 4: number of long vertical runs with greyscale differences less than 20
Descriptor 5: number of descriptor 3 runs > 16 pixels long with mean greyscale > 200
Descriptor 6: number of horizontal locally dark pixels with greyscale > 230
Descriptor 7: number of pixels with greyscale less than 24
Descriptor 8: number of pixels with greyscale equal to a horizontal neighbour
Descriptor 9: number of pixels with greyscale equal to their right neighbour
Descriptor 10: number of pixels with greyscale less than their left and right next-but-one neighbours with contrast greater than 16.

When applied to the twenty surfaces in Fig. 4.42 the following table of values is obtained. This table has a number of striking features, including the large differences in the lengths of the scales. The shortest scale is D2 which has values 16 to 55 while the longest is D3 with range 3953 to 19282.

Surface	D1	D2	D3	D4	D5	D6	D7	D8	D9	D10
S1	149	16	4002	7420	0	0	0	3496	1900	1947
S2	150	16	3953	6414	0	0	0	3752	1950	1583
S3	141	19	7110	10263	0	0	0	2829	1475	2542
S4	144	16	4126	8436	0	0	0	3375	1872	2425
S5	166	56	16425	19517	6	5410	136	2154	1179	2928
S6	156	60	18072	20166	1	4680	536	2114	1121	3143
S7	167	57	16874	19188	11	6147	202	2523	1434	2538
S8	159	63	19282	20618	1	5620	723	2151	1144	3446
S9	136	27	9477	10016	0	135	0	2766	1296	1538
S10	170	30	14559	13693	3	1253	0	2162	1013	2073
S11	169	29	9668	13711	0	1567	0	2482	1313	3275
S12	157	32	15083	17081	0	1207	0	1864	983	4495
S13	164	54	12658	14498	66	3906	317	2993	1569	1097
S14	159	55	13640	14689	40	4095	327	2751	1404	1310
S15	164	56	12958	14908	88	4287	444	3064	1686	914
S16	170	54	12971	14135	86	5547	183	3715	2044	873
S17	143	53	13733	18219	14	1643	1128	2161	1207	3505
S18	135	55	15908	19315	0	1132	2212	1939	1124	4000
S19	152	61	14999	20124	3	2587	1848	1920	1128	3310
S20	149	55	16320	19550	0	2063	1397	1944	1089	4130

Table 4.8 The relation between the surfaces and the descriptors

These descriptors may seem rather arbitrary, and indeed they are. When a programmer tries to discriminate objects in images they try to find descriptors that characterise one or more classes. Our descriptors are typical of such an attempt. For example, a small greyscale variance indicates the paving slabs, while a large number is associated with the small stones.

One of the problems in pattern recognition is that the samples of a particular class may vary considerably. For example, paving slab S9 (mean greyscale 136) is darker than S10, and S11(mean greyscales 170 and 169) while S12 is between them (greyscale 157). As shown in Fig. 4.45 S9 is at the far left of the D1 scale while D10 and D11 are at the far right, so this particular dimension is not useful for clustering the concrete samples, D9, D10, D11 and D12. In general, as Fig. 4.45 shows, there is no single descriptor that obviously characterises a particular class, but combinations of descriptors can.

If one has prior knowledge of the classes it is relatively easy to partition the scales so that surfaces of the same type appear in the same sub-intervals. For example, paving slabs S1 S2, S3 and S4 are in the D2 interval 16–19, while concrete surfaces S9, S10, S11 and S12 are all in the interval 27–32. All the other surfaces are mixed

Simplicial Complexes and Q-analysis

up higher up the scale. Less obviously, the small stones surfaces S13, S14, S15 and S16 all uniquely occupy the D5 interval 40–68, while the tarmac road surfaces S17, S18, S19 and S20 uniquely occupy the D7 interval 1128–2212.

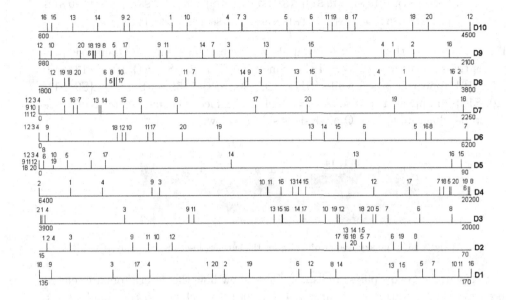

Fig. 4.45 Surface numbers arranged along the descriptor

The set of subinterval descriptors in the table below was abstracted from Fig. 4.45. For example D1_140–155 means the interval [140, 155] for descriptor D1. The table shows the relationship between these intervals and the surfaces.

	S1	S2	S3	S4	S5	S6	S7	S8	S9	S10	S11	S12	S13	S14	S15	S16	S17	S18	S19	S20
D1_140–155	1	1	1	1	0	0	0	0	0	0	0	0	0	0	0	0	1	0	1	1
D2_0–19	1	1	1	1	0	0	0	0	0	0	0	0	0	0	0	0	0	0	0	0
D2_20–32	0	0	0	0	0	0	0	0	1	1	1	1	0	0	0	0	0	0	0	0
D3_0–7500	1	1	1	1	0	0	0	0	0	0	0	0	0	0	0	0	0	0	0	0
D3_7501–10000	0	0	0	0	0	0	0	1	0	0	0	0	0	0	0	0	0	0	0	0
D3_16400–20000	0	0	0	0	1	1	1	1	0	0	0	0	0	0	0	0	0	0	0	0
D4_0–9000	1	1	0	1	0	0	0	0	0	0	0	0	0	0	0	0	0	0	0	0
D4_10301–14000	0	0	0	0	0	0	0	0	0	1	1	0	0	0	0	0	0	0	0	0
D4_14001–15000	0	0	0	0	0	0	0	0	0	0	0	1	1	1	1	0	0	0	0	0
D4_18000–21000	0	0	0	0	0	0	0	0	0	0	0	0	0	0	0	0	1	1	1	1
D5_16–100	0	0	0	0	0	0	0	0	0	0	0	0	1	1	1	1	0	0	0	0
D6_0–0	1	1	1	1	0	0	0	1	0	0	0	0	0	0	0	0	0	0	0	0
D6_100–2600	0	0	0	0	0	0	0	1	1	1	1	0	0	0	0	1	1	1	1	1
D6_3000–7000	0	0	0	0	1	1	1	0	0	0	0	1	1	1	1	0	0	0	0	0
D7_0–0	1	1	1	1	0	0	0	0	1	1	1	0	0	0	0	0	0	0	0	0
D7_100–750	0	0	0	0	1	1	1	1	0	0	0	0	1	1	1	1	0	0	0	0
D7_1000–2000	0	0	0	0	0	0	0	0	0	0	0	0	0	0	0	0	1	1	1	1
D8_0–2750	0	0	0	0	1	1	1	1	0	1	1	1	0	0	0	0	1	1	1	1
D8_2751–4000	1	1	1	1	0	0	0	0	1	0	0	0	1	1	1	1	0	0	0	0
D9_0–1210	0	0	0	0	1	1	1	1	0	1	1	1	0	0	0	0	1	1	1	1
D9_1400–2100	1	1	1	1	0	0	0	0	0	0	0	1	1	1	1	0	0	0	0	0
D10_100–1320	0	0	0	0	0	0	0	0	0	0	0	1	1	1	1	0	0	0	0	0

Table 4.9 The relation between the surfaces and the subinterval descriptors

80 Hypernetworks in the Science of Complex Systems

$q = 7$: {S1 S2 S4}, {S3}

$q = 6$: {S1 S2 S3 S4}, {S13 S14 S15 S16}

$q = 5$: {S1 S2 S3 S4}, {S13 S14 S15 S16}, {S17 S19 S20}, {S5 S6 S8} {S7} {S9} {S10}, {S11}

$q = 4, 3$: {S1 S2 S3 S4}, {S13 S14 S15 S16}, {S17 S19 S20 S18} {S5 S6 S7 S8}, {S9 S11 S10 S12}

$q = 2, 0$: {S1 S2 S3 S4 S9 S10 S11 S12 S17 S18 S19 S20 S5 S6 S8 S7 S13 S14 S15 S16}

The Q-analysis of this matrix is given above. As can be seen, the Q-analysis partitions the surfaces into classes of the same type, {S1, S2, S3, S4} and {S13, S14, S15, S16} appear at $q = 6$ while {S17, S19, S20} {S18} {S5, S6, S7, S8}, {S9, S11, S10, S12} all appear at $q = 4$. At $q = 2$ all the surfaces become 2-connected. The skyscraper diagram for the Q-analysis is shown below.

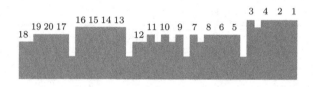

Fig. 4.46 The skyscraper diagram for the surface-descriptor Q-analysis

Below, the q-Galois pairs are listed. These show how the descriptors are related to clusters of the surfaces, e.g at $q = 6$ D6_3001+ is related to the small stones, S5 S6 S7 S8, and the large stones, S13 S14 S15 S16. At $q = 5$ all the descriptors ⟨D4_18001+ D8_0–2750 D9_0–1210 D6_100–2600⟩ are related to all of the tarmac road surfaces ⟨S17 S18 S19 S20⟩. These Galois pairs give insights into which descriptors are most effective at identifying the classes.

$q = 10$ ⟨D8_0–2750⟩ ↔ ⟨S5 S6 S7 S8 S10 S11 S12 S17 S18 S19 S20⟩
$q = 8$ ⟨D8_2751+ ⟩ ↔ ⟨S1 S2 S3 S4 S9 S13 S14 S15 S16⟩
$q = 8$ ⟨D9_1401+⟩ ↔ ⟨S1 S2 S3 S4 S7 S13 S14 S15 S16⟩
$q = 8$ ⟨D8_0–2750 D9_0–1210⟩ ↔ ⟨S5 S6 S8 S10 S12 S17 S18 S19 S20⟩
$q = 7$ ⟨D7_0–0⟩ ↔ ⟨S1 S2 S3 S4 S9 S10 S11 S12 ⟩
$q = 7$ ⟨D8_2751+ D9_1401+⟩ ↔ ⟨S1 S2 S3 S4 S9 S13 S14 S15 S16 ⟩
$q = 7$ ⟨D4_18000 D8_0–2750 D9_0–1210⟩ ↔ ⟨ S5 S6 S8 S17 S18 S19 S20⟩
$q = 7$ ⟨D6_3001+ D7_100–750⟩ ↔ ⟨S5 S6 S7 S8 S13 S14 S15 S16⟩
$q = 7$ ⟨D6_100–2600 ⟩ ↔ ⟨S9 S10 S11 S12 S17 S18 S19 S20⟩
$q = 6$ ⟨ D1_140–155⟩ ↔ ⟨S1 S2 S3 S4 S17 S19 S20⟩
$q = 6$ ⟨D7_0–0⟩ ↔ ⟨S1 S2 S3 S4 S9 S10 S11 S12⟩
$q = 6$ ⟨D8_2751+ D9_1400+} ↔ ⟨S1 S2 S3 S4 S13 S14 S15 S16 ⟩
$q = 6$ ⟨D4_18000 D8_0–2750 D9_0–1210 D6_100–2600⟩ ↔ ⟨S17 S18 S19 S20⟩
$q = 6$ ⟨D6_3001+ D7_100–750⟩ ↔ ⟨S5 S6 S7 S8 S13 S14 S15 S16⟩
$q = 5$ ⟨D1_140–155⟩ ↔ ⟨S1 S2 S3 S4 S17 S19 S20⟩
$q = 5$ ⟨D7_0–0⟩ ↔ ⟨S1 S2 S3 S4 S9 S10 S11 S12⟩
$q = 5$ ⟨D8_2751+ D9_1401+⟩ ↔ ⟨S1 S2 S3 S4 S13 S14 S15 S16⟩
$q = 5$ ⟨D4_18001+ D8_0–2750 D9_0–1210 D6_100–2600⟩ ↔ ⟨S17 S18 S19 S20⟩
$q = 5$ ⟨D6_3001+ D7_100–750⟩ ↔ ⟨S5 S6 S7 S8 S13 S14 S15 S16⟩
$q = 4$ ⟨D1_140–155⟩ ↔ ⟨S1 S2 S3 S4 S17 S19 S20 1⟩
$q = 4$ ⟨D4_18001+ D8_0–2750 D9_0–1210 D6_100–2600⟩ ↔ ⟨S17 S18 S19 S20 ⟩

4.18.3 Example: Random Q-analysis

It is interesting to consider the Q-analysis of $H_A(B,R)$ when the relation R between sets A and B is random with no underlying structure. To investigate this, Fig. 4.47 shows the hypernetwork $H_A(B,R)$ for a random relation R between sets A and B which each have 100 elements. Here the probability of a being related to b is $1/2$.

dimension q	23	24	25	26	27	28	29	30	31	32	33	34	35	36	37	38	39	40	41
$\check{q}$ frequency	1	1	3	2	3	8	6	9	16	16	12	9	6	4	4	0	0	0	0
$\hat{q}$ frequency	0	0	0	0	0	0	0	0	0	0	0	0	0	1	1	1	3	2	0

dimension q	42	43	44	45	46	47	48	49	50	51	52	53	54	55	56	57	58	59	60
$\check{q}$ frequency	0	0	0	0	0	0	0	0	0	0	0	0	0	0	0	0	0	0	0
$\hat{q}$ frequency	2	4	5	8	8	4	14	4	10	5	7	5	6	6	3	0	0	0	1

The list above shows the frequency of the dimensions, $\hat{q}$, of the random simplices and the frequencies of the dimensions of their largest shared faces, $\check{q}$. Figure 4.47(a) shows these frequencies as a histogram and Fig. 4.47 shows the associated skyscraper diagram for the Q-analysis of $H_A(B,R)$.

(b) the skyscraper diagram for the Q-analysis of $H_A(B,R)$

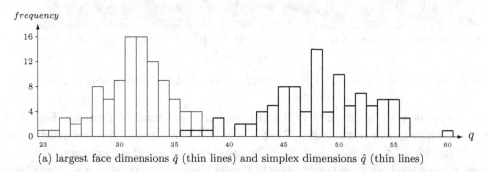

(a) largest face dimensions $\check{q}$ (thin lines) and simplex dimensions $\hat{q}$ (thin lines)

Fig. 4.47 Q-analysis of random hypernetwork, $H_A(B,R)$, $|A|=100$, $|B|=100$ $\mathrm{pr}(a_i\,R\,b_j)=1/2$

The overall pattern is that all the simplices have eccentricity between 0.3 and 0.4, and that a little structure appears at $q = 37$ with many simplices collapsing into a large component at $q = 34$ until the whole structure becomes connected as one component at $q = 23$. In other words, the random Q-analysis does not show the structural features found in real systems.

Figure 4.48 shows the skyscraper diagrams for $H_A(B; R_n)$ where $a_i\ R_n\ b_j$ with probability $1/n$ for $n = 5, 10, 20, 30, 40$ and 50. The dimensions drop as the probability $1/n$ decreases. However in all cases the simplices are relatively eccentric but form a large component for a relatively low value of q.

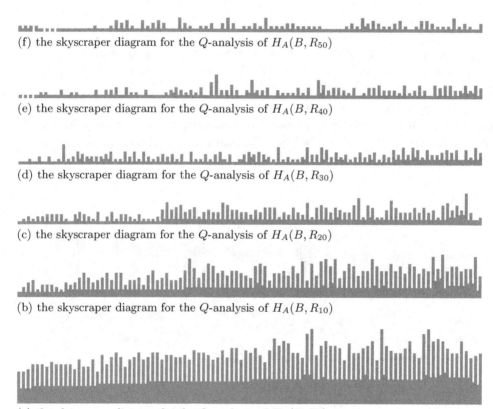

(f) the skyscraper diagram for the Q-analysis of $H_A(B, R_{50})$

(e) the skyscraper diagram for the Q-analysis of $H_A(B, R_{40})$

(d) the skyscraper diagram for the Q-analysis of $H_A(B, R_{30})$

(c) the skyscraper diagram for the Q-analysis of $H_A(B, R_{20})$

(b) the skyscraper diagram for the Q-analysis of $H_A(B, R_{10})$

(a) the skyscraper diagram for the Q-analysis of $H_A(B, R_5)$

Fig. 4.48   Random Q-analysis: $|A| = 100$, $|B| = 100$, and $a_i\ R_n\ b_j$ with probability $1/n$

In the previous examples the sets A and B each had one hundred elements. Figure 4.49 shows a Q-analysis in which B has a thousand elements and the probability of a_i being related to b_j is et to $1/10$. The same general structure emerges. All the simplices have high eccentricity, mostly greater than 0.8, associated with them having high dimensions but being relatively disconnected.

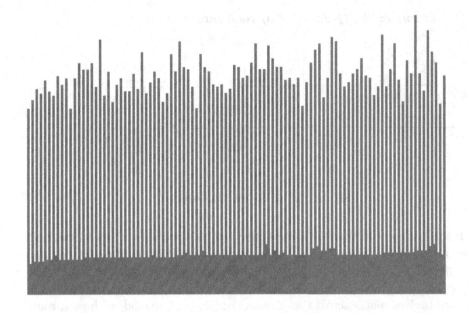

Fig. 4.49 Random Q-Analysis: $|A| = 100$, $|B| = 1000$, and $a_i \; R_n \; b_j$ with probability $1/10$

Figure 4.50 shows the structure vector of $H_A(B, R)$. The components above $q=30$ contain just one simplex. The highest dimensional component appears at $q = 128$ and the number of components grows to 100 as the dimension decreases to $q = 85$. This number of single-simplex components remains until two of the simplices become connected at $q = 22$. As the dimension decreases the number of components rapidly collapses until the simplices all form a single component at $q = 13$.

Generally the structure of random hypernetworks follows some simple rules. Unsurprisingly, as the relational probability increases the dimensions of the simplices and shared faces. This is related to the number of elements in the sets A and B, with $H_A(B, R)$ being less highly connected as the number of elements in B increases. The most striking feature of random hypernetworks is that pairs or subsets of simplices do not become highly connected for relatively high values of q.

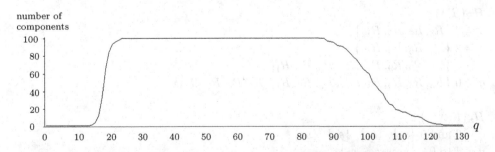

Fig. 4.50 Random Q-analysis: $|A| = 100$, $|B| = 1000$, and $a_i \; R_n \; b_j$ with probability $1/10$

4.18.4 Example: the Q-analysis of road intersections

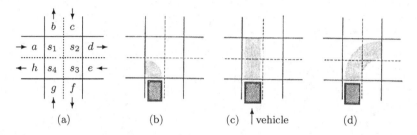

Fig. 4.51 A narrow road intersection

All the examples in this section assume vehicles travel on the left of the road as they do in the UK. Figure 4.51(a) shows a narrow road intersection divided into four *segments*, s_1, s_2, s_3, and s_4. There are four places where vehicles can enter the intersection, a, c, e and g, and four places where vehicles can leave, b, d, f and h. There are twelves *routes* across this intersection ab, ac, ... gb, gd, with each route defined to be λ-related to the segments it crosses. For example, route R_1 between a and b traverses just road segment s_1, while route R_2 between a and d traverses s_2 and s_3. This route can be considered to be a simplex $\langle s_2, s_3; \lambda \rangle$. Because the intersection is narrow, right turn vehicles occupy all of the segments, for example the route R_3 between a and f is associated with the simplex $\langle s_1, s_2, s_3, s_4; \lambda \rangle$.

The incidence matrices of the relation λ and its Q-analysis are shown below.

Route	O-D pair	s_1	s_2	s_3	s_4	Route	O-D pair	s_1	s_2	s_3	s_4
R_1	ab	1	0	0	0	R_7	ef	0	0	1	0
R_2	ad	1	1	0	0	R_8	eh	0	0	1	1
R_3	af	1	1	1	1	R_9	eb	1	1	1	1
R_4	cd	0	1	0	0	R_{10}	gh	0	0	0	1
R_5	cf	0	1	1	0	R_{11}	gb	1	0	0	0
R_6	ch	1	1	1	1	R_{12}	gd	1	1	1	1

$H_R(S, \lambda)$
$q = 3 \ \{R_3, R_6, R_9, R_{12}\}$
$q = 2 \ \{R_3, R_6, R_9, R_{12}\}$
$q = 1 \ \{R_3, R_6, R_9, R_{12}, R_2, R_5, R_8, R_{11}\}$
$q = 0 \ \{R_3, R_6, R_9, R_{12}, R_2, R_5, R_8, R_{11}, R_1, R_4, R_7, R_{10}\}$

$H_S(R, \lambda^{-1})$
$q = 6$ and 5 $\{s_1\}, \{s_2\}, \{s_3\}, \{s_4\}$
$q = 4$ to 0 $\{s_1, s_2, s_3, s_4\}$

Figure 4.52 shows a wider intersection. In the previous case only one vehicle could use the intersection at a time, but in this case it is possible, for example, for a vehicle to turn right from g to d at the same time a vehicle is turning left from e to f. This is reflected in the connectivity structure.

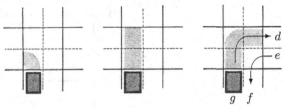

Fig. 4.52 A wide road intersection

The incidence matrices of the relation λ, its Q-analysis are shown below.

Route	O-D pair	s_1	s_2	s_3	s_4
R_1	ab	1	0	0	0
R_2	ad	1	1	0	0
R_3	af	1	1	1	0
R_4	cd	0	1	0	0
R_5	cf	0	1	1	0
R_6	ch	0	1	1	1

Route	O-D pair	s_1	s_2	s_3	s_4
R_7	ef	0	0	1	0
R_8	eh	0	0	1	1
R_9	eb	1	0	1	1
R_{10}	gh	0	0	0	1
R_{11}	gb	1	0	0	0
R_{12}	gd	1	1	0	1

$H_R(S, \lambda)$
$q = 2 \; \{R_3\}, \{R_6\}, \{R_9\}, \{R_{12}\}$
$q = 1 \; \{R_3, R_6, R_9, R_{12}, R_2, R_5, R_8, R_{11}\}$
$q = 0 \; \{R_3, R_6, R_9, R_{12}, R_2, R_5, R_8, R_{11}, R_1, R_4, R_7, R_{10}\}$

$H_S(R, \lambda^{-1})$
$q = 5 \text{ to } 3 \; \{s_1\}, \{s_2\}, \{s_3\}, \{s_4\}$
$q = 2 \text{ to } 0 \; \{s_1, s_2, s_3, s_4\}$

Figure 4.53 compares the skyscraper diagrams for the narrow and wide intersections and shows that the structures for the wide intersection are less highly connected than for those of the narrow intersection.

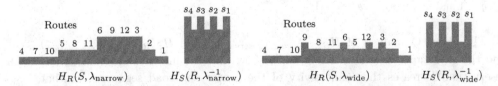

Fig. 4.53 Comparison of the narrow and wide intersections

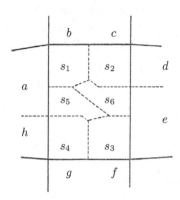
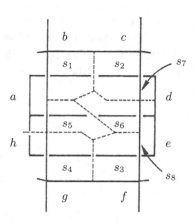

(a) intersection with central reservation   (b) intersection with underpass

Fig. 4.54 Junction designs with a central reservation and an underpass

Figure 4.54(a) shows a road intersection with a central reservation. The incidence matrices of the relation $\lambda_{central\_reservation}$ and its Q-analysis are shown below.

Route	O-D pair	s_1	s_2	s_3	s_4	s_5	s_6		Route	O-D pair	s_1	s_2	s_3	s_4	s_5	s_6
R_1	ab	1	0	0	0	0	0		R_7	ef	0	0	1	0	0	0
R_2	ad	1	1	0	0	0	0		R_8	eh	0	0	1	1	0	0
R_3	af	0	0	1	0	1	0		R_9	eb	1	0	0	0	0	1
R_4	cd	0	1	0	0	0	0		R_{10}	gh	0	0	0	1	0	0
R_5	cf	0	1	1	0	0	1		R_{11}	gb	1	0	0	1	1	0
R_6	ch	0	1	1	1	0	1		R_{12}	gd	1	1	0	1	1	0

$H_R(S, \lambda_{central\_reservation})$

$q = 3$: $\{R_6\}, \{R_{12}\}$
$q = 2$: $\{R_6, R_5\}, \{R_{12}, R_{11}\}$
$q = 1$: $\{R_6, R_5, R_8\}, \{R_{12}, R_{11}, R_2\}, \{R_3\}, \{R_9\}$
$q = 0$: $\{R_8, R_5, R_8, R_{12}, R_{11}, R_2, R_3, R_9, R_1, R_4, R_7, R_{10}\}$

$H_S(R, \lambda^{-1}_{central\_reservation})$

$q = 4$ and 3: $\{s_1\}, \{s_2\}, \{s_3\}, \{s_4\}$
$q = 2$: $\{s_1\}, \{s_2\}, \{s_3\}, \{s_4\}, \{s_5\}, \{s_6\}$
$q = 1$ and 0: $\{s_1, s_2, s_3, s_4, s_5, s_6\}$

The above Q-analysis of $H_R(S, \lambda_{central\_reservation})$ and $H_S(R, \lambda^{-1}_{central\_reservation})$ and the skyscraper diagrams in Fig. 4.55 show that the introduction of a central reservation decreases the connectivity of the routes and road segments, so that vehicles on them interact less and flow is improved.

Figure 4.54(b) shows a road intersection with a central reservation The incidence matrices of the relation $\lambda_{\text{underpass}}$ and its Q-analysis are as follows.

Route	O-D pair	s_1	s_2	s_3	s_4	s_5	s_6	s_7	s_8
R_1	ab	1	0	0	0	0	0	0	0
R_2	ad	0	0	0	0	0	0	1	0
R_3	af	0	0	1	0	1	0	0	0
R_4	cd	0	1	0	0	0	0	0	0
R_5	cf	0	1	1	0	0	1	0	0
R_6	ch	0	1	1	1	0	1	0	0

Route	O-D pair	s_1	s_2	s_3	s_4	s_5	s_6	s_7	s_8
R_7	ef	0	0	1	0	0	0	0	0
R_8	eh	0	0	0	0	0	0	0	1
R_9	eb	1	0	0	0	0	1	0	0
R_{10}	gh	0	0	0	1	0	0	0	0
R_{11}	gb	1	0	0	1	1	0	0	0
R_{12}	gd	1	1	0	1	1	0	0	0

$H_R(S, \lambda_{\text{underpass}})$
$q = 3$: $\{R_6\}, \{R_{12}\}$
$q = 2$: $\{R_6, R_5\}, \{R_{12}, R_{11}\}$
$q = 1$: $\{R_6, R_5, R_{12}\}, \{R_3\}, \{R_2\}$
$q = 0$: $\{R_8, R_5, R_8, R_{12}, R_{11}, R_3, R_9, R_4, R_7, R_{10}\} \{R_2\}, \{R_1\}$

$H_S(R, \lambda_{\text{underpass}}^{-1})$
$q = 3$: $\{s_1\}, \{s_2\}, \{s_3\}, \{s_4\}$
$q = 2$: $\{s_1\}, \{s_2\}, \{s_3\}, \{s_4\}, \{s_5\}, \{s_6\}$
$q = 1$: $\{s_1, s_2, s_3, s_4, s_5, s_6\}$
$q = 0$: $\{s_1, s_2, s_3, s_4, s_5, s_6\}, \{s_7\}, \{s_8\}$

The introduction of an underpass disconnects the traffic even more so that $H_R(S, \lambda_{\text{underpass}})$ and $H_S(R, \lambda_{\text{underpass}}^{-1})$ each have three disconnected components. The traffic on one of these components is locally independent of the traffic of the others and again flow is improved by this configuration.

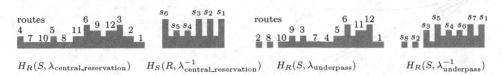

$H_R(S, \lambda_{\text{central\_reservation}}) \quad H_S(R, \lambda_{\text{central\_reservation}}^{-1}) \quad H_R(S, \lambda_{\text{underpass}}) \quad H_R(S, \lambda_{\text{underpass}}^{-1})$

Fig. 4.55 Skyscraper diagrams for junctions with a central reservation and an underpass

The skyscraper diagrams in Fig. 4.55 complete the picture of disconnection from the most constrained narrow intersection, the wider intersection, the intersection with central reservation to the most disconnected intersection with underpass. In all cases the structure of the road intersection backcloth constrains the flow of traffic through the intersection. The generality is that designing intersections to be as disconnected as possible will increase their capacity.

Multidimensional Spaghetti

Figure 4.56(a) shows a design for a road system combining a motorway T-junction with a large roundabout, and Fig. 4.56(b) shows an alternative. Which design is best?

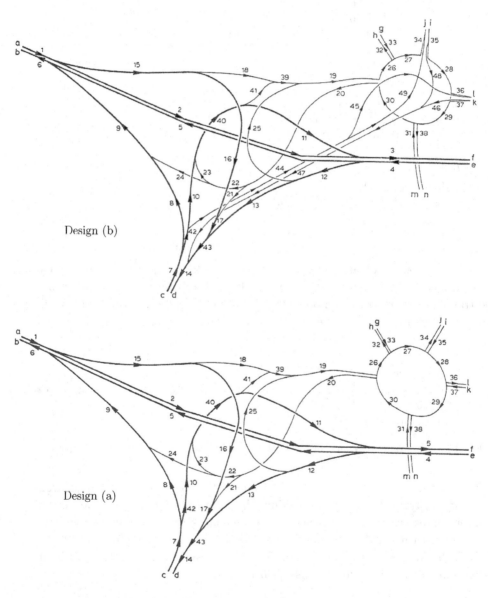

Fig. 4.56 A motorway T-junction combined with a large roundabout

Design (a) is more complicated than Design (a) but at first sight it is not clear if this will be an advantage or not. In this case the road system is represented by sets of numbered *links* and forms a network. A *route* between an origin O and destination D is a contiguous sequence of links, or road segments, starting at O and ending at D. Note that a route is more than a set of road segments. The sequential relationship, λ, makes it a *structured* set of links. Thus a route is a simplex, *e.g.* in both designs there is a simplex $\sigma(R_{cb}) = \langle s_7, s_8, s_9, s_6; \lambda \rangle$ representing the route between c and b.

The set of all routes for Design (a) forms the hypernetwork $H_R(S; \lambda_a)$ with conjugate $H_S(R; \lambda_a^{-1})$. Similarly, the set of all routes for Design (b) forms the hypernetwork $H_R(S; \lambda_b)$ with conjugate $H_S(R; \lambda_b^{-1})$. These hypernetworks have the following structure vectors:

	$q=0$					$q=5$					$q=11$		
$H_R(S; \lambda_a)$	(1	3	8	7	11	8	8	11	11	7	3	1	)
$H_R(S; \lambda_b)$	(1	4	11	11	13	8	8	10	10	6	2	1	)

	$q=0$					$q=5$						$q=11$						$q=17$
$H_S(R; \lambda_b^{-1})$	(1	1	1	1	6	8	2	2	2	1	1	1	5	5	5	5	5	5)
$H_S(R; \lambda_b^{-1})$	(1	1	2	3	7	12	2	1	1	1	3	5	5	5	5	5	0	0)

Inspection of these structure vectors shows Design (a) has more components at relatively high q values and fewer components at relatively low q values. This reflects the simplices of Design (a) having higher dimensions (having more road segments) and being more highly connected for relatively high values of q. In other words Design (b) is more highly disconnected than Design (b) and this, in Atkin's terminology, advantageously *obstructs* the interaction between flows on the different routes.

To make this more formal, let $(c_0, c_1, ..., c_{\max-q})$ and $(c'_0, c'_1, ..., c'_{\max-q})$ be two structure vectors. Let a *flipover value*, be defined to be a value, q_f, such that $c_i \leq c'_i$ for $i < q_f$ and $c_i \geq c_k$ for $i \geq q_f$.

Flipover values do not always exist, but by inspection, $q_f = 5$ is a flipover value for the structure values of $H_R(S, \lambda_a)$ and $H_R(S, \lambda_b)$. This shows that the routes in Design (b) are lower dimension and less highly connected than the routes of Design (b), and this structural feature suggests that Design (b) will perform best.

Although there is no clear flipover value for the conjugate structure vectors, it can be seen that Design (a) has higher dimension road segments (more routes traversing them) which are more highly connected (*e.g.* at $q = 11$ five simplices connect up to become one component) than Design (b). In contrast, Design (b) has higher numbers of low dimensional components at $q = 4$ and $q = 5$ showing the road segments carry less routes and there is less route-interaction on them.

Magic Roundabouts

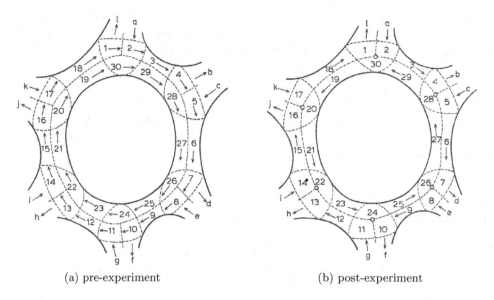

(a) pre-experiment (b) post-experiment

Fig. 4.57   A magic roundabout

Road intersections are a major cause of bottlenecks in urban traffic systems and many ways have been tried to improving the flow through them. Roundabouts have been a major feature of road systems in the UK for many years, and they can provide an excellent self-organising way of directing traffic flows. However, when they become highly congested their performance can be significantly regraded.

In the nineteen seventies a large complicated roundabout (Fig. 4.57(a)) in the English town of Hemel Hempstead caused significant tailbacks and the UK Road Research Laboratory tried an experimental configuration (Fig. 4.57(b)).

The idea at the heart of the new configurations is idea that traffic can go round the roundabout in both directions. Although this may seem highly counter-intuitive, the new design can give very good results because it significantly disconnects the routes and the road segments. To see the potential benefit of the new design consider the routes R_{aj} and R_{gd}. In design (a) two of these routes are the simplices

$$\sigma(R_{aj}) = \langle s_2, s_3, s_4, s_5, s_6, s_7, s_8, s_9, s_{10}, s_{11}, s_{12}, s_{13}, s_{14}, s_{15}, s_{16}; \lambda_a \rangle$$
$$\sigma(R_{gj}) = \langle s_{11}, s_{12}, s_{13}, s_{14}, s_{15}, s_{16}, s_{17}, s_{18}, s_1 s_2, s_3, s_4, s_5, s_6, s_7; \lambda_a \rangle$$

which are 11-near each other, and therefore there is significant interaction between the traffic on them. In comparison the routes for design (b) are:

$$\sigma'(R_{ab} = \langle s_2, s_{30}, s_{19}, s_{20}, s_{16}; \lambda_b \rangle$$
$$\sigma'(R_{gd} = \langle s_{11}, s_{24}, s_{25}, s_{26}, s_7; \lambda_b \rangle$$

which are completely disconnected. The designs have the following structure vectors

$H_R(S; \lambda_a)$ $\qquad Q = (\overset{q=0}{1} \quad 1 \quad 1 \quad 1 \quad \overset{q=4}{2} \quad 2 \quad 7 \quad 13 \quad \overset{q=8}{13} \quad 12 \quad 12 \quad 6 \quad \overset{q=12}{6})$

$H_R(S; \lambda_b)$ $\qquad Q = (1 \quad 1 \quad 2 \quad 2 \quad 8 \quad 2 \quad 18 \quad 12 \quad 12 \quad 0 \quad 0 \quad 0 \quad 0)$

$H_S(R; \lambda_a)$ $\qquad Q = (\overset{q=0}{1} \quad 1 \quad 1 \quad 1 \quad \overset{q=4}{1} \quad 1 \quad 7 \quad 7 \quad \overset{q=8}{7} \quad 12 \quad 12 \quad \overset{q=11}{12})$

$H_S(R; \lambda_a)$ $\qquad Q = (1 \quad 1 \quad 1 \quad 7 \quad 7 \quad 7 \quad 18 \quad 18 \quad 18 \quad 0 \quad 0 \quad 0)$

It can be seen that $q = 7$ is a flipover value for $H_R(S; \lambda_a)$ and $H_R(S; \lambda_a)$, while $q = 8$ is a flipover value for $H_S(R; \lambda_a^{-1})$ and $H_S(R; \lambda_b^{-1})$. The structures show a significant decrease in connectivity for the experimental design.

These structural results suggest that the flows through the intersection should be improved by the new design, and this was the case:

> The current experiment being undertaken at the Plough has, in terms of capacity and reduction of delays, proved a success and a new lease of life has been given to the junction, *West Herts Transportation Study, Hertfordshire County Council, 1973*

The experimental design implemented in the early nineteen seventies remains in place today some forty years later. The reason is clear – changing the structure of the roundabout significantly disconnected the traffic that competes for space on the roundabout. Similar roundabout designs can now be found in other places in the UK.

One drawback to the new design is that people unfamiliar with using such roundabout may find them confusing, and there is the possibility of accidents related to this.

An important insight from this study of road intersections is that infrastructure such as road layouts can be *designed* to give better performance. Here the underlying design heuristic is that disconnecting the route-segment structure of roads generally increases the capacity.

This insight applies to road systems in general. Disconnecting routes reduces the interaction between the vehicles they carry, and can decouple their dynamics with beneficial effects. However, road systems must have a minimal level of connectivity if vehicles are to be able to travel freely. A better understanding of the way connectivity constrains traffic dynamics could lead to more effective investments in building new roads and intersections.

4.18.5 Example: The Wisdom of Crowds

Groups of people often collectively give reliable answers to questions, even when some are uncertain. To investigate this, consider a mathematics test given to forty five students $\{s_1, s_2, ..., s_{45}\}$. Each member q_j of the set of questions, $\{q_1, q_2, ..., q_{20}\}$, had seven possible answers denoted $\mathtt{A}_j$, $\mathtt{B}_j$, $\mathtt{C}_j$, $\mathtt{D}_j$, $\mathtt{E}_j$, $\mathtt{F}_j$ and $\mathtt{G}_j$ in Table 4.10.

	q_1	q_2	q_3	q_4	q_5	q_6	q_7	q_8	q_9	q_{10}	q_{11}	q_{12}	q_{13}	q_{14}	q_{15}	q_{16}	q_{17}	q_{18}	q_{19}	q_{20}
s_1	C_1	B_2	A_3	G_4	C_5	E_6	C_7	D_8	E_9	A_{10}	F_{11}	C_{12}	B_{13}	D_{14}	F_{15}	D_{16}	G_{17}	C_{18}	C_{19}	F_{20}
s_2	C_1	D_2	A_3	G_4	C_5	E_6	C_7	D_8	E_9	A_{10}	F_{11}	C_{12}	B_{13}	D_{14}	C_{15}	D_{16}	G_{17}	D_{18}	C_{19}	F_{20}
s_3	C_1	B_2	A_3	G_4	C_5	F_6	C_7	C_8	E_9	A_{10}	F_{11}	C_{12}	B_{13}	D_{14}	F_{15}	D_{16}	G_{17}	D_{18}	C_{19}	F_{20}
s_4	C_1	B_2	A_3	G_4	C_5	E_6	C_7	D_8	E_9	A_{10}	F_{11}	C_{12}	B_{13}	D_{14}	F_{15}	D_{16}	G_{17}	D_{18}	C_{19}	F_{20}
s_5	C_1	D_2	A_3	G_4	C_5	F_6	C_7	C_8	E_9	C_{10}	F_{11}	C_{12}	B_{13}	D_{14}	B_{15}	B_{16}	G_{17}	D_{18}	E_{19}	G_{20}
s_6	C_1	B_2	A_3	G_4	C_5	F_6	C_7	D_8	E_9	A_{10}	F_{11}	C_{12}	B_{13}	D_{14}	A_{15}	D_{16}	G_{17}	D_{18}	C_{19}	F_{20}
s_7	C_1	D_2	A_3	C_4	C_5	F_6	C_7	D_8	D_9	A_{10}	E_{11}	C_{12}	G_{13}	B_{14}	F_{15}	D_{16}	G_{17}	D_{18}	B_{19}	F_{20}
s_8	C_1	B_2	A_3	G_4	C_5	E_6	C_7	D_8	E_9	A_{10}	F_{11}	C_{12}	B_{13}	C_{14}	F_{15}	D_{16}	G_{17}	C_{18}	C_{19}	F_{20}
s_9	C_1	D_2	A_3	A_4	C_5	F_6	C_7	B_8	G_9	G_{10}	F_{11}	C_{12}	B_{13}	B_{14}	E_{15}	D_{16}	G_{17}	D_{18}	C_{19}	F_{20}
s_{10}	C_1	D_2	A_3	G_4	C_5	E_6	C_7	C_8	D_9	A_{10}	F_{11}	C_{12}	B_{13}	D_{14}	F_{15}	D_{16}	G_{17}	D_{18}	D_{19}	F_{20}
s_{11}	C_1	B_2	A_3	D_4	C_5	F_6	C_7	D_8	E_9	A_{10}	F_{11}	C_{12}	B_{13}	D_{14}	E_{15}	D_{16}	G_{17}	D_{18}	C_{19}	F_{20}
s_{12}	C_1	B_2	A_3	B_4	C_5	F_6	A_7	A_8	F_9	F_{10}	A_{11}	E_{12}	G_{13}	C_{14}	E_{15}	A_{16}	C_{17}	D_{18}	B_{19}	A_{20}
s_{13}	C_1	B_2	A_3	G_4	C_5	F_6	C_7	D_8	E_9	A_{10}	F_{11}	C_{12}	B_{13}	D_{14}	F_{15}	D_{16}	G_{17}	D_{18}	C_{19}	F_{20}
s_{14}	C_1	D_2	A_3	E_4	C_5	E_6	C_7	C_8	E_9	A_{10}	F_{11}	C_{12}	G_{13}	D_{14}	F_{15}	B_{16}	G_{17}	B_{18}	A_{19}	F_{20}
s_{15}	C_1	B_2	A_3	G_4	C_5	F_6	D_7	F_8	C_9	C_{10}	F_{11}	C_{12}	G_{13}	D_{14}	F_{15}	D_{16}	G_{17}	D_{18}	C_{19}	F_{20}
s_{16}	C_1	B_2	A_3	G_4	C_5	E_6	C_7	D_8	E_9	A_{10}	F_{11}	C_{12}	B_{13}	D_{14}	F_{15}	D_{16}	G_{17}	D_{18}	C_{19}	F_{20}
s_{17}	C_1	B_2	A_3	A_4	C_5	F_6	C_7	D_8	E_9	C_{10}	F_{11}	B_{12}	B_{13}	C_{14}	E_{15}	D_{16}	G_{17}	D_{18}	C_{19}	F_{20}
s_{18}	C_1	D_2	A_3	G_4	C_5	F_6	C_7	D_8	E_9	A_{10}	F_{11}	C_{12}	B_{13}	C_{14}	E_{15}	D_{16}	G_{17}	E_{18}	F_{19}	D_{20}
s_{19}	C_1	B_2	A_3	G_4	C_5	E_6	C_7	D_8	E_9	A_{10}	F_{11}	C_{12}	B_{13}	D_{14}	F_{15}	D_{16}	G_{17}	D_{18}	C_{19}	F_{20}
s_{20}	C_1	B_2	A_3	G_4	C_5	F_6	C_7	A_8	E_9	A_{10}	F_{11}	C_{12}	B_{13}	D_{14}	F_{15}	F_{16}	G_{17}	D_{18}	C_{19}	F_{20}
s_{21}	C_1	B_2	A_3	G_4	C_5	E_6	C_7	D_8	E_9	C_{10}	B_{11}	C_{12}	B_{13}	C_{14}	F_{15}	D_{16}	G_{17}	B_{18}	C_{19}	G_{20}
s_{22}	C_1	D_2	A_3	G_4	C_5	F_6	D_7	D_8	E_9	G_{10}	F_{11}	C_{12}	B_{13}	C_{14}	F_{15}	A_{16}	G_{17}	A_{18}	C_{19}	F_{20}
s_{23}	C_1	D_2	A_3	C_4	C_5	F_6	C_7	D_8	E_9	A_{10}	F_{11}	C_{12}	B_{13}	D_{14}	F_{15}	D_{16}	G_{17}	A_{18}	C_{19}	F_{20}
s_{24}	C_1	B_2	A_3	G_4	C_5	F_6	C_7	B_8	E_9	A_{10}	F_{11}	C_{12}	B_{13}	D_{14}	E_{15}	B_{16}	G_{17}	C_{18}	C_{19}	F_{20}
s_{25}	C_1	B_2	A_3	G_4	C_5	E_6	C_7	D_8	E_9	A_{10}	F_{11}	C_{12}	B_{13}	D_{14}	F_{15}	D_{16}	G_{17}	D_{18}	G_{19}	F_{20}
s_{26}	C_1	D_2	A_3	B_4	C_5	F_6	C_7	B_8	F_9	G_{10}	D_{11}	C_{12}	G_{13}	G_{14}	B_{15}	B_{16}	E_{17}	C_{18}	B_{19}	F_{20}
s_{27}	C_1	B_2	A_3	G_4	C_5	E_6	C_7	C_8	E_9	A_{10}	F_{11}	C_{12}	B_{13}	D_{14}	F_{15}	D_{16}	G_{17}	D_{18}	E_{19}	F_{20}
s_{28}	B_1	B_2	A_3	G_4	C_5	F_6	D_7	D_8	D_9	A_{10}	F_{11}	C_{12}	G_{13}	D_{14}	F_{15}	D_{16}	G_{17}	D_{18}	C_{19}	E_{20}
s_{29}	C_1	B_2	A_3	G_4	C_5	E_6	C_7	D_8	E_9	A_{10}	F_{11}	C_{12}	B_{13}	D_{14}	E_{15}	D_{16}	G_{17}	D_{18}	B_{19}	F_{20}
s_{30}	C_1	B_2	A_3	G_4	C_5	F_6	C_7	D_8	E_9	A_{10}	F_{11}	C_{12}	B_{13}	A_{14}	E_{15}	B_{16}	G_{17}	D_{18}	C_{19}	F_{20}
s_{31}	C_1	B_2	A_3	G_4	C_5	F_6	C_7	D_8	E_9	A_{10}	F_{11}	C_{12}	B_{13}	C_{14}	F_{15}	D_{16}	G_{17}	D_{18}	D_{19}	F_{20}
s_{32}	C_1	B_2	A_3	G_4	C_5	E_6	C_7	D_8	E_9	A_{10}	F_{11}	C_{12}	B_{13}	D_{14}	F_{15}	D_{16}	G_{17}	D_{18}	C_{19}	F_{20}
s_{33}	C_1	B_2	A_3	G_4	C_5	E_6	X_7	B_8	D_9	A_{10}	A_{11}	C_{12}	E_{13}	D_{14}	E_{15}	B_{16}	G_{17}	D_{18}	C_{19}	F_{20}
s_{34}	C_1	B_2	A_3	G_4	C_5	E_6	C_7	C_8	E_9	A_{10}	F_{11}	C_{12}	B_{13}	C_{14}	F_{15}	D_{16}	G_{17}	D_{18}	C_{19}	F_{20}
s_{35}	C_1	B_2	A_3	G_4	C_5	F_6	C_7	B_8	B_{10}		F_{11}	C_{12}	B_{13}	D_{14}	F_{15}	A_{16}	G_{17}	D_{18}	B_{19}	F_{20}
s_{36}	C_1	B_2	A_3	B_4	C_5	E_6	C_7	D_8	E_9	A_{10}	F_{11}	C_{12}	B_{13}	D_{14}	F_{15}	D_{16}	G_{17}	D_{18}	C_{19}	F_{20}
s_{37}	C_1	D_2	A_3	G_4	C_5	E_6	A_7	A_8	E_9	A_{10}	B_{11}	C_{12}	B_{13}	B_{14}	E_{15}	D_{16}	G_{17}	C_{18}	D_{19}	F_{20}
s_{38}	C_1	D_2	A_3	D_4	C_5	F_6	C_7	F_8	E_9	B_{10}	B_{11}	A_{12}	D_{13}	G_{14}	B_{15}	B_{16}	G_{17}	C_{18}	C_{19}	A_{20}
s_{39}	C_1	B_2	A_3	G_4	C_5	D_6	E_9	A_{10}			F_{11}	C_{12}	B_{13}	D_{14}	G_{15}	D_{16}	G_{17}	D_{18}	E_{19}	G_{20}
s_{40}	C_1	B_2	A_3	G_4	C_5	F_6	C_7	E_8	E_9	A_{10}	F_{11}	C_{12}	B_{13}	B_{14}	F_{15}	B_{16}	G_{17}	D_{18}	B_{19}	A_{20}
s_{41}	C_1	D_2	A_3	D_4	C_5	F_6	C_7	B_8	D_9	C_{10}	X_{11}	A_{12}	G_{13}	D_{14}	E_{15}	A_{16}	G_{17}	D_{18}	C_{19}	A_{20}
s_{42}	C_1	B_2	A_3	G_4	C_5	E_6	C_7	D_8	E_9	A_{10}	F_{11}	C_{12}	B_{13}	D_{14}	F_{15}	D_{16}	G_{17}	D_{18}	C_{19}	F_{20}
s_{43}	C_1	B_2	A_3	G_4	C_5	E_6	C_7	D_8	E_9	A_{10}	F_{11}	C_{12}	B_{13}	D_{14}	F_{15}	D_{16}	G_{17}	D_{18}	C_{19}	F_{20}
s_{44}	C_1	B_2	A_3	G_4	C_5	F_6	C_7	A_8	E_9	A_{10}	F_{11}	C_{12}	B_{13}	D_{14}	E_{15}	B_{16}	D_{17}	C_{18}	C_{19}	F_{20}
s_{45}	D_1	B_2	A_3	G_4	C_5	E_6	C_7	B_8	E_9	A_{10}	F_{11}	C_{12}	G_{13}	D_{14}	F_{15}	D_{16}	G_{17}	D_{18}	C_{19}	F_{20}

Table 4.10. The relation between students and their answers

Can the correct answers be abstracted from this table with no further information? To test this the relation R is defined between the students and their answers. Student s_i is R-related to answer a_j if this is the answer they give to q_j. For example, student s_1 is R-related to C_1, B_2, A_3, G_4, and so on.

Answer	Students	Answer	Students	Answer	Students	Answer	Students
q_1 - C_1	43	q_6 - F_6	24	q_{11} - F_{11}	37	q_{16} - D_{16}	31
q_2 - B_2	32	q_7 - C_7	40	q_{12} - C_{12}	41	q_{17} - G_{17}	42
q_3 - A_3	45	q_8 - D_8	26	q_{13} - B_{13}	35	q_{18} - D_{18}	33
q_4 - G_4	34	q_9 - E_9	36	q_{14} - D_{14}	30	q_{19} - C_{19}	30
q_5 - C_5	45	q_{10} - A_{10}	34	q_{15} - F_{15}	26	q_{20} - F_{20}	36

Table 4.11. The most popular answers selected by the 45 students.

For each question the most frequently given answers are shown in Table 4.11. The first column shows that all students except two gave the answer C_1 to question q_1, making it highly likely that this is the correct answer. In general one would expect the majority response to be correct.

At first sight the answers in Table 4.11 are correct, since in all cases more than half the students gave these responses. For most of the questions the students overwhelmingly agree, but for some the agreement is not so clear. For example, for question q_6 the answer F_6 was selected by 24 students (53%). The answer D_8 to questions q_8 was given by 26 students (58%), and the answer F_{15} to question q_{15} was also selected by 26 students (58%). How certain can one be that the most popular answers are really correct in these cases?

Of particular interest is 21 students answering E_6 (47%) to q_6, compared to 24 (53%) for F_6. Is the majority correct? To answer this question, consider the students viewed as relational simplices, e.g. $\sigma(s_1) = \langle C_1, B_2, A_3, G_4, C_5, E_6, C_7, D_8, E_9, A_{10}, F_{11}, C_{12}, B_{13}, D_{14}, F_{15}, D_{16}, G_{17}, C_{18}, C_{19}, F_{20}\rangle$. $H_S(Q;R)$ is the family of the relational simplices $\sigma(s_1), ..., \sigma(s_{45})$.

The Q-analysis of the hypernetwork $H_S(Q,R)$ in Fig. 4.58 shows the component $\{s_{42}, s_{16}, s_{43}, s_4, s_{32}, s_{19}\}$ at $q = 19$, meaning that each of these students gave exactly the same answers to all twenty questions, i.e. the simplex $\sigma = \langle C_1, B_2, A_3, G_4, C_5, E_6, C_7, D_8, E_9, A_{10}, F_{11}, C_{12}, B_{13}, D_{14}, F_{15}, D_{16}, G_{17}, D_{18}, C_{19}, F_{20}\rangle$. Its vertices are exactly the same as the list of most frequently occurring answers given in Table 4.11, with the exception of E_6, instead of F_6. Have these six students answered all the questions correctly with the exception of q_6?

Fig. 4.58 Q-analysis of the student-questions relation, $H_S(Q,R)$ showing connected students

In this system, if more than two students get all the answers right then they will have identical answer simplices. This is the first indication that E_6 is correct rather than the more popular answer F_6. Since there is only one non-trivial 19-component, all the other students have at least one vertex different to any other, which means that if F_6 is correct only one student, s_{13}, got all the answers right.

Consider the weaker students in this cohort. They will get a number of the answers wrong, not only missing the correct answer but also giving a scattering of incorrect answers, and they will tend to be more eccentric in the answers they give. In other words, one expects the better students to be more highly connected and the weaker students to be less highly connected.

To investigate the lower level connectivities, students $s_4, s_{16}, s_{19}, s_{32}, s_{42}$, and s_{43} were removed from the system, and the Q-analysis rerun (Figure 4.59).

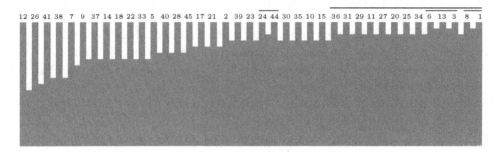

Fig. 4.59   Q-analysis of $H_S(Q, R)$ with students $s_{42}, s_{16}, s_{43}, s_4, s_{32}$ and s_{19} removed

At $q = 18$ there are three components, $\{s_{24}, s_{44}\}$, $\{s_6, s_{13}, s_3\}$ and $\{s_8, s_1\}$. Of these students, $s_{24}, s_{44}, s_6, s_{13}$, and s_3 gave the answer F_6 while s_8 and s_1 gave the answer E_6. Thus, five of the most highly connected students favoured F_6 while, including the six removed for this analysis, eight favoured E_6 (62%).

A larger component emerges at $q = 17$, with students $s_1, s_3, s_6, s_8, s_{11}, s_{13}, s_{20}, s_{25}, s_{27}, s_{29}, s_{31}, s_{34}$, and s_{36}. Eight of these students favour E_6 while five favour F_6. Combined with the previous six, this means that 14 of the most highly connected students favour E_6 (74%) while 5 favour F_6.

What about the most disconnected students at the left of Fig. 4.59? Examination of Table 4.10 shows that $s_{12}, s_{26}, s_{41}, s_{38}$ and s_7 all gave the answer F_6. Assuming these are the weakest students, this is another strong indication that F_6 is wrong. This is a strong indication that E_6 is the correct answer to q_6.

Thus, although F_6 is the most popular answer for q_6, the most highly connected students overwhelmingly prefer E_6. Assuming that the most highly connected students will be the best, this is a very strong indication that E_6 is the correct answer.

Having reached this conclusion without any information about the questions or answers other than that given in Table 4.10, the conclusion can be tested by reference to the examination paper. Question 6 reads as follows: "A body moves in such a way that its speed (in miles per hour) after t hours is $4t^3$. How far has it travelled after 3 hours?" It gives the options (A_6) 16 miles, (B_6) 27 miles, (C_6) 54 miles, (D_6) 64 miles, (E_6) 81 miles, (F_6) 108 miles, and (G_6) 243 miles. The stronger students correctly realised that they had to integrate $4t^3$ and substitute 3 into t^4 to give 81 miles (E_6), while the weaker student incorrectly substituted 3 directly into $4t^3$ to obtain 108 miles (F_6).

This example shows how multidimensional connectivity can be used to reason about systems. It also illustrates that the "wisdom of crowds" may be more subtle than majority decision making, and that the way individuals cluster together through their connectivity can be significant.

4.18.6 Example: Multidimensional Structure in Road Networks

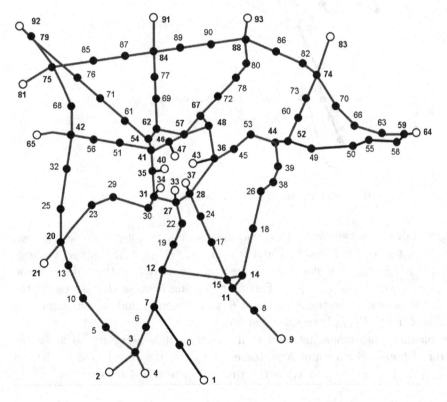

Fig. 4.60 The network of major roads in Cambridge

Figure 4.60 shows the network of major roads in and around the city of Cambridge in England. Let a *route*, $R_{0,n}$ in the network between origin v_0 and destination v_n be defined to be a sequence of vertices $v_0, v_1, \ldots v_{n-1}, v_n$, where there is a road between v_{i-1} and v_i for $i = 1, \ldots n$. As a structured sequence of vertices a route can be written as a simplex, $\sigma(R_{0,n}) = \langle v_0, v_1, \ldots, v_n \rangle$.

It is of course more usual to define a route as a set of links $\ell_{0,1}, \ell_{1,2}, \ldots \ell_{n-1,n}$ where link $\ell_{i-1,i}$ has vertices v_{i-1} and v_i for $i = 1, \ldots, n$, but for the moment using vertices makes the discussion clearer.

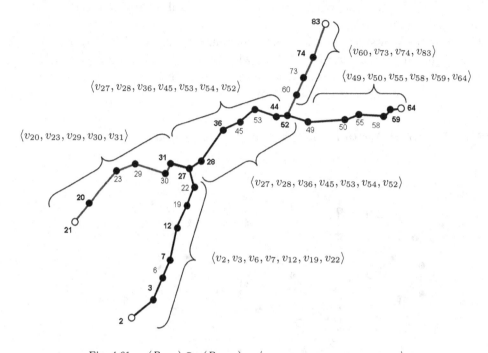

Fig. 4.61 $\sigma(R_{2,83}) \cap \sigma(R_{21,64}) = \langle v_{27}, v_{28}, v_{36}, v_{45}, v_{53}, v_{54}, v_{52} \rangle$

Figure 4.61 shows two routes $\langle v_{21}, v_{20}, v_{23}, v_{29}, v_{30}, v_{31}, v_{27}, v_{28}, v_{36}, v_{45}, v_{53}, v_{54}, v_{52}, v_{49}, v_{50}, v_{55}, v_{58}, v_{59}, v_{64} \rangle$ and $\langle v_2, v_3, v_6, v_7, v_{12}, v_{19}, v_{22}, v_{27}, v_{28}, v_{36}, v_{45}, v_{53}, v_{44}, v_{52}, v_{60}, v_{73}, v_{74}, v_{83} \rangle$. These are connected by the 6-dimensional face $\langle v_{27}, v_{28}, v_{36}, v_{45}, v_{53}, v_{44}, v_{52} \rangle$. This is interesting because this is exactly the substructure on which the traffic on route $R_{1,83}$ between v_1 and v_{83} interacts with the traffic on route $R_{21,64}$ between v_{21} and v_{64}.

For planning and managing road traffic systems it is necessary to know the *route travel times*. For a route $R = \langle v_0, v_1, \ldots, v_n \rangle$ let the travel time be defined $\Sigma_{i=1}^{n} t(v_{i-1,i}, v_i)$ where $t(v_{i-1,i}, v_i)$ is the time taken to travel between nodes v_{i-1} and v_i.

Let the *flow* of vehicles across a route be the number of vehicles entering and leaving that route in unit time. In the simplest case, as the flow of vehicles on any particular route increases the number of vehicles on the links it traverses increases.

There are many complications in the relationship between traveltimes, densities, speeds and flows, depending how the variables are defined. Nonetheless it is generally the case that vehicle speeds reduce and travel times increase as the density of traffic on multilane roads increases as illustrated in Figure 7.68.

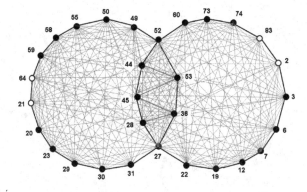

Fig. 4.62 Routes $R_{2,28}$ and $R_{21,64}$ as intersecting polyhedra

The shared face of two routes is exactly the site of interaction, with the traffic in each route slowing down the traffic on the other. As the demand to travel either route increases, so the traveltime on each increases. Also, the more highly two routes are connected by a q-dimensional face, the greater is the effect of the interaction.

When routes intersect their travel times become *coupled*, *i.e.* they are no longer independent. This is because when two routes share link vertices they both contribute vehicles to the density with vehicles on each route slowing down the vehicles on the other.

Figure 4.63 shows three routes, $R_{91,2}$, $R_{21,46}$ and $R_{9,83}$. Routes $R_{91,2}$ and $R_{9,83}$ share no vertices, but each shares a 1-dimensional face with $R_{21,46}$. Thus, $R_{2,91}$ and $R_{9,83}$ are disjoint but 1-connected through $R_{21,46}$. Suppose the number of vehicles travelling on $R_{91,2}$ increases as $\delta f(R_{91,2}) > 0$. Then the number of vehicles on $\langle v_{31}, v_{27}\rangle$ increases with $\delta k\langle v_{31}, v_{27}\rangle > 0$ and the traveltime on $\langle v_{31}, v_{27}\rangle$ increases as $\delta t\langle v_{31}, v_{27}\rangle > 0$.

This increase in traveltime will be experienced by the vehicles on route $R_{21,46}$ as $\delta t(R_{21,46}) > 0$ precisely because it shares the link $\langle v_{31}, v_{27}\rangle$. The increased travel time on $R_{21,46}$ may result in less vehicles travelling that route, with $\delta f(R_{21,46}) < 0$, resulting in less vehicles traversing $\langle v_{44}, v_{52}\rangle$ so that $\delta k\langle v_{44}, v_{52}\rangle < 0$ and $\delta t\langle v_{44}, v_{52}\rangle < 0$. The decrease in time across $\langle v_{44}, v_{52}\rangle$ results in the time on $R_{9,83}$ decreasing as $\delta t(R_{9,83}) < 0$. Thus, an increase in traffic flow on $R_{91,2}$ can cause a decrease in travel time on $R_{9,83}$, even though these routes shared no links or vertices.

Let V be the set of vertices in Figure 4.60 and R be the set of shortest routes between the white vertices. Let $H_R(V)$ and $H_V(R)$ be the simplicial families of the relation between R and V. Figure 4.64 shows the skyscraper diagram of the Q-analysis of $H_R(V)$.

In cities the route travel times are determined as the combined travel time in free flow and the time spent waiting in queues to traverse intersections. Often the queuing time is much greater that the free flow time, and it is common to include a *junction penalty* for the nodes representing intersections. When a junction penalty

98   *Hypernetworks in the Science of Complex Systems*

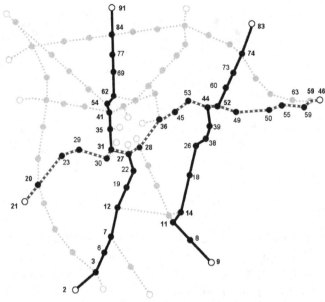

(a) Routes $R_{2,91}$ and $R_{9,83}$ share no vertices but are connected through $R_{21,46}$.

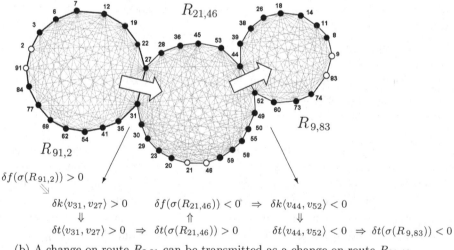

$\delta f(\sigma(R_{91,2})) > 0$

$\delta k\langle v_{31}, v_{27}\rangle > 0 \qquad \delta f(\sigma(R_{21,46})) < 0 \;\Rightarrow\; \delta k\langle v_{44}, v_{52}\rangle < 0$
$\Downarrow \qquad\qquad\qquad \Uparrow \qquad\qquad\qquad \Downarrow$
$\delta t\langle v_{31}, v_{27}\rangle > 0 \;\Rightarrow\; \delta t(\sigma(R_{21,46})) > 0 \qquad \delta t\langle v_{44}, v_{52}\rangle < 0 \;\Rightarrow\; \delta t(\sigma(R_{9,83})) < 0$

(b) A change on route $R_{2,91}$ can be transmitted as a change on route $R_{21,46}$.

Fig. 4.63 The dynamics on disjoint routes are coupled through a chain of connection

is included the minimum time routes change and a different structure appears, as shown in Fig. 4.64(b).

When there are no junction penalties the nodes in the centre have the highest dimensions and are highly connected to each other. When junction penalties are introduced the dimensions of the city centre nodes decrease because trips through

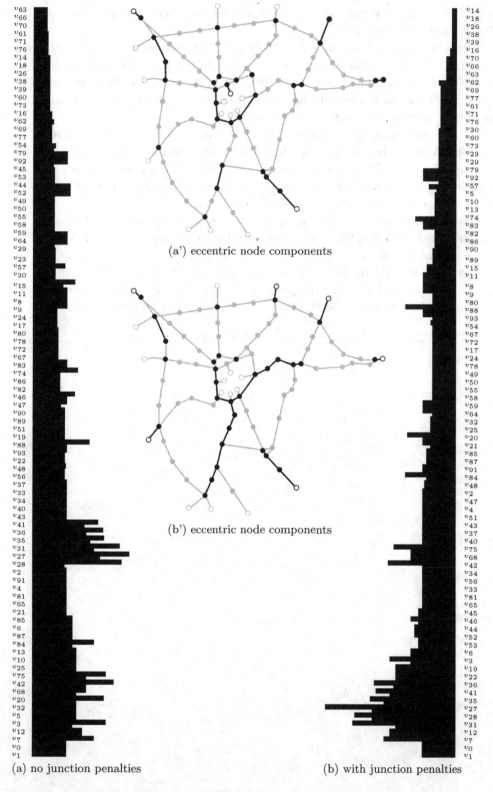

(a') eccentric node components

(b') eccentric node components

(a) no junction penalties

(b) with junction penalties

Fig. 4.64 Q-analysis for $K_V(R)$, without and with junction penalties

the city centre are slower so that spatially longer routes become attractive in terms of their lower travel times.

Simply assigning a junction a travel penalty is not very realistic because the travel costs across junction depend on the route taken across the junction and the design of the junction. For example, junctions with inappropriately set traffic signals may cause some routes to have longer delays than others, and may cause longer than necessary delays to all the routes.

The structures revealed in Figure 4.64 can be relevant for the design of road networks. When a sequence of links is highly connected the roads and intersections are acting as a kind of 'corridor' bringing together the traffic on many routes. The significance of this is that such corridors are emergent higher level structures that can be analysed and potentially changed for the better.

The analysis of road intersections in the next section illustrates the use of Q-analysis in understanding how the connectivity between the parts of road systems can be used in their design.

Figures 4.64(a) and (b) show the parts of the road network associated with the most highly connected and eccentric parts of the network. When junction penalties are included the network connectivity increases with distinct corridors emerging.

Chapter 5

Backcloth and traffic: dynamics constrained by topology

Atkin's theory of Q-analysis makes a distinction between the relatively fixed structure of a system that changes in a relatively slow way, and the values of mappings or functions defined on that structure that can change over much shorter timescales.

For example, it takes many years to build a cathedral. Once it is built there are many associated numbers that can change over much shorter time periods, *e.g.* the number of people in the cathedral. Another example is a transportation system made up of roads, intersections, bridges, car parks, *etc.* This infrastructure is expensive, changes relatively slowly, and it exists whether or not it is used. In comparison, the traffic of vehicles it supports can be represented by numbers such as the flow, density and travel time which vary over short time scales. Atkin suggested that the relatively fixed structure of systems be called the *backcloth* and that activity represented by patterns of relatively fast changing numbers be called the *traffic*.

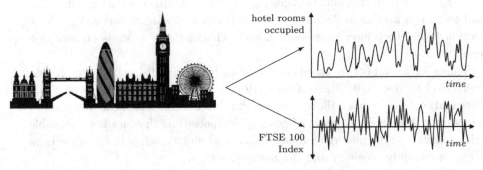

Fig. 5.1 Examples of traffic on the London backcloth

Figure 5.1 illustrates how a city such as London supports many diverse activity patterns such as hotel room occupancy and the FTSE 100 Index. There are many other kinds of traffic in a city, *e.g.* its shopping traffic, the traffic of people in and out of hospital, property tax traffic, air quality traffic, legislation traffic, the traffic of crime, and so on. Much of this traffic is important for planning and commercial purposes and it is captured as time series. The time scale for this traffic is relatively short compared to the decades and centuries it takes to create the backcloth.

5.1 The space-time backcloth constrains physical and social traffic

Figure 5.2 shows how the space-time backcloth constrains physical and social motion. In Fig. 5.2(a) the speed of light constrains the future trajectories of physical objects. In general, starting from any initial position, there are parts of space-time, $\mathbb{R}^3 \times \mathbb{R}$, that can never be reached.

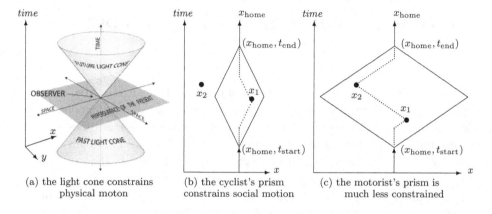

(a) the light cone constrains physical moton

(b) the cyclist's prism constrains social motion

(c) the motorist's prism is much less constrained

Fig. 5.2 Space-time geometry constrains system trajectories.

Figures 5.2(b) and (c) show how social systems are also constrained by their space-time geometry. These diagrams show a cyclist and a motorist leaving home at time t_{start} and returning home at time t_{end}. In this two-dimensional time-distance world there is a maximum distance they can travel backward or forward in the x direction before they have to go back home. This distance depends on how fast they can travel.

Figures 5.2(b) and (c) are based on work done by the Swedish geographer Torsen Hägerstrand in the 1970s [Hägerstrand (1970)]. There are clear issues of social justice and equality between those who can afford to use cars and those who cannot. The motorist has access to parts of geographic space-time that are not accessible to the cyclist, *e.g.* the motorist can access both x_1 and x_2, while in the same time the cyclist can only access x_1 and can never access x_2.

Let R be a relation between the set of people P and the set of located activities A, and let $\sigma(p) = \langle a_1, ..., a_n \rangle$ where person p can access activity a_i, $i = 1, ..., n$. Then

$$\sigma(p_{\text{pedestrian}}) \lesssim \sigma(p_{\text{cyclist}}) \lesssim \sigma(p_{\text{motorist}}).$$

Also the motorist has much greater opportunities in *combining* activities in space-time trajectories. For example, in western countries the simplex $\langle a_{\text{home}}, a_{\text{nursery}}, a_{\text{school}}, a_{\text{workplace}}, a_{\text{shops}}, a_{\text{workplace}}, a_{\text{school}}, a_{\text{nursery}}, a_{\text{home}} \rangle$ is part of many parents' working day.

5.2 Structuring the social backcloth to constrain the traffic

Physical space and physical processes put constraints on human beings that make some things inevitable and some things impossible. As we have just seen, there are parts of the space-time backcloth that are impossible for some people to reach. However, much of the social backcloth structure is created by ourselves in order to impose structure and predictability where otherwise there would be too many possibilities and potential chaos.

Figure 5.3 shows a small car park with parking bays marked out by lines painted on the ground. These lines create the parking bays as emergent objects, $\langle \ell_1, \ell_2, \ell_3 \rangle$, $\langle \ell_3, \ell_4, \ell_5 \rangle$, and so on. These simplices form the backcloth for the traffic of parking, represented by a mapping f with $f(\langle \ell_3, \ell_4, \ell_5 \rangle) = 1$ if the bay is occupied and $f(\langle \ell_3, \ell_4, \ell_5 \rangle) = 0$ otherwise.

In Fig. 5.3(a) the cars are parking normally within the bays. However although the painted bays normally constrain parking in this way, it is possible for drivers to break the rules and ignore the bay markings, as shown in Fig. 5.3(b).

A person leaving their car across the bay markings effectively changes the backcloth, in this case making two parking spaces unusable for others. Generally people do not violate the rules, possibly because they do not want to incur penalties, but often because they do not want to be antisocial and disrupt the social backcloth.

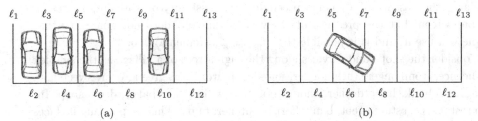

Fig. 5.3 Space structured and used by social convention

This example illustrates how social space can be artificially structured to create objects such as parking bays. In fact a very large part of the social world is artificial, *designed* to carry traffic in socially desired ways. Other examples include the relationships between people that constrain how they will act to each other, *e.g.* most bridge players accept structuring relationships that preclude cheating. Similarly most business people accept the rule that once they have shaken hands on a deal they will not try to change the terms. Generally we human beings *make* structures which socially constrain the traffic of future behaviour, and make it more predictable and manageable.

5.3 The backcloth allows and forbids but does not require

Road network infrastructure is the relatively fixed backcloth structure supporting the dynamic traffic of patterns of vehicle movements. Let $f(\langle a,b \rangle)$ be the flow on a road link $\langle a,b \rangle$. Then there is difference between $\langle a,b \rangle$ not existing and $f(\langle a,b \rangle)$ being zero. For example, in Fig. 5.4(b) the bridge may carry no traffic, $f(\langle a,b \rangle) = 0$, but the link $\langle a,b \rangle$ still exists with the potential to carry traffic in the future.

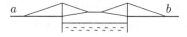

(a) there is no link $\langle a,b \rangle$ across the river, so 0 vehicles flow between a and b

(b) The bridge link $\langle a,b \rangle$ exists but carries no vehicles, so $f\langle a,b \rangle = 0$ vehicles flow between a and b

Fig. 5.4 The absence of a link forbids traffic, the presence of a link allows but does not require.

In Figure 5.4(a) the absence of a link in the backcloth *forbids* vehicles from travelling from a to b across the river, while in Fig. 5.4(b) the presence of the bridge *allows* traffic to flow from a to b, but the presence of the bridge does not *require* that any vehicles actually do travel from a to b in any particular time interval.

In general the absence of infrastructure makes particular activities impossible while the presence of infrastructure enables activity but does not require it to happen. For example, in the previous section the absence of a car park forbids parking, while the existence of a car park allows but does not require cars to park there. Similarly, the absence of a cinema precludes the possibility of watching movies on the big screen, but the presence of the vertex $\langle \text{cinema} \rangle$ does not guarantee that people will use it, and it is possible that $f_{\text{cinema goers}} \langle \text{cinema} \rangle = 0$.

Consider the shops and services along the high street of a village as the simplex $\langle$off-licence, ironmongers, fishmonger, jeweller, hairdresser, greengrocer, sweetshop, restaurant, bank, haberdasher, shoe shop, barber, butcher, baker, dress shop, Indian restaurant, estate agent, bank, florist, butcher, pub$\rangle$. This simplex and its faces provide backcloth infrastructure for village life. Although is is unlikely that anyone uses all these facilities at any particular time, some faces are well trafficked, *e.g.* a family might frequently contribute to the spending traffic on the simplex $\langle$butcher, baker, greengrocer$\rangle$. Of course they do not go to the $\langle$butcher$\rangle$, return home, go to the $\langle$baker$\rangle$, return home, go to the $\langle$greengrocer$\rangle$ and return home. They use $\langle$butcher, baker, greengrocer$\rangle$ as a *simplex*, taking advantage of the spatial relationships and synergies that bind these activities together, allowing just one trip. If any of the vertices of this simplex did not exist, the simplex would cease to exist and the family would have to travel outside the village to find a simplex $\langle$butcher, baker, greengrocer$\rangle$ able to support their desired multidimensional spending traffic. This illustrates a fundamental idea in Q-analysis [Gould *et al* (1984)]:

The multidimensional backcloth *allows* and *forbids* but does *not require*.

5.4 Traffic as patterns of numbers on the backcloth simplices

Let $\mathcal{F}$ be a simplicial family, let $\mathbb{N}$ be a set of numbers, and let ϕ be a mapping assigning a number $\phi(\sigma)$ to each σ in $\mathcal{F}$, $\phi : \mathcal{F} \to \mathbb{N}$.

To illustrate this consider a fruitshop as a simplex with vertices the fruit they sell, $\sigma_{\text{fruitshop}} = \langle \text{oranges, pears, apples, bananas} \rangle$. Various mappings can be defined on this simplex such as the money taken in a week for each of the fruit, *e.g.* $\phi : \langle \text{apples} \rangle \to \phi(\text{apples})$, as shown in Fig. 5.5. Then the range of fruit sold and the simplex σ may remain unchanged over long periods of time, while the value of $\phi(\text{apples})$ changes weekly as a time series.

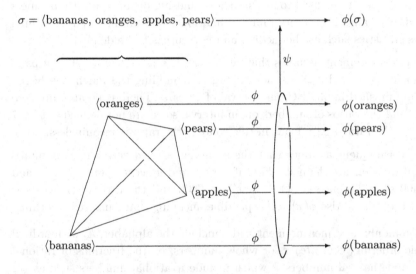

Fig. 5.5 Functions on a simplex as traffic

It is common for values to be associated with simplices that depend on the values on the vertices, so that there is a function $\psi : (\phi\langle v_0 \rangle, ..., \phi\langle v_p \rangle) \to \phi(\sigma)$, as illustrated in Fig. 5.5. Although the traffic in social systems has many examples of non-linear functions ψ from vertex values to simplex values, frequently they are linear. For example, if $\phi(\sigma)$ is the total takings in the shop, then $\phi(\sigma) = \psi(\phi(\text{bananas}), \phi(\text{apples}), \phi(\text{pears}), \phi(\text{oranges})) = \phi(\text{bananas}) + \phi(\text{apples}) + \phi(\text{pears}) + \phi(\text{oranges}) = \sum_{\langle v \rangle \lesssim \sigma} \phi\langle v \rangle$.

An entrepreneurial fruit seller might try to increase trade by offering a combination of three fruits for the price of the least expensive, for example $\phi\langle \text{apples, pears, oranges} \rangle = \psi(\phi(\text{apples}), \phi(\text{pears}), \phi(\text{oranges})) = 3 \times \min\{\phi(\text{apples}), \phi(\text{pears}), \phi(\text{oranges})\}$. This results in a non-linear relationship between the quantities of fruit sold and the amount of money taken.

5.5 Measuring the traffic

Traffic includes measurements made on vertices and simplices, and it is important that the measurement scales are defined appropriately.

In 1946 Stanley Smith Stevens suggested four scales of measurement: *nominal, ordinal, interval* and *ratio* [Stevens (1946)]. Suppose X is a set of things to be measured, that S is a set called a *scale* and ϕ is a mapping from X to S.

<u>Nominal measurement</u> is effectively classification. All that is assumed of S is that it is a finite set, $S = \{s_1, ..., s_n\}$. Each s_i determines a class, $\{x \mid \phi(x) = s_i\}$. These classes are disjoint and form a partition of X.

<u>Ordinal measurement</u> assumes that S has an order relation so that $\phi x \leq \phi y$. For example, S could be a scale used to measure how amusing people are with members deadpan < droll < amusing < very funny < hilarious. Measurement on an ordinal scales allows statistics such as the median and percentiles to be defined.

<u>Interval scale measurement</u> assumes that the interval, $\phi(x) - \phi(x')$, between pairs of measurements $\phi(x)$ and $\phi(x')$ is meaningful, *e.g.* building the Parthenon began in 447 BC and finished in 438 BC, an interval of 9 years. There is no absolute zero on this scale and the ratios of numbers on an interval scale are not defined, *e.g.* 447 and 438 are in the ratio 1.02 : 1 but in this context the ratio is meaningless.

<u>Ratio scale measurement</u> assumes that the scale has a fixed zero and that multiplication and division are defined. *E.g.* if cars x and x' costs $\phi(x) = £800$ and $\phi(x') = £400$ the difference is $\phi(x) - \phi(x') = £400$ and the ratio of the costs is $\phi(x)/\phi(x') : 1$, or 2 : 1. Also $\phi(x) = 0$ is possible, meaning that car x costs nothing.

The most commonly used measurement scales include the alphabet, $\mathbb{A}$, with ordinal lexicographic ordering, the *integers* or whole numbers, $\mathbb{N}$, the fractions or rational numbers $\mathbb{Q}$, and the real numbers $\mathbb{R}$ which include irrational numbers such as $\sqrt{2}$ and π.

To avoid confusion, the scale with the weakest properties should be used, *e.g.*, let $\phi(x)$ be a measure of how much person x likes opera on a five point scale, $\{1, 2, 3, 4, 5\}$. Suppose $\phi(x) = 2$ and $\phi(x') = 4$. Then it is tempting, but incorrect, to say that x' likes opera twice as much as x. Here it would be better to use the scale $\{A, B, C, D, E\}$ since the ratio D/B is not defined and the fallacious interpretations could be avoided.

Steven's scheme can be used to show which statistics can be meaningfully computed:

Legitimately computable statistics	Nominal	Ordinal	Interval	Ratio	$\mathbb{A}$	$\mathbb{N}$	$\mathbb{Q}$	$\mathbb{R}$
Frequency distributions	Yes	Yes	Yes	Yes	Yes	Yes	Yes	Yes
Medians and percentiles	No	Yes	Yes	Yes	No	Yes	Yes	Yes
Addition and subtraction	No	No	Yes	Yes	No	Yes	Yes	Yes
Means, standard deviations	No	No	Yes	Yes	No	Yes	Yes	Yes
Ratios	No	No	No	Yes	No	No	Yes	Yes

It can be argued that it is impossible to make measurements using the scale $\mathbb{R}$ or, even if a measurement could be made as an irrational number, in general it could not be recorded. Certainly, *all* measurements stored in digital computers are finite rational numbers in $\mathbb{Q}$, often being representations of truncated decimal fractions. As the table above shows this is not a practical constraint since the meaningful computable statistics for the rationals, $\mathbb{Q}$, are the same as for the reals, $\mathbb{R}$.

Ordinal measurements and order relations

Some subtle distinctions can be made between different types of ordinal measurements depending on the following *order axioms*:

For all s, s' and s'' in S,

(i) *reflexivity* $s \leq s$

(ii) *antisymmetry* $s \leq s'$ and $s' \leq s$ if and only if $s = s'$

(ii') *asymmetry* $s < s'$ implies $s' \not< s$

(iii) *transitivity* $s \leq s'$ and $s' \leq s''$ imply $s \leq s''$

(iv) *total comparability*: $s \leq s'$ or $s' \leq s$

The relation "less than or equal" on common number systems has all four properties, and is called a *total order*. In comparison the relation "less than" is not reflexive but is antisymmetric and transitive. It is called a *strict order*.

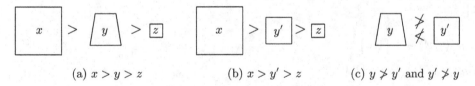

(a) $x > y > z$ (b) $x > y' > z$ (c) $y \not> y'$ and $y' \not> y$

Fig. 5.6 y does not fit inside y', and y' does not fit inside y

Figure 5.6 shows an ordering on the set of shapes defined as $x > x'$ if x' can fit inside x. This relationship is reflexive and it is transitive, but it is not antisymmetric and as y and y' show it is not totally comparable. A reflexive transitive relation is called a *quasi order*.

What is the scale for the "fits inside" measurement? One possibility is to use the measurements $\phi(x) = (\text{width}(x), \text{height}(x))$ with $x > x'$ if $\text{width}(x) > \text{width}(x')$ and $\text{height}(x) > \text{height}(x')$. As Fig. 5.6 shows, this does not work because y is both wider and taller than y' but y' does not fit inside y.

One possible scale for the "fits inside" measurement would be a standard set of shapes such as squares, rectangles, triangles, circles, *etc.*, possibly parameterised by measurements such as height, width, radius. This could be complicated to implement and a better approach could be to make a comparison between two shapes, establishing the relationship directly.

Measurement scales with a fixed highest value can suffer from the *ceiling effect* or *floor effect* in which the top or bottom parts of the scale are under-used. For example, a university course might allocate five marks for the quality of a homework essay with the markers giving only scores of two, three or four marks, arguing that none of the essays was excellent and none of them was completely lacking in quality. This can be overcome to some extent by extending the scale but this does not completely solve the problem, *e.g.* in some subjects essays are marked on a scale of zero to one hundred, but markers rarely give a score higher than eighty which has evolved as the measure of almost unattainable excellence.

Apart from the scale being too short, it can be misleading if it is too long, *e.g.* how much do you like opera on a seven thousand point scale?

5.6 Measuring the backcloth

At its simplest the backcloth is made up of simplicial families and the traffic is defined to be mappings on the simplicial families associating patterns of numbers with the simplices, their vertices and their other faces. Let V be a set of vertices and R a relation. The relatedness of vertices can be measured by the function ϕ_R as follows

<u>Nominal relations</u>: things are either "not related" or "related" on a binary scale $\{0,1\}$.

$$\phi_R : \langle v, v' \rangle \longrightarrow \{1, 0\}$$

$$\phi_R : \langle v_0, ..., v_p \rangle \longrightarrow \{1, 0\}$$

For example $\phi\langle$Elizabeth II, Queen of England$\rangle = 1$ since, at the time of writing, there is no doubt that this binary relation holds. Similarly, at present $\phi\langle$Wales, Scotland, Northern Ireland, England$\rangle = 1$ since there is no doubt that these regions form the United Kingdom. The backcloth dynamics for nominal measurement involve existing relations breaking $(1 \rightarrow 0)$ and new relations forming $(0 \rightarrow 1)$. For example, Scots nationalists would like independence, removing Scotland from the UK simplex, while Turkey would like to become a vertex in the European Union simplex.

<u>Ordinal relations</u>: $v \, R \, v'$ if $\phi\langle v \rangle < \phi\langle v' \rangle$

$$\phi\langle v \rangle < \phi\langle v' \rangle \text{ implies } \phi_R : \langle v, v' \rangle = 1$$

Backcloth and traffic: dynamics constrained by topology 109

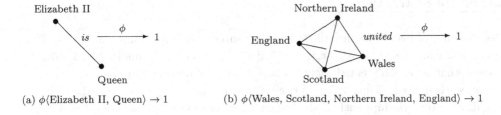

(a) $\phi\langle\text{Elizabeth II, Queen}\rangle \to 1$ (b) $\phi\langle\text{Wales, Scotland, Northern Ireland, England}\rangle \to 1$

Fig. 5.7 Backcloth structures defined by nominal relations

Ordinal relations are very common in social systems. For example, you might like one person more than another, someone might be considered to be more knowledgable than another, or a soccer team might be considered to be more entertaining than another.

The grading of people in many organisations uses an ordinal scale {Grade 1, Grade 2,...} so that $\phi\langle p\rangle$ is the grade of person p. This ordering on the vertices can induce relations beween people according to whether one has a higher grade than the other, as illustrated in Fig. 5.8(a).

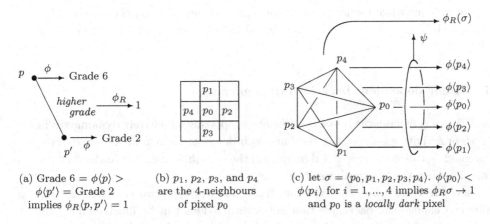

(a) Grade 6 = $\phi\langle p\rangle >$ (b) p_1, p_2, p_3, and p_4 (c) let $\sigma = \langle p_0, p_1, p_2, p_3, p_4\rangle$. $\phi\langle p_0\rangle <$
$\phi\langle p'\rangle$ = Grade 2 are the 4-neighbours $\phi\langle p_i\rangle$ for $i = 1,...,4$ implies $\phi_R\sigma \to 1$
implies $\phi_R\langle p, p'\rangle = 1$ of pixel p_0 and p_0 is a *locally dark* pixel

Fig. 5.8 Ordinal values on the vertices can determine relational structure

In digital images the 4-neighbours of a pixel are the four pixels above, to the right, below and to the right, as illustrated in Fig. 5.8(b). Let a *locally dark pixel* be one that has lower greyscale than its 4-neighbours, where the greyscales are interpreted as ordinal numbers. Then all five pixels are required to decide if a pixel is locally dark. The traffic on the pixels as vertices are the greyscales, and these can be used to determine that $\phi_R : \langle p_0, p_1, p_2, p_3, p_4\rangle \to 1$, as shown in Fig. 5.8(c). As will be seen, it is surprising how useful ordinal relations on pixels can be.

Interval scale relationships

The mapping of people to their birthdays measures them on an interval scales of dates. For example, $\phi_{\text{older than}}\langle x, x'\rangle = 1$ if $\phi\langle x\rangle < \phi\langle x'\rangle$. As another example, suppose that a person's entitlement to social housing depends on the date they registered their need. Then $\phi_{\text{priority}}\langle x, x'\rangle = 1$ if $\phi_{\text{registered}}\langle x\rangle < \phi_{\text{registered}}\langle x'\rangle$. In everyday life relationships such as these seem obvious and easy to handle. However, when there are many of them, and when they interact, they can become complicated and hard to manage. Before we can understand the complex it is essential to make the simple well defined.

Ratio scale relationships

Consider the mapping of people to the amount they are paid, $\phi\langle x\rangle$. This is a ratio scale. It supports the ordinal "paid less than" relation $\phi_{\text{paid less than}}\langle x, x'\rangle = 1$ if $\phi\langle x\rangle < \phi\langle x'\rangle$. It also supports the interval "pay differential" traffic, so that $\phi_{\text{pay differential}}\langle x, x'\rangle = \phi\langle x\rangle - \phi\langle x'\rangle$. Perhaps more interesting is that the scale supports the ratio $\phi_{\text{pay ratio}}\langle x, x'\rangle = \phi\langle x\rangle \div \phi\langle x'\rangle$. In some organisations this ratio between the highest and lowest paid is of the order 10:1, with the bosses earning about ten times as much as the lowest paid. However, in some organisations bosses are paid millions while the lowest paid get tens of thousands, and the ratio exceeds 100:1, possibly associated with organisational volatility or vulnerability to extreme events.

5.7 Traffic-dependent backcloth dynamics

The dichotomy of a relatively static backcloth supporting a relatively dynamic traffic of system activity assumes that the backcloth does not change in the short term. This could be called the *type-one* dynamics of the system, in contrast to the *type-two* dynamics when the backcloth itself changes.

There are cases when the type-one traffic induces type-two changes, and the traffic can change the backcloth that supports it. This can be illustrated by the rare cases in which railway trains (traffic) become derailed and damage the track (backcloth), requiring engineers to spend many days restoring the backcloth (type-two change) so that the normal traffic of trains can be restored (type-one change).

Consider an electrical circuit such as the wiring in a house. Normally the circuit carries the traffic of electricity but occasionally switching on another device (type one change) causes an overload and the wiring can burn out destroying part of the backcloth (type-two change). This is well understood and most circuits are *designed* to fail in predictable ways by the introduction of fuses into the backcloth. Then when failure occurs it is relatively simple to repair the backcloth so that it can again carry the traffic.

At larger scales this backcloth-traffic interaction is not so well understood and despite power grids protecting themselves locally by tripping out, this type-two change can cascade in ways that are not understood to cause cascading blackouts, including a massive blackout across Europe in November 2006:

> One of the worst and most dramatic power failures in three decades plunged millions of Europeans into darkness over the weekend, halting trains, trapping dozens in lifts and prompting calls for a central European power authority. The blackout, which originated in north-western Germany, also struck Paris and 15 French regions, and its effects were felt in Austria, Belgium, Italy and Spain. ... The power loss came about when Germany's network became overloaded probably as a result of a routine shut down of a high-voltage transmission line under the Ems river to allow a ship to pass by safely. The fallout from the incident, said to be one of the worst since the 1970s, left engineers and politicians aghast, and underlined the interdependence of European countries' electricity grids. ... Work was under way yesterday to try to identify why such a routine operation provoked such a massive power failure.'
>
> Stephen Castle, *The Independent*, 6 Nov 2006.

The many other examples of the traffic interacting with backcloth structure include traffic accidents which can close roads, and disagreement traffic that can break relationships between people.

Sometimes part of the traffic is required to maintain the backcloth. A house that supports the traffic of family life must be maintained, requiring part of the family income traffic to repair leaks in the roof and replace worn out appliances. Similarly, road traffic includes the vehicles that repair the road, rail traffic includes specialised locomotives and trucks to repair the track, organisational traffic includes the cost of the human relations department that maintains employment structures, and part of a government's tax traffic is needed to maintain the government backcloth.

Although excessive traffic can damage or destroy the backcloth, too little maintenance traffic can degrade the backcloth and its ability to carry traffic. It is essential to invest in the backcloth in order to maintain its long term ability to carry system traffic, but in social systems this adds expenditure to the bottom line. Politicians with a short mandate are reluctant to spend tax revenue traffic on the long term health of the backcloth when the benefit will only be felt after their mandate ends. Similarly business people may be reluctant to invest in the long term backcloth when their bonus traffic depends on short term shareholder dividend traffic.

On the other hand, there can be excessive use of the short term traffic to support the long term backcloth. Despots may build themselves palaces while their people starve, and company executives may set up lavish offices as sales are falling.

As the above examples show, while the backcloth is essential to support the traffic, there is inevitable tension between using the traffic from one part of the system to maintain backcloth structure, or to create new backcloth structure in another part of the systems.

5.8 Matrices as backcloth supporting flow traffic

	1 EU	2 OW	3 CE	4 NA	5 LA	6 ME	7 AS	8 CH	9 IN	10 OC	11 AF
1. EU-27	0.0	12.0	15.2	22.0	6.3	8.9	23.2	7.8	2.6	2.3	10.1
2. Other W. Europe	69.4	0.0	2.2	9.8	1.9	2.9	11.7	2.6	0.9	0.8	1.3
3. Central & E. Europe	62.8	4.0	0.0	4.5	1.3	7.4	15.7	6.4	1.7	0.1	4.1
4. North America	27.7	2.6	1.8	0.0	27.0	5.4	30.0	9.0	2.0	2.5	3.0
5. Latin America	15.9	2.3	0.8	57.0	0.0	1.9	19.1	10.2	1.6	0.5	2.6
6. Middle East	22.6	3.5	4.2	33.7	3.2	0.0	26.9	5.1	7.3	1.0	4.8
7. Asia	32.0	1.6	4.0	32.5	8.5	10.0	0.0	27.6	4.5	5.0	6.4
8. China	20.1	0.5	3.4	20.4	4.8	4.4	40.4	0.0	2.5	1.9	4.1
9. India	21.8	0.6	1.7	12.2	3.1	22.5	29.3	6.4	0.0	1.0	8.0
10. Oceania	10.8	0.3	0.3	5.2	1.5	2.7	77.8	26.5	9.2	0.0	1.4
11. Africa	40.6	4.1	2.3	21.2	4.6	4.9	21.7	8.8	5.5	0.6	0.0

Table 5.1 Manufactured products world trade matrix in 2009 (%).

Table 5.1 shows the 2009 world trade matrix for manufactured products based on Eurostat figures [EC (2011)]. Two Asian countries, China and India, are also shown separately. This matrix expresses relationships between the regions of the world according to the percentage of a country's exports. For example, Asia has approximately the same relationship between Europe (EU27) and North America, 32.0% and 32.5% of Asia's exports flowing to each respectively.

It is interesting to compare the trading patterns for the regions. The flows in the matrix imply the existence of backcloth structure such as trucking, shipping and airline routes between countries and the infrastructure that supports this.

A lot can be gained from the matrices by inspection. For example, column 1 for Europe (EU-27) has relatively large numbers, showing that Europe imports a relatively large proportion of the exports of the other regions. Column 1 shows Europe's exporting pattern is more limited, with the majority of its exports going to Asia (23.2%) and North America (22.0%), followed by Central and Eastern Europe (15.2%) and Other Western Europe (12.0%). These observations give an insight into the physical and cultural backcloth that supports this traffic. For example, the countries of Europe have a history or maritime exploration, discovery, conquest and trade with well established commercial and physical infrastructure to support the import of exotic and inexpensive goods and raw materials. The first column of the matrix reflects this. The other countries of Western, Central and Eastern Europe had more restricted access to seaways but good overland routes to Western Europe. Presumably these backcloth constraints are reflected in the majority of their exports (> 60%) going to the relatively rich Western Europe.

A more systematic comparative analysis of the traffic data can act as surrogate for the underlying backcloth connectives. There are many ways this could be done, but one of the simplest is *slicing*. For example, let the matrix in Table 5.1 be converted to a binary incidence matrix by *slicing out* values below 2% to become zero and *slicing in* all other values as 1 as shown in Table 5.2.

	EU	OW	CE	NA	LA	ME	AS	CH	IN	OC	AF
1.EU-27	1	1	1	1	1	1	1	1	1	1	1
2.Other W. Europe	1	1	1	1	0	1	1	1	0	0	0
3.Central & E. Europe	1	1	1	1	0	1	1	1	0	0	1
4.North America	1	1	0	1	1	1	1	1	1	1	1
5.Latin America	1	1	0	1	1	0	1	1	0	0	1
6.Middle East	1	1	1	1	1	1	1	1	1	0	1
7.Asia	1	0	1	1	1	1	1	1	1	1	1
8.China	1	0	1	1	1	1	1	1	1	0	1
9.India	1	0	0	1	1	1	1	1	1	0	1
10.Oceania	1	0	0	1	0	1	1	1	1	1	0
11.Africa	1	1	1	1	1	1	1	1	1	0	1

Table 5.2 Manufactured products world trade exports $\geq 2\%$ in 2009.

In Table 5.2 the diagonal entries have all been set to 1 since each country exports and imports to itself. Let E be the set of countries as exporters and I the set of countries as importers. Let the conjugate simplicial families associated with Table 5.2 be denoted $F_E(I, R_{2\%})$ and $F_I(E, R_{2\%})$. Then Europe dominates both $F_E(I, R_{2\%})$ and $F_I(E, R_{2\%})$ as exporter and importer, being related to all countries. More interesting at this level are the imports (columns) of the lower dimensional countries in $F_I(E, R_{2\%})$:

$$\sigma_{I,2\%}(\text{OW}) = \langle \text{EU, OW, CE, NA, LA, ME, AF} \rangle$$
$$\sigma_{I,2\%}(\text{CE}) = \langle \text{EU, OW, CE, ME, AS, CH, AF} \rangle$$
$$\sigma_{I,2\%}(\text{OC}) = \langle \text{EU, NA, AS, OC} \rangle$$

where $\sigma_{I,2\%}(\text{OW})$ shares the face $\langle \text{EU, OW, CE, AF} \rangle$ with $\sigma_{I,2\%}(\text{CE})$, but $\sigma_{I,2\%}(\text{OC})$ is relatively disconnected, sharing only the vertex $\langle \text{EU} \rangle$ with the other two, $\langle \text{OW, CE, OC} \rangle \leftrightarrow \langle \text{EU} \rangle$ meaning that these three relatively weak importing countries have only Europe as their common source of manufactured products. At the 10% level OC leaves the picture as an importer while OW and CE reman related to only EU (Table 5.3).

	EU	OW	CE	NA	LA	ME	AS	CH	IN	OC	AF
1.EU-27	1	1	1	1	0	0	1	0	0	0	1
2.Other W. Europe	1	1	0	0	0	0	1	0	0	0	0
3.Central & E. Europe	1	0	1	0	0	0	1	0	0	0	0
4.North America	1	0	0	1	1	0	1	0	0	0	0
5.Latin America	1	0	0	1	1	0	1	1	0	0	0
6.Middle East	1	0	0	1	0	1	1	0	0	0	0
7.Asia	1	0	0	1	0	1	1	1	0	0	0
8.China	1	0	0	1	0	0	1	1	0	0	0
9.India	1	0	0	1	0	1	1	0	1	0	0
10.Oceania	1	0	0	0	0	0	1	1	0	1	0
11.Africa	1	0	0	1	0	0	1	0	0	0	1

Table 5.3 Manufactured products world trade exports $\geq 10\%$ in 2009.

114 Hypernetworks in the Science of Complex Systems

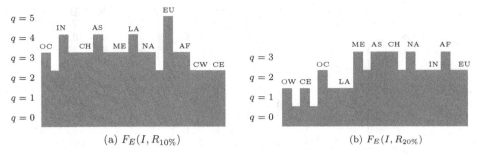

Fig. 5.9 Sky scraper diagrams Q-analysis of the export relations $R_{10\%}$ and $R_{20\%}$.

Figure 5.9(a) shows a Q-analysis of $K_E(I, R_{10\%})$ at the 10% level. Europe has the highest dimension at $q = 5$ with Africa as a 3-dimensional face. India, Asia and Latin America all have relatively high dimension at $q = 4$. $\sigma_{I,10\%}(\text{EU})$ and $\sigma_{I,10\%}(\text{AS})$ dominate $K_I(E, R_{10\%})$, both being related to all the other countries.

$K_E(I, R_{10\%})$ has two simplices, $\langle\text{EU, AS, MA}\rangle$ and $\langle\text{EU, AS, CH}\rangle$ forming significant Galois pairs:

$\langle\text{EU, AS, NA}\rangle \leftrightarrow \langle\text{EU, AF, NA, LA, ME, AS, CH, IN}\rangle$

$\langle\text{EU, AS, CH}\rangle \leftrightarrow \langle\text{LA, AS, CH, OC}\rangle$

	EU	OW	CE	NA	LA	ME	AS	CH	IN	OC	AF
1. EU-27	1	0	0	1	0	0	1	0	0	0	1
2. Other W. Europe	1	1	0	0	0	0	0	0	0	0	0
3. Central & E. Europe	1	0	1	0	0	0	0	0	0	0	0
4. North America	1	0	0	1	1	0	1	0	0	0	0
5. Latin America	0	0	0	1	1	0	0	0	0	0	0
6. Middle East	1	0	0	1	0	1	1	0	0	0	0
7. Asia	1	0	0	1	0	0	1	1	0	0	0
8. China	1	0	0	1	0	0	1	1	0	0	0
9. India	1	0	0	0	0	1	1	0	1	0	0
10. Oceania	0	0	0	0	0	0	1	1	0	1	0
11. Africa	1	0	0	1	0	0	1	0	0	0	1

Table 5.4 Manufactured products world trade exports $\geq 20\%$ in 2009.

Table 5.4 shows the matrix sliced at 20%. Fig. 5.9(b) shows five regions with dimension $q = 3$ at the 20% level, the Middle East, Asia, China, North America, and Africa. The structure has the Galois pair:

$\langle\text{EU, NA, AS}\rangle \leftrightarrow \langle\text{EU, ME, AS, CH, AF}\rangle$

with Europe, North America and Asia receiving twenty percent or more of the imports from these five regions.

The analysis above shows that slicing data matrices can give interesting insights. However, the selection of the slicing value affects the structure and its interpretation.

5.9 Matrices as backcloth supporting descriptor traffic

Figure 5.10(a) shows an images of a set of hand drawn vowels, the letters $A_1, ..., A_5, E_1, ..., E_5, I_1, ..., I_5, U_1, ..., U_5$. Figure 5.10(b) shows another image of hand drawn vowels, the letters $A'_1, ..., A'_5, E'_1, ..., E'_5, I'_1, ..., I'_5, U'_1, ..., U'_5$.

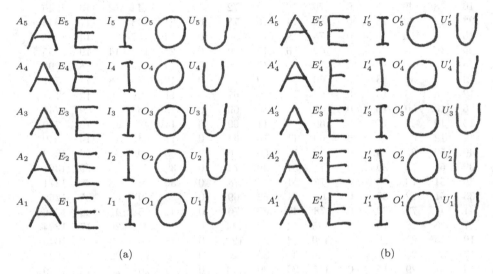

Fig. 5.10 Images of two sets of hand drawn vowels

At the top of Table 5.5 are fifteen 2×2 arrays of black and white pixels. Every block of four pixels in the image either has all white pixels (d_0, not shown) or one of these configurations. The blocks of pixels are denoted d_1 to d_{15} because they are what will be called *descriptor dimensions*. The table shows the frequency of occurrence of the pixel configurations for each of the vowel characters in the image shown in Fig. 5.10(a).

As can be seen, the E characters have relatively high values for the configurations d_3, ▀▀, and d_{12}, ▄▄, corresponding to the three horizontal strokes in the E characters. Similarly, the U characters have relatively high values for the configurations d_6, ▐▐, and d_9, ▌▌, corresponding to the two vertical strokes in the U characters. The configurations d_5, ▜▜, and d_{10}, ▙▙ do not occur in the images for any of the characters in these images, shown by the columns of zeros.

Thus each character is mapped to a point in 14-dimensional space. What kind of scales are these, what is the backcloth, and what is the traffic? In general the backcloth is formed by objects and relations between them. In this example there are two sets of objects, the letters and pixel configuration descriptors. They are called *descriptors* because the numbers associated with the characters describe them.

	d_1	d_2	d_3	d_4	d_5	d_6	d_7	d_8	d_9	d_{10}	d_{11}	d_{12}	d_{13}	d_{14}	d_{15}
A_1	51	40	37	39	0	65	40	31	62	0	51	58	31	39	1188
A_2	42	39	32	36	0	67	39	29	71	0	42	48	29	36	1098
A_3	40	39	37	32	0	64	39	30	65	0	40	54	30	32	1108
A_4	41	37	33	32	0	73	37	26	75	0	41	53	26	32	1114
A_5	36	52	44	32	0	84	52	40	92	0	36	60	40	32	1357
E_1	16	8	149	15	0	79	6	14	72	0	14	144	12	13	1345
E_2	22	12	143	18	0	86	11	16	78	0	21	143	15	17	1344
E_3	15	10	164	15	0	73	9	17	66	0	14	157	16	14	1393
E_4	25	17	131	20	0	62	16	15	59	0	24	138	14	19	1255
E_5	15	12	139	9	0	76	11	17	65	0	14	140	16	8	1295
I_1	13	11	36	13	0	69	10	9	71	0	12	38	8	12	757
I_2	12	6	45	13	0	67	5	7	67	0	11	43	6	12	751
I_3	6	10	50	10	0	65	9	10	69	0	5	46	9	9	748
I_4	15	10	44	16	0	69	9	11	69	0	14	42	10	15	779
I_5	12	12	67	11	0	69	11	16	64	0	11	64	15	10	850
O_1	40	50	68	39	0	77	50	50	76	0	40	69	50	39	1373
O_2	39	51	73	39	0	70	51	49	72	0	39	75	49	39	1363
O_3	36	51	73	38	0	73	51	48	78	0	36	74	48	38	1371
O_4	33	51	72	35	0	68	51	52	69	0	33	69	52	35	1258
O_5	41	47	73	39	0	78	47	47	76	0	41	75	47	39	1346
U_1	17	25	28	18	0	145	24	27	144	0	16	25	26	17	1224
U_2	19	25	22	19	0	120	24	24	121	0	18	23	23	18	1045
U_3	17	31	26	17	0	125	30	30	126	0	16	27	29	16	1095
U_4	14	27	29	15	0	115	26	29	114	0	13	26	28	14	1009
U_5	20	28	28	18	0	121	27	28	119	0	19	30	27	17	1072

Table 5.5 Pixel configuration counts for the characters in Fig. 5.10

Let c_i represent the ith vowel character and $d_j(c_i)$ be the number of times configuration d_j occurs in vowel character c_i. It could be said that a descriptor d_i is related to a character c_j if it occurs in the image at least once, $d_i(c_j) > 0$. In this case, since $d_5(c_j) = 0$ and $d_{10}(c_j) = 0$ for all j, every character is associated with the *descriptor simplex* $\langle d_1, d_2, d_3, d_4, d_6, d_7, d_8, d_{10}, d_{11}, d_{12}, d_{13}, d_{14}, d_{15} \rangle$. This single simplex acts as the backcloth for all the descriptor configuration counts, a pattern of fifteen numbers for each vowel.

Inspection of Table 5.5 shows that some characters have relatively low or high counts. For example, the characters E_1, E_2, E_3, E_4, E_5, I_1, I_2, I_3, I_4, I_5 have the lowest counts for configuration, d_7, namely 6, 11, 9, 16, 11, 10, 5, 9, 9 and 11. The next lowest value is 24 for U_1 and U_2, so there is a gap on this dimension between the values 16 and 24 separating the E and I characters from the others. Similarly, of the remaining configuration counts, U_1, U_2, U_3, U_4 and U_5 all have counts of 30 or less. Thus the d_7 dimension could be divided into d_1-Low, d_1-Medium (abbreviated to d_1-Mid) and d_1-High corresponding to the intervals [0, 20], [21,33] and [34, ∞).

d_1	d_2	d_3	d_4	d_6	d_7	d_8	d_9	d_{11}	d_{12}	d_{13}	d_{14}	d_{15}
6	6	22	9	62	5	7	59	5	23	6	8	748
12	8	26	10	64	6	9	62	11	25	8	9	751
12	10	28	11	65	9	10	64	11	26	9	10	757
13	10	28	13	65	9	11	65	12	27	10	12	779
14	10	29	13	67	9	14	65	13	<u>30</u>	12	12	<u>850</u>
15	11	32	15	67	10	15	66	14	38	14	13	1009
15	12	33	15	68	11	16	67	14	42	15	14	1045
15	12	36	15	69	11	16	69	14	43	15	14	1072
16	12	37	16	69	11	17	69	14	46	16	15	1095
17	17	37	17	69	<u>16</u>	17	69	16	48	16	16	1098
17	25	44	18	70	24	24	71	16	53	23	17	1108
19	25	44	18	73	24	26	71	18	54	26	17	1114
20	27	45	18	73	26	27	72	19	58	26	17	1188
22	28	50	19	73	27	28	72	21	60	27	18	1224
<u>25</u>	<u>31</u>	67	<u>20</u>	76	<u>30</u>	29	75	<u>24</u>	<u>64</u>	28	<u>19</u>	1255
33	37	<u>68</u>	32	77	37	29	76	33	69	29	32	1258
36	39	72	32	78	39	30	76	36	69	29	32	1295
36	39	73	32	79	39	30	78	36	74	30	32	1344
39	40	73	35	84	40	31	78	39	75	31	35	1345
40	47	73	36	<u>86</u>	47	<u>40</u>	<u>92</u>	40	<u>75</u>	40	36	1346
40	50	131	38	115	50	47	114	40	138	47	38	1357
41	51	139	39	120	51	48	119	41	140	48	39	1363
41	51	143	39	121	51	49	121	41	143	49	39	1371
42	51	149	39	125	51	50	126	42	144	50	39	1373
51	52	164	39	145	52	52	144	51	157	52	39	1393

Table 5.6 Dividing the descriptor dimensions into intervals for Q-analysis

Table 5.6 shows the pixels counts for the configurations in increasing order, and underlines entries that could be used as thresholds for dividing the scales. Then the set of intervals is $D = \{d_1$-Low, d_1-High, d_2-Low, d_2-High, d_3-Low, d_3-High, d_4-Low, d_4-High, d_5-Low, d_6-High, d_7-Low, d_7-Mid, d_7-High, d_8-Low, d_8-High, d_9-Low, d_9-High, d_{11}-Low, d_{11}-High, d_{12}-Low, d_{12}-LMid, d_{12}-HMid, d_{12}-High, d_{13}-Low, d_{13}-High, d_{14}-Low, d_{14}-High, d_{15}-Low, d_{15}-High$\}$.

This establishes a binary relation between the set of vowel characters, C, and the interval descriptors D. For example,

$$\sigma(A_1) = \langle d_1\text{-High}, d_2\text{-High}, d_3\text{-Low}, d_4\text{-High}, d_6\text{-Low}, d_7\text{-High}, d_8\text{-Low},$$
$$d_9\text{-Low}, d_{11}\text{-High}, d_{12}\text{-LMid}, d_{13}\text{-High}, d_{14}\text{-High}, d_15\text{-High}\rangle.$$

The Skyscraper diagram for a Q-analysis of this relation is shown in Fig. 5.11, where the Q-analysis partitions the characters into three groups corresponding to A, E, I, O and U.

118   Hypernetworks in the Science of Complex Systems

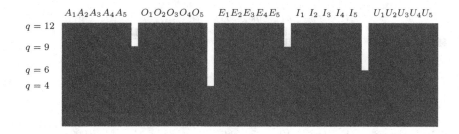

Fig. 5.11   The skyscraper diagram for the Q-analysis of vowel characters versus intervals

As can be seen in the Q-analysis, all the simplices have the same dimension, $q = 12$, since they are all related to thirteen of the intervals. The simplices for each of the A characters are all the same, as are those for each of the E, I, O and U characters. This is why they form blocks at $q = 12$ in Fig. 5.11.

The skyscraper diagram also shows that the A characters are most similar to the O characters at $q = 9$, and that the E characters are most similar to the I characters, also at $q = 9$. However, the A and O characters are clearly separated from the others becoming connected only at $q = 4$. In contrast, the U characters form a relatively disconnected component until $q = 6$.

In this case the numbers in the matrix can be both viewed as traffic, and be used to defined backcloth structures. Here the backcloth structure provides a way of classifying or recognising the characters, and it is interesting to ask if the process generalises. When the vowel characters in Fig. 5.10(b) are added the Q-analysis shows again the that vowels are highly connected by type (Fig. 5.12).

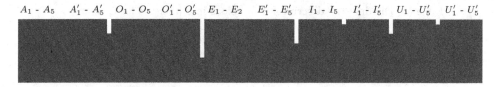

Fig. 5.12   Generalisation: the A, E, I, O and U characters form clusters in the Q-analysis

Although it is interesting that these simple pixel configurations can discriminate the handwritten vowels, this example is presented for illustration rather than building a character recognition system. Although the method generalised for the examples given here, it is unlikely that it would work in all cases.

5.10 Q-transmission in simplicial complexes

The concept of q-transmission is based on the idea that q-connectedness constrains the dynamics of system behaviour. In particular it is assumed that that changes will be transmitted differently between simplices if they are q-connected as opposed to be being $(q-1)$-connected.

Let σ_1 and σ_2 be two q-near simplices in a simplicial family supporting a traffic mapping f. Suppose that a change $\delta f \sigma_1$ of the value of f on σ_1 causes a change $\delta f \sigma_2$ of the value of f on σ_2. Then the change δf is said to be q-*transmitted* from σ_1 to σ_2, as illustrated in Fig. 5.13.

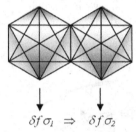
(a) q-transmission from σ_1 to σ_2

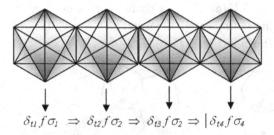
(b) q-transmission along a chain of q-connection

Fig. 5.13 q-transmission, for $q = 1$

A mapping on a simplicial family, f, is said to have a q-*transmission property* if

$$\left.\begin{array}{ll}\text{(i)} & \sigma_q \lesssim \sigma_p \\ \text{(ii)} & \sigma'_q \lesssim \sigma_p \\ \text{(iii)} & \delta f_t \sigma_q \neq 0 \\ \text{(iv)} & \delta f_t \sigma'_q = 0\end{array}\right\} \text{ together imply that } \delta f_{t+1}\, \sigma'_q \neq 0 \text{ for time } t+1 > t$$

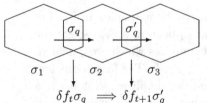
(a) the q-transmission property

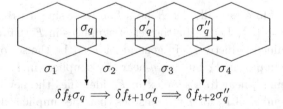
(b) propagation of change due to q-transmission

Fig. 5.14 The q-transmission property and the propagation of traffic changes across the backcloth

The q-transmission property is illustrated in Fig. 5.14(a) where the simplices σ_1, σ_2, are σ_3 are pairwise q-near. (i) Let $\sigma_q = \sigma_1 \cap \sigma_2$, so $\sigma_q \lesssim \sigma_2$. (ii) Let $\sigma'_q = \sigma_2 \cap \sigma_3$, so $\sigma'_q \lesssim \sigma_2$. (iii) Suppose there is a change in f at time t on σ_1, $\delta f_t \sigma_q \neq 0$. (iv) Then there will be a change $\delta f_{t+1} \sigma'_q \neq 0$ on σ'_q.

Consider Fig. 5.14(b) which extends Fig. 5.14(a) by another simplex, σ_4 with σ_3 and σ_4 also q-near. (i) as before $\sigma'_q = \sigma_2 \cap \sigma_3$, so $\sigma'_q \lesssim \sigma_3$. (ii) let $\sigma''_q = \sigma_3 \cap \sigma_4$, so $\sigma''_q \lesssim \sigma_3$. (iii) we have just showed that given a change $\delta f_t \sigma_q \neq 0$ there is a change $\delta f_{t+1} \sigma'_q \neq 0$. (iv) therefore there will be a change $\delta f_{t+2} \sigma'' q \neq 0$ on σ''_q. Thus the q-transmission property results in changes being propagated along chains of q-connection. Here change on σ_1 is q-transmitted through σ_2 and σ_3 to σ_4.

q-transmission fronts

The general idea behind q-transmission is that change will be propagated from a source throughout a simplicial family in a way conditioned by its q-connectivity. Figure 5.15(a) shows a chain of q-connection, and Fig. 5.15(b) shows how such chains can make up q-transmission fronts, through which changes are transmitted, as shown in Fig. 5.15(c).

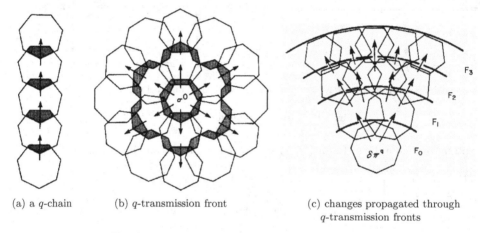

(a) a q-chain   (b) q-transmission front   (c) changes propagated through q-transmission fronts

Fig. 5.15 q-transmission through q-transmission fronts

More formally, let F be a family of simplices. Let σ be a simplex in F. Let $F_0 = \{\sigma\}$. Let F_1 be the set of simplices in F that are q-near σ_0. F_2, F_3, etc. be defined inductively. In general let F_{k+1} be the set of simplices which are q-near to a simplex in F_k but not q-near any simplices in F_j for , $j < k$. The first condition, being q-near to a simplex in F_k identifies the new q-near simplices. Not being a member of F_j for $j < k$, ensures that the simplices do not belong to previous transmission fronts. Thus the transmission fronts are disjoint sets of simplices.

q-transmission and q-percolation

The idea of q-transmission contrasts with percolation in networks. The q-transmission property requires that q-dimensional simplices are required for change to be transmitted rather than vertices. A system will be said to q-percolate when the simplices form a single q-connected component.

q-transmission, time, and dynamics

Let f have the q-transmission property, and let F_0, F_1, F_2, be the q-transmission fronts associated with simplex σ, as shown in Fig. 5.15. Then a change δf will be q-transmitted to a q-face of every simplex in transmission front F_τ at time $t = \tau$. This follows from the definitions and the construction of the transmission fronts. It gives an insight into the dynamics of simplicial families, and how the backcloth topology constrains those dynamics. It is however a special case.

Pure chains of q-connection as shown in Fig. 5.15(a) are rare. Generally chains of simplices are connected by mixed q-values. The lowest q-value will define q-chains, but the combinatorial picture is more complicated than this.

The time interval is assumed to be one tick of the clock, but the nature of the clock is not specified. For applications, one tick of the clock underlies computer simulation which can be used to investigate the consequences of the connectivity of the backcloth.

q-transmission and the q-graph

Let S be a simplicial family and let $S_q = \{\sigma_{p,i} \mid \text{ for } i \in I_q\}$ where I_q is the index set for simplices in S with dimension $p \geq q$.

For each $\sigma_{p,i}$ in S let there be a vertex $v(\sigma_{p,i})$ in a new structure called the q-graph of S, $G_q(S)$. Let $v(\sigma_{p,i})$ be called a *simplex vertex*. Let there be an edge in $G_q(S)$ between $v(\sigma_{p,i})$ and $v(\sigma_{r,j})$ if and only if $\sigma_{p,i}$ and $\sigma_{r,i}$ are q-near. This is illustrated in Fig. 5.16.

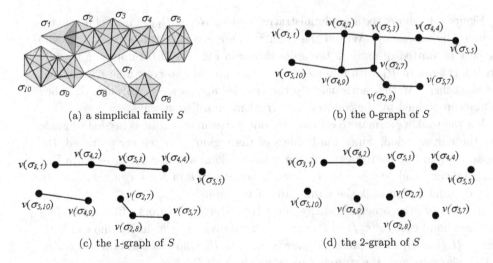

Fig. 5.16 q-transmission through q-transmission fronts

The q-graph constrains q-transmission. A necessary condition for q-transmission is that two simplex vertices are connected in the q-graph. Thus 0-transmission is possible between all simplices in Fig. 5.16, but 1-transmission is more constrained while 2-transmission is only possible between σ_1 and σ_2.

Although it is possible to interpret q-transmission as change over the q graph, there is an ambiguity in the q-graph between three q-connected simplices having a q-dimensional hub or a q-dimensional hole, as shown in Fig. 4.29.

5.11 Examples

Example: q-transmission in social-geographical space

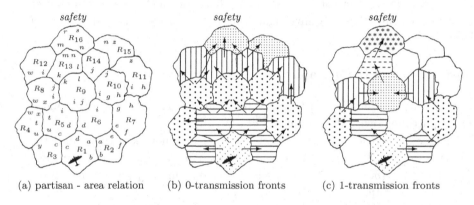

(a) partisan - area relation (b) 0-transmission fronts (c) 1-transmission fronts

Fig. 5.17 q-transmission wounded aircrew behind enemy lines

Figure 5.17 shows sixteen administrative regions $R_1, ..., R_{16}$ in part of a country occupied by an enemy. Within the country there are partisans $a, b, ... , z$ who are only permitted to move in the areas shown in Fig. 5.17(a). Suppose a friendly aircraft crashes in R_1 with one member of the injured two-crew able to walk but not the other. As the partisans help the crew escape to safety through R_{16} they will be constrained by the connectivity structure in different ways.

For the walking airman to escape only one partisan at a time is needed to guide him through a region. Since the borders of the region are intensely patrolled, the handover from one partisan to another must be done within a region, e.g. in Fig. 5.17(b) d can hand over to i in R_6, i can hand over to k in R_9, k can hand over to m in R_{13} and m can walk the airman to safety through R_{16}.

In contrast, the other airman requires two stretcher bearers. In Fig. 5.17(c) $\langle a,b \rangle$ can hand over to $\langle e,f \rangle$ in R_3 who can hand over to $\langle g,h \rangle$ in R_7 who can hand over to $\langle i,j \rangle$ in R_{10} who can hand over to $\langle k,l \rangle$ in R_9 who can hand over to $\langle m,n \rangle$ in R_{13} who can carry the stretcher to safety through R_{16}.

This shows how the backcloth constrains q-transmission, with the dynamics of systems dependent on q. Here 0-paths are shorter and more numerous than 1-paths.

Example: q-transmission through departmental secretaries

Consider a department with five secretaries serving twenty members of authoring staff in a publishing company. Each secretary formats the work for six authors according to the skills needed for their specialisms, *e.g.* a_1 writes on sport and photographs are very important while a_{12} writes on economic matters with lots of tables and graphs. The secretaries share some skills, but can only support the authors as shown in Table 5.6.

	a_1	a_2	a_3	a_4	a_5	a_6	a_7	a_8	a_9	a_{10}	a_{11}	a_{12}	a_{13}	a_{14}	a_{15}	a_{16}	a_{17}	a_{18}	a_{19}	a_{20}
s_1	1	1	1	1	1	1	0	0	0	0	0	0	0	0	0	0	0	0	0	0
s_2	0	0	0	0	1	1	1	1	1	1	0	0	0	0	0	0	0	0	0	0
s_3	0	0	0	0	0	0	0	1	1	1	1	1	1	0	0	0	0	0	0	0
s_4	0	0	0	0	0	0	0	0	0	0	1	1	1	1	1	1	0	0	0	0
s_5	1	1	0	0	0	0	0	0	0	0	0	0	0	0	0	0	1	1	1	1

Table 5.7 The relation between authors and secretaries

Suppose all the secretaries are busy and authors, a_3 and a_4 bring a coauthored piece of work their secretary s_1. This work is associated with the face $\langle a_3, a_4 \rangle$ of $\sigma(s_1)$. In this case the only way s_1 can do the work is to shuffle some of her other work on to s_2 through the shared faces $\langle a_5, a_6 \rangle$, $\langle a_5 \rangle$ and $\langle a_6 \rangle$, or to S_5 through the shared faces $\langle a_1, a_2 \rangle$, $\langle a_2 \rangle$ and $\langle a_1 \rangle$. In order to do this s_2 or s_3 may need to shuffle some work onto s_3 and s_4 through their shared face. In this case the work can be 1-transmitted through the transmission fronts $F_0(s_1) = \{s_1\}$, $F_1(s_1) = \{s_1, s_5\}$, $F_2(s_1) = \{s_3, s_4\}$, and the structure has transmission number 2.

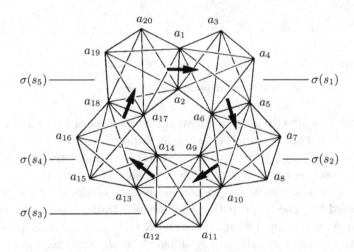

Fig. 5.18 The author-secretary structure

Example: Wage dynamics on the job-company backcloth

The wages traffic in the employment market is subject to many forces for change, *e.g.* the use of incentives and rewards to the workforce when a company does well, the notion of the "rate for the job", and the maintenance of "pay differentials" to reflect qualifications, skills and effort required.

For example, consider the companies c_1, c_2 and c_3 a set of jobs $\{j_1, j_2, ... j_{14}\}$. Let job j_k be related to company c_i if job k_k can be found in that company, as shown in Fig. 5.19. Suppose that company c_1 is being successful and gives its employees an increase of $x\%$ in their wages.

Then the employees with jobs j_5 and j_6 in company c_2 will be paid less than employees with the same job in c_1. Suppose the "rate for the job" lets them successfully demand an increase $x'\%$ from their management.

Then the other workers in company c_2 with jobs $j_7, ..., j_{10}$ may use the "wage differential" principle to successfully demand the $x'\%$ increase.

Now workers in company c_3 with job j_{10} will be paid less than employees in c_2 with job in j_{10}. Then the "rate for the job" principle applies and these workers may successfully press the c_3 employers to increase their wages by $x''\%$.

After this the "wages differential" principle kicks in, and the other workers in c_3 may successfully demand an an increase of $x''\%$ in their wages.

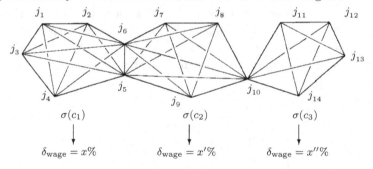

Fig. 5.19 Transmission of wage increases

This is a simplistic description of the dynamic adjustment of wages and salaries, and many other forces are at work. For example, if company c_2 resists the pressure to increase wages on $\langle j_5, j_6 \rangle$ there may be an undesirable traffic of employees from c_2 to c_1 as employees seek to switch to the better paid job in company c_1.

Alongside the wages paid traffic, people have "expectations" traffic of what their wages should be. The "expected rate for the job" traffic depends on absolute criteria such as qualifications and relative criteria such as what other people are paid. When the actual pay rate is very different to the expected rate workers experience dissatisfaction traffic. When many workers experience this there may be "low morale" and "alienation" traffic which may damage or even destroy the backcloth.

Backcloth and traffic: dynamics constrained by topology

Example: q-transmission of jokes through public houses

Figure 5.20 shows a set of British pubs with vertices the people who visit them regularly to socialise. Suppose there is an elaborate joke that requires two people to tell properly, one as the "female-funny-guy" and the other as the "straight-man", with the pair forming the simplex ⟨female-funny-guy, straight-man⟩.

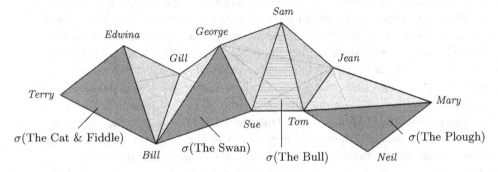

Fig. 5.20 A family of pubs and clients

Suppose Terry and Edwina make up a joke to tell to their friends in The Cat & Fiddle pub. Jokes spread like wildfire within a pub and so within a short time everyone in the pub has heard Terry and Edwina's joke, including Gill and Bill. Since ⟨Gill, Bill⟩ is a face of both σ(The Cat & Fiddle) and σ(The Swan) these two people can learn the joke in The Cat & Fiddle and tell it in The Swan. Everyone in the The Swan pub soon learns the new joke including the pair ⟨Sue, George⟩ which is also a face of The Bull. The 1-simplex ⟨Jean, Tom⟩ allows the joke to be transmitted from The Bull to The Plough, and to the pair ⟨Mary, Neil⟩. In this case the chain of 1-connection between The Cat & Fiddle and The Plough is necessary for the joke to be transmitted.

In this example the joke can be transmitted through the 1-connected components of the pub-people structure. It cannot be transmitted to components that are only 0-connected, and it cannot be transmitted to components that are disconnected from the origin of the joke.

In cities pubs are highly connected by their clients, and jokes are transmitted very quickly. Even pubs in remote rural areas share customers with more urban pubs and it seems that, eventually, every joke reaches every pub. No doubt there are "long-linking" polyhedra and "small world" 1-connectivities aiding this kind of transmission.

Most jokes can be told by just one person and the minimum structure needed to transmit them is single shared vertex. One-person jokes certainly travel very quickly indeed. Perhaps the need for a transmission structure inhibits the formulation of high dimensional jokes? A joke that took, say, ten people to tell properly would have a low chance of being transmitted from pub to pub.

5.12 Dimensions, probabilities and q-transmission

The idea underlying q-transmission is that q-nearness is sufficient for change to be transmitted while $(q-1)$-nearness not. Although there are non-trivial examples of q-transmission, it seems that in the great majority of cases sharing just one vertex is sufficient for change to be transmitted between two simplices. For example, if just one person sits on two committees, this is sufficient for information to flow between them. Similarly, if two pubs are connected by just one person, this is sufficient for a joke to be transmitted from one to the other. On the other hand, sharing these people does not guarantee the information or jokes will be transmitted. More precisely, in the terms of Section 5.3, being disconnected *forbids* transmission, while sharing just one vertex *allows* transmission, but *does not require* transmission.

Let $prob(\sigma_i, x, t)$ be the probability that simplex σ_i has property x at time t. Let the probability of change being transmitted from one simplex to another through a zero-dimensional face $\langle v_i \rangle$ be p. Then in Fig. 5.21(a) the probability of x being transmitted from σ to $\langle v_1, v_2 \rangle$ in $[t, t+1]$ is p, the probability of x being transmitted from $\langle v_1, v_2 \rangle$ to $\langle v_2, v_3 \rangle$ in $[t+1, t+2]$ is p^2. In general the probability of x being transmitted from $\langle v_{t-1}, v_t \rangle$ to $\langle v_t, v_{t+1} \rangle$ between times t and $t+1$ is p^t. For example, $prob(\sigma', x, t+4) = .0001$ for $p = 0.1$. In comparison, $prob(\sigma', x, t+4) \simeq 16p^4 - 32p^3 \simeq .0013$ for $p = 0.1$ in Fig. 5.21(b) for a 1-connected chain. This is an order of magnitude higher.

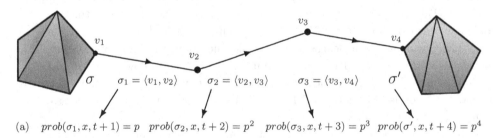

(a) $prob(\sigma_1, x, t+1) = p$ $prob(\sigma_2, x, t+2) = p^2$ $prob(\sigma_3, x, t+3) = p^3$ $prob(\sigma', x, t+4) = p^4$

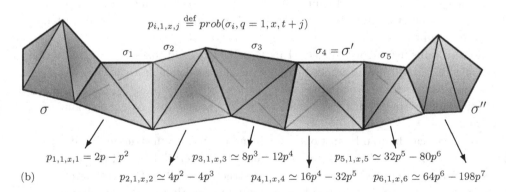

(b) $p_{1,1,x,1} = 2p - p^2$ $p_{3,1,x,3} \simeq 8p^3 - 12p^4$ $p_{5,1,x,5} \simeq 32p^5 - 80p^6$
   $p_{2,1,x,2} \simeq 4p^2 - 4p^3$ $p_{4,1,x,4} \simeq 16p^4 - 32p^5$ $p_{6,1,x,6} \simeq 64p^6 - 198p^7$

Fig. 5.21 Probabilistic q-transmission

The probabilities in Fig. 5.21(b) are calculated as follows. Let σ_1 and σ_2 be 1-near sharing the 1-dimensional face $\langle v_1, v_2 \rangle$, where σ_1 has property x. It is assumed that the probabilities of transmission of x across the vertices are independent, and equal to p for both v_1 and v_2. Then there are two chances for x to be transmitted from σ_1 to σ_2, first across v_1 and second across v_2. However the probability of transmission from σ_1 to σ_2 will be less than $2p$. This is because the probability of x not being transmitted to σ_2 across $\langle v_1 \rangle$ is $(1-p)$ and the probability of x not being transmitted to σ_2 across $\langle v_2 \rangle$ is $(1-p)$. Thus the probability of x *not* being transmitted from σ_1 to σ_2 is $(1-p) \times (1-p) = 1 - 2p + p^2$, and the probability of it being transmitted is $1 - (1 - 2p + p^2) = 2p - p^2$. The term p^2 is the probability of x being transmitted across both v_1 and v_2. Thus the probability of x being transmitted to σ_n at time $t+n$ in Fig. 5.21(b) is $(2p - p^2)^n$.

By a similar argument the probability of x being transmitted to σ_n at time $t+n$ along a 2-dimensional chain of connection is $(3p - 3p^2 + p^3)^n$, and the probability of x being transmitted to σ_n at time $t+n$ along a 3-dimensional chain of connection is $(4p - 6p^2 + 4p^3 - p^4)^n$. Table 5.8 shows some illustrative calculations for $p = 0.1$ and $p = 0.5$. As can be seen the dimension of the connectivity has a significant impact on the probability of x being transmitted.

		$t+1$	$t+2$	$t+3$	$t+4$	$t+5$	$t+6$
$p=0.1$	$q=0$	0.100	0.0100	0.00100	0.0001	0.00001	0.000001
$p=0.1$	$q=1$	0.190	0.0361	0.00686	0.0013	0.00025	0.000047
$p=0.1$	$q=2$	0.271	0.0734	0.01990	0.0054	0.00147	0.000396
$p=0.1$	$q=3$	0.364	0.1324	0.04819	0.0175	0.00638	0.002322
$p=0.5$	$q=0$	0.500	0.2500	0.12500	0.0625	0.03125	0.015625
$p=0.5$	$q=1$	0.750	0.5625	0.42188	0.3164	0.23731	0.177979
$p=0.5$	$q=2$	0.875	0.7656	0.66992	0.5862	0.51291	0.448795
$p=0.5$	$q=3$	0.938	0.8789	0.82397	0.7725	0.72420	0.678934

Table 5.8 Illustrative probabilities for q-transmission

Of course these calculations assume isolated chains of connection that are unlikely to be encountered in practice. However they make the point that higher dimensional connectivities facilitate transmission considerably.

Heterogeneous transmission

The calculations above assume that the probabilities of transmission through the vertices are independent. If this were the case there is no need to restrict the analysis to q-connected simplices and one could consider chains of connection with different dimensions between the simplices. In Chapter 8 the process of a simplex gaining a property x will be called an *event* and transmission between events with heterogeneous dimensions will be the norm.

5.13 Supertraffic

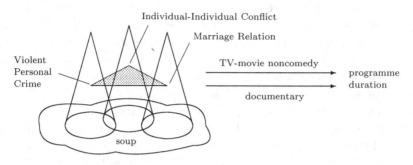

Fig. 5.22 Supertraffic on the programme content backcloth makes ([Gould *et al* (1984)])

A study of the international flow of television programs [Gould *et al* (1984)] found that typical classifications had classes such as *News, Current Affairs, Drama, Documentary, Sport* and *Children's programmes*. The first two of these reflect the *topicality* of the content, *News* being "now" and *Current Affairs* being "recent". *Drama* and *Documentary* reflect the fictional or factual *form* of the programme. *Sport* reflects the *content*, while *Children's Programmes* reflects the expected *audience*. As Fig. 5.22 shows, the same content backcloth with different form traffic can make completely different programmes, *e.g.* ⟨Marriage Relation, Individual-Individual Conflict, Violent Personal Crime; $R_{\text{TV movie}}$⟩ could be a classic Kojak episode, while ⟨Marriage Relation, Individual-Individual Conflict, Violent Personal Crime; $R_{\text{documentary}}$⟩ could be a factual programme on domestic violence.

This study suggested that the different ways of presenting the content backcloth could be combined into *supertraffic* as illustrated in Fig. 5.23 where the traffic combinations ⟨child education, ethical concern, documentary⟩ and ⟨ethical concern, documentary, aesthetic enrichment⟩ map the programme content to transmission times. These combinations of mappings can also be associated with the relations $R_{\text{child education}} \wedge R_{\text{ethical concern}} \wedge R_{\text{documentary}}$ and $R_{\text{ethical concern}} \wedge R_{\text{documentary}} \wedge R_{\text{aesthetic enrichment}}$.

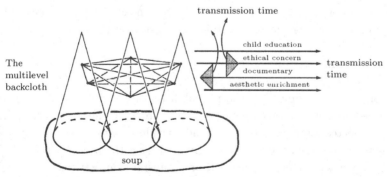

Fig. 5.23 Structured traffic on the structured backcloth ([Gould *et al* (1984)])

5.14 The topology of the backcloth

Intuitively two structures are topologically equivalent if they can be transformed into each other by stretching without tearing, for example a teacup is topologically equivalent to a wedding ring. The term topology is used ambiguously in network theory, often meaning the network structure rather than the topological structure.

Intuitively, two networks $N = (V, E)$ and $N' = (V', E')$ are structurally equivalent if they can be transformed into each other preserving their connectivity properties. More formally let ϕ be a one-to-one onto mapping between N and N' assigning ϕv in V' to each v in V, and $\phi(v_1, v_2)$ in E' to each edge (v_1, v_2) in E. Then N and N' are *structurally equivalent* if such a mapping ϕ exists with $\phi(v_1, v_2) = (\phi v_1, \phi v_2)$. Apart from the number of vertices and edges being the same, this definition ensures that the edges are connected to the vertices in the same way.

Figures 5.24(a) and (b) show two simple networks. Each has four vertices and four edges. However v_5 has degree 3 while the vertices v_1, v_2, v_3 and v_4 all have degree 2 and none can be mapped to v_5 in a way that preserves edge structure.

Figure 5.24(c) contains a loop or cycle but is not structurally equivalent to any of the other networks with cycles because it has only three vertices.

In comparison, the networks in Fig. 5.24(d) and (e) are structurally equivalent as demonstrated by the mapping $\phi v_{12} = v_{18}$, $\phi v_{13} = v_{17}$, $\phi v_{14} = v_{16}$ and $\phi v_{15} = v_{19}$. This case illustrates the generality that the the structure preserving mapping need not be unique since $\phi v_{14} = v_{19}$ and $\phi v_{15} = v_{16}$ would also be structure preserving.

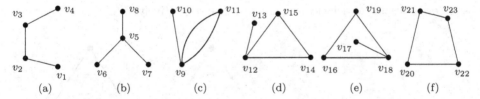

Fig. 5.24 Structural equivalence versus topological equivalence

In contrast to structural equivalence, topology is concerns the way that objects can be deformed into each other without tearing, *e.g.* a cylinder can be squashed to half its height without tearing, and the squashed and unsquashed cylinders are topologically equivalent. The squashed cylinder can be squashed again and again until it becomes a circle and, perhaps surprisingly, cylinders and circles are topologically equivalent in terms of having a single "hole" and the same homotopy type.

A curved line between two vertices is topologically equivalent to a straight line between those vertices. Thus each curved line in Fig. 5.24(c) is topologically equivalent to a straight line between v_9 and v_{10}. A line between two vertices can also be squashed to half it length, with the short and long line being topologically equivalent. Repeated squashing will transform the line to a single vertex. Thus an isolated line between two vertices is topologically equivalent to the vertices.

However this is a *local* perspective. A *global* perspective can reveal other properties. For example, the two edges between v_9 and v_{11} together form a *loop*, and together they cannot be distorted to a point without breaking one of the edges. Loops are examples of holes and generally structures with different numbers of holes have different topological structures. For example, the capital letters of the alphabet form three sets of topologically equivalent objects: {C, E, F, G, H, I, J, K, L, M, N, R, S, T, U, V, W, X, Y, Z}, {A, D, O, P, Q}, and {B}. These classes correspond to a point, a disk with a hole, and a disk with two holes.

From this topological perspective, in Fig. 5.24 networks (a) and (b) are equivalent to a point, while networks (c), (d), (e) and (f) are equivalent to a circle.

5.15 Algebraic topology

Q-analysis and hypernetwork theory have their origins in algebraic topology, a branch of mathematics that allows topological properties to be expressed algebraically, making them easier to compute.

Simplex Orientation

The difference between the set $\{v_0, v_1, ..., v_p\}$ and the simplex $\langle v_0, v_1, ..., v_p \rangle$ is that the order of the vertices matters. Simplices are said to be *oriented* by the order of their vertices. For example, below $\langle a, b \rangle$ and $\langle b, a \rangle$ have opposite orientations, the former from a to b and the latter from b to a. Similarly $\langle a, b, c \rangle$ and $\langle c, b, a \rangle$ have "opposite" orientations, anticlockwise and clockwise respectively.

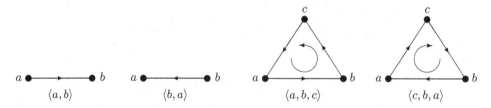

Fig. 5.25 The order of the vertices induces an orientation on a simplex

If the edge $\langle a, b \rangle$ in Fig. 5.25 carries a flow of, say, $f\langle a, b \rangle$ from a to b it could be said that $\langle b, a \rangle$ carries a flow of *minus* $f\langle a, b \rangle$ from b to a. This suggests the association $f\langle a, b \rangle \stackrel{\text{def}}{=} -f\langle b, a \rangle$, and giving simplices a positive or negative *orientation* as $\langle a, b \rangle \stackrel{\text{def}}{=} -\langle b, a \rangle$.

This generalises to the triangles with $\langle a, b, c \rangle \stackrel{\text{def}}{=} -\langle c, b, a \rangle$. Given this, what are the orientations of $\langle a, c, b \rangle$, $\langle b, a, c \rangle$, $\langle b, c, a \rangle$ and $\langle c, a, b \rangle$? In Fig. 5.25 it can be seen that $\langle a, c, b \rangle$ "goes round" the triangle with the same clockwise orientation as $\langle c, a, b \rangle$, effectively starting at c instead of a. Similarly, $\langle b, a, c \rangle$ also has a clockwise orientation, effectively starting at b. A similar argument suggests that $\langle b, c, a \rangle$ and

$\langle c, a, b \rangle$ have an anticlockwise orientation. Suppose triangles with the same orientation are defined to be equal. Then:

$$\langle a,b,c\rangle = -\langle a,c,b\rangle = \langle b,c,a\rangle = -\langle b,a,c\rangle = \langle c,a,b\rangle = -\langle c,b,a\rangle.$$

This sequence is obtained by starting with clockwise $\langle a, b, c \rangle$, swapping b and c, and changing the sign to get anticlockwise $-\langle a, c, b \rangle$. Swapping a and b and changing the sign gives the clockwise $\langle b, c, a \rangle$. Swapping a and c and changing the sign gives the anticlockwise $-\langle b, a, c \rangle$. Swapping a and c and changing the sign gives the clockwise $\langle b, c, a \rangle$. Finally, swapping b and c gives the anticlockwise $-\langle c, b, a \rangle$. This suggests the general rule that "swapping a pair of vertices in a simplex changes the sign of the resulting simplex":

$$\langle ..., v, ..., v', ... \rangle \stackrel{\text{def}}{=} -\langle ..., v', ..., v, ... \rangle$$

Figure 5.26(a) shows a tetrahedron oriented by the order of its vertices. Its faces are also oriented by the order of their vertices. For higher dimensional simplices there is no obvious intuition for the orientaton such as the direction of a link or the clockwise-anticlockwise orientation of the triangles. However, the above rule determines which vertex orderings give the same orientation.

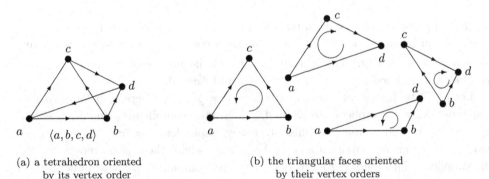

(a) a tetrahedron oriented by its vertex order

(b) the triangular faces oriented by their vertex orders

Fig. 5.26 Simplices with orientation induced by their vertex ordering

Chains of simplices

Simplicial families and simplicial complexes are *sets* of simplices. It can be useful to define "chains" of simplices linked together by addition and subtraction. An expression of the form $n_1\sigma_1 + n_2\sigma_2 + ... + n_m\sigma_m$ is called a *chain* of simplices where $n_1...n_m$ are numbers. This can be written as

$$\sum_{i=1}^{m} n_i\sigma_i \stackrel{\text{def}}{=} n_1\sigma_1 + n_2\sigma_2 + ... + n_m\sigma_m$$

Chains can be added and subtracted, e.g. $(n_1\sigma_1 + n_2\sigma_2 + ... + n_m\sigma_m) + (n'_1\sigma_1 + n'_2\sigma_2 + ... + n'_m\sigma_m) = (n_1 + n'_1)\sigma_1 + (n_2 + n'_2)\sigma_2 + ... + (n_m + n'_m)\sigma_m$. Thus in general

$$\sum_{i=1}^{m} n_i\sigma_i + \sum_{i=1}^{m} n'_i\sigma_i = \sum_{i=1}^{m} (n_i + n'_i)\sigma_i$$

Under appropriate conditions a set of chains under the operation of addition form what is call a "group", as explained in the next section.

Chain groups

A set S forms a *commutative group* or *abelian group* under the operator $+$ if the following axioms are satisfied:

(i) Closure: $s + s'$ belongs to S for all s and s' belonging to S.
(ii) Associativity: $s + (s' + s'') = (s + s') + s''$ for all s, s' and s'' belonging to S.
(iii) Identity: there exists an element e of S, often denoted 0, such that
$s + e = e + s = s$ for all s in S.
(iv) Inverses: for every s in S there exists an element s' in S, often denoted $-s$, such that $s + s' = s' + s = e$.
(v) Commutativity: $s + s' = s' + s$ for all s and s' in S.

Let $F = \{\sigma_1, \sigma_2, ..., \sigma_N\}$ be a simplicial family. For σ a simplex in F let $C(\sigma) = \{n\sigma \,|\, n \text{ is an integer}\}$. Then $C(\sigma)$ is a commutative group under addition, where $n\sigma + n'\sigma \stackrel{\text{def}}{=} (n + n')\sigma$. The identity is 0σ and the inverse of $n\sigma$ is $-n\sigma$. For simplicity of notation let $1\sigma = \sigma$, $-1\sigma = -\sigma$ and $0\sigma = 0$.

Let $C(F)$ be the set of chains of the form $c = \sum n_i\sigma_i$ where σ_i belongs to F, and repetitions of simplices are allowed, i.e. c may contain $n_i\sigma_i$ and $n_j\sigma_j$ with $\sigma_i = \sigma_j$ but n_i and n_j may be different. For example, let $c_1 = 7\sigma_1 - 3\sigma_2 - 4\sigma_1 + 4\sigma_2$. Then c can be rewritten as $c = \sum_{i=1}^{N} n_i\sigma_i$ where there is no repetition of the simplices. This follows from associativity and commutativity which allows the terms in each σ_i to be shuffled together and reduced to a single term $n_i\sigma_i$. For example $c_1 = 7\sigma_1 - (3\sigma_2 - 4\sigma_1) + 2\sigma_2 \stackrel{\text{commutativity}}{=} 7\sigma_1 + (4\sigma_1 - 3\sigma_2) + 2\sigma_2 \stackrel{\text{associativity}}{=} (7\sigma_1 + 4\sigma_1) + (-3\sigma_2 + 2\sigma_2) = 11\sigma_1 - 1\sigma_2 = 11\sigma_1 - \sigma_2$. $C(F)$ is a commutative group with identity $0 \,(= \sum_{i=1}^{N} 0\,\sigma_i)$.

Cycles as chains

Chains of simplices can be used to represent cycles, e.g. in Fig. 5.27 $c_1 = \langle x_1, x_2 \rangle + \langle x_2, x_3 \rangle + \langle x_3, x_1 \rangle$ and $c_2 = \langle x_2, x_4 \rangle + \langle x_4, x_3 \rangle + \langle x_3, x_2 \rangle$ are cycles. Since $\langle x_3, x_2 \rangle = -\langle x_2, x_3 \rangle$ the chain c_2 can be rewritten as $c_2 = \langle x_2, x_4 \rangle + \langle x_4, x_3 \rangle - \langle x_2, x_3 \rangle$. Thus $c_1 + c_2 = [\langle x_1, x_2 \rangle + \langle x_2, x_3 \rangle + \langle x_3, x_1 \rangle] + [\langle x_2, x_4 \rangle + \langle x_4, x_3 \rangle - \langle x_2, x_3 \rangle]$.

In this sum the edges $\langle x_2, x_3 \rangle$ of c_1 and $\langle x_3, x_2 \rangle = -\langle x_2, x_3 \rangle$ of c_2 cancel out giving $c_1 + c_1 = \langle x_1, x_2 \rangle + \langle x_2, x_4 \rangle + \langle x_4, x_3 \rangle + \langle x_3, x_1 \rangle$.

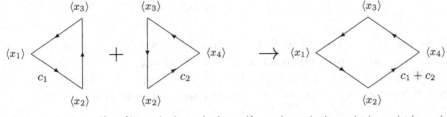

$[\langle x_1, x_2 \rangle + \langle x_2, x_3 \rangle + \langle x_3, x_1 \rangle] + [\langle x_2, x_4 \rangle + \langle x_4, x_3 \rangle + \langle x_3, x_2 \rangle] \rightarrow \langle x_1, x_2 \rangle + \langle x_2, x_4 \rangle + \langle x_4, x_3 \rangle + \langle x_3, x_1 \rangle$

Fig. 5.27 The addition of two appropriately oriented cycle chains creates another cycle chain.

Until now we have used an intuitive approach to cycles based on cycles of 1-simplices. The general notion of cycle can made more precise using chains, as will now be explained.

Boundaries of Simplices

The *boundary* of a p-simplex is made up of its $(p-1)$-faces. For example, the boundary of the tetrahedron (3-simplex) $\langle v_0, v_1, v_2, v_3 \rangle$ in Fig. 5.28 is made up of the triangles (2-dimensional faces) $\langle v_1, v_2, v_3 \rangle$, $\langle v_0, v_2, v_3 \rangle$, $\langle v_0, v_1, v_3 \rangle$ and $\langle v_0, v_1, v_2 \rangle$.

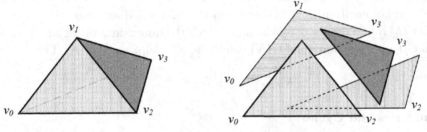

(a) a solid 3-dimensional tetrahedron   (b) the 2-dimensional faces of the tetrahedron

Fig. 5.28 The 2-dimensional triangular faces form the boundary of a 3-dimensional tetrahedron

The *boundary operator*, denoted ∂, maps a p-simplex to its *boundary*, i.e. a chain made up of its $(p-1)$-dimensional faces. It goes through the vertices of the simplex one by one, removing each vertex to obtain a face and flipping the sign (and hence the orientation) to make a chain:

$$\partial \langle v_0, v_1, ..., v_p \rangle \stackrel{\text{def}}{=} \Sigma_{i=0}^{p} (-1)^i \langle v_0, ..., \hat{v}_i, ..., v_p \rangle$$

where $\langle v_0, ..., \hat{v}_i, ..., v_p \rangle$ is the face of $\langle v_0, ..., v_p \rangle$ obtained by removing the vertex v_i.

For example, the triangle in Fig. 5.29(a) has the boundary $\langle v_0, v_1 \rangle + \langle v_1, v_2 \rangle + \langle v_2, v_0 \rangle$. By definition $\partial \langle v_0, v_1, v_2 \rangle$ is the chain of 1-simplices:

$$\partial \langle v_0, v_1, v_2 \rangle = \langle v_1, v_2 \rangle - \langle v_0, v_2 \rangle + \langle v_0, v_1 \rangle$$

obtained as follows: remove $\langle v_0 \rangle$ from $\langle v_0, v_1, v_2 \rangle$ to get $+\langle v_1, v_2 \rangle$; remove $\langle v_1 \rangle$ to get $-\langle v_0, v_2 \rangle$; and remove $\langle v_2 \rangle$ to get $+\langle v_0, v_1 \rangle$. Rewriting $-\langle v_0, v_2 \rangle$ as $\langle v_2, v_0 \rangle$ and rearranging terms gives $\partial \langle v_0, v_1, v_2 \rangle = \langle v_0, v_1 \rangle + \langle v_1, v_2 \rangle + \langle v_2, v_0 \rangle$ which corresponds to the intuitive notion of a boundary as a path around the outside of the simplex.

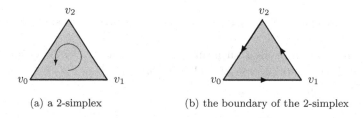

(a) a 2-simplex   (b) the boundary of the 2-simplex

Fig. 5.29 $\langle v_0, v_1, v_2 \rangle$ has boundary $\langle v_1, v_2 \rangle - \langle v_0, v_2 \rangle + \langle v_1, v_2 \rangle = \langle v_0, v_1 \rangle - \langle v_1, v_2 \rangle + \langle v_2, v_0 \rangle$

In order for $\partial : C(F) \to C(F)$ to be well defined it is necessary that, for any simplex σ, the chain group $C(F)$ contains all the faces of σ. In other words it is necessary that F is a simplicial complex. For consistency with the literature, F and $C(F)$ will be rewritten as K and $C(K)$ in the following discussion.

Let $C_p(K)$ be the set of all chains in which the dimensions of the simplices is p. Each $C_p(K)$ is a subgroup of $C(K)$ called the pth *chain group* of K. Then

$$\partial : C_p(K) \to C_{p-1}(K)$$

Boundaries and Cycles

The boundary operator gives a way of making precise the intuitive concept of "cycle". Let c_p be a p-chain in $C_p(K)$. Then c_p is a p-*cycle* in K if

$$\partial c_p = 0$$

The boundary operator is *nilpotent, i.e.* when it is applied twice the result is always zero. This means that the boundary operator applied twice to *any* simplex, or chain of simplices, will result in zero. For example, in Fig. 5.29 $\partial(\partial\langle v_0, v_1, v_2\rangle) = \partial(\langle v_1, v_2\rangle - \langle v_0, v_2\rangle + \langle v_0, v_1\rangle) = \langle v_2\rangle - \langle v_1\rangle - (\langle v_2\rangle - \langle v_0\rangle) + \langle v_1\rangle - \langle v_0\rangle = 0$. In this case the first application of ∂ gives the boundary of the triangle which is a cycle. The second application maps the cycle to zero.

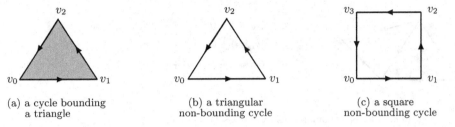

(a) a cycle bounding a triangle   (b) a triangular non-bounding cycle   (c) a square non-bounding cycle

Fig. 5.30   Bounding and non-bounding cycles

The boundary of the "filled in" triangle in Fig. 5.30(a) could be called a *bounding cycle*. In contrast the chain in Fig. 5.30(b) goes round a triangular hole. It is a cycle because $\partial(\langle v_0, v_1 \rangle + \langle v_1, v_2 \rangle + \langle v_2, v_3 \rangle$ is the chain $\langle v_1 \rangle - \langle v_0 \rangle + \langle v_2 \rangle - \langle v_1 \rangle + \langle v_3 \rangle - \langle v_0 \rangle = 0$. However it is not a bounding cycle since it's not the boundary of anything. It is a *non-bounding* cycle. Similarly the chain in Fig. 5.30(c) is a non-bounding cycle since $\partial(\langle v_0, v_1 \rangle + \langle v_1, v_2 \rangle + \langle v_2, v_3 \rangle + \langle v_3, v_0 \rangle)$ is the chain $\langle v_1 \rangle - \langle v_0 \rangle + \langle v_2 \rangle - \langle v_1 \rangle + \langle v_3 \rangle - \langle v_2 \rangle + \langle v_3 \rangle - \langle v_0 \rangle$ which is zero. This cycle is not a bounding cycle since the rectangle is not filled in. Non-bounding cycles provide an algebraic way of characterising "holes" in topological spaces.

Homology groups

Let $C_{p+1}(K)$ and $C_p(K)$ be the $(p+1)$ and p dimensional chain groups of the complex K. Let $Z_p(K)$ be defined as the set of all cycles in $C_p(K)$:

$$Z_p(K) \stackrel{\text{def}}{=} \{\partial c_p \mid \partial c_p = 0\}$$

This set forms a commutative subgroup of $C_p(K)$. Let the set of *bounding cycles*, $B_p(K)$, be defined as

$$B_p(K) \stackrel{\text{def}}{=} \{\partial c_{p+1} \mid \text{ for all } c_{p+1} \text{ in } C_{p+1}(K)\}$$

This set is also a commutative subgroup of $C_p(K)$, and $B_p(K) \subseteq Z_p(K)$. The pth *homology group* of K is defined to be the members of $Z_p(K)$ not in $B_p(K)$, which is defined to be the *factor group*

$$H_p(K) \stackrel{\text{def}}{=} Z_p(K)/B_p(K)$$

$H_p(K)$ is the set of equivalence classes of p-cycles where two p-cycles are considered equivalent if their difference is the boundary of a $(p+1)$ chain. These p cycles are associated with p-dimensional holes. p-cycles associated with the same hole are defined to be "homologous". For example, if c_p is a non-bounding cycle then $n \, c_p$ goes round the same hole n times. The equivalence set of cycles $\{nc_p \mid n \text{ in } \mathbb{Z}\}$ has the same cardinality as the integers, which in this context is often denoted by $\mathbb{Z}$.

136   Hypernetworks in the Science of Complex Systems

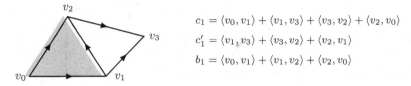

$$c_1 = \langle v_0, v_1 \rangle + \langle v_1, v_3 \rangle + \langle v_3, v_2 \rangle + \langle v_2, v_0 \rangle$$
$$c_1' = \langle v_1, v_3 \rangle + \langle v_3, v_2 \rangle + \langle v_2, v_1 \rangle$$
$$b_1 = \langle v_0, v_1 \rangle + \langle v_1, v_2 \rangle + \langle v_2, v_0 \rangle$$

Fig. 5.31 c_1 is homologous to c_1', $c_1 \sim c_1'$, since $c_1 = c_1' + b_1$

Two cycles c_p and c_p' in $Z_p(K)$ are said to be *homologous*, written $c_p \sim c_p'$, if there exists b_p in $B_p(K)$ such that $c_p = c_p' + b_p$. This is illustrated in Fig. 5.31(a) where $c_1 + b_1 = \langle v_0, v_1 \rangle + \langle v_1, v_3 \rangle + \langle v_3, v_2 \rangle + \langle v_2, v_0 \rangle - (\langle v_0, v_1 \rangle + \langle v_1, v_2 \rangle + \langle v_2, v_0 \rangle) = \langle v_1, v_2 \rangle + \langle v_2, v_3 \rangle + \langle v_3, v_1 \rangle = c_1'$, Thus $c_1 = c_1' + b_1$ and $c_1 \sim c_1'$.

The homology of the cylinder

Figure 5.32(a) shows a cylinder. The cylinder has a hole you could put your arm through. In Fig. 5.32 it is cut down the long edge from vertex v_0 to vertex v_3 and opened out to form a sheet and *triangulated*, *i.e.* divided into triangles so that each triangle unambiguously covers a part of the sheet, and all the triangles together make up the sheet. The triangles are uniquely defined by their vertices.

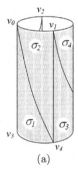 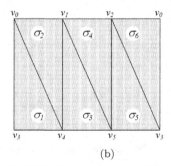

$$\sigma_1 = \langle v_0, v_3, v_4 \rangle$$
$$\sigma_2 = \langle v_0, v_1, v_4 \rangle$$
$$\sigma_3 = \langle v_1, v_4, v_5 \rangle$$
$$\sigma_4 = \langle v_1, v_2, v_5 \rangle$$
$$\sigma_5 = \langle v_2, v_3, v_5 \rangle$$
$$\sigma_6 = \langle v_0, v_2, v_2 \rangle$$

(a)   (b)

Fig. 5.32   A triangulated cylinder

The chain $c_1 = \langle v_0, v_1 \rangle + \langle v_1, v_2, \rangle + \langle v_2, v_0 \rangle$ is a cycle since $\partial(\langle v_0, v_1 \rangle + \langle v_1, v_2, \rangle + \langle v_2, v_0 \rangle) = \langle v_1 \rangle - \langle v_0 \rangle + \langle v_2 \rangle - \langle v_1 \rangle + \langle v_0 \rangle - \langle v_2 \rangle = 0$. The chain $c_2 = \langle v_3, v_4 \rangle + \langle v_4, v_5, \rangle + \langle v_5, v_3 \rangle$ is also cycle. The cycles c_1 and c_2 are both non-bounding. They are homologous because $c_2 = c_1 + \partial(\sigma_1 - \sigma_2 + \sigma_3 - \sigma_4 + \sigma_5 + \sigma_6)$. $H_1(K)$ is the free group of non-bounding cycles homologous to c_1. It characterises the single hole through the cylinder, written as $H_1(K) \cong \mathbb{Z}$. In general $H_0(K) \cong \mathbb{Z}$ when K is connected. For the cylinder $H_0(K) \cong \mathbb{Z}$, $H_1(K) \cong \mathbb{Z}$, and $H_p(K) \cong 0$ for $p > 1$.

The homology of the torus

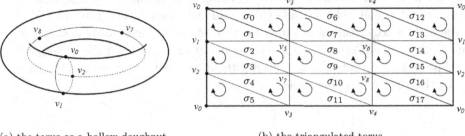

(a) the torus as a hollow doughnut   (b) the triangulated torus

Fig. 5.33 Relations links and relational simplices

Figure 5.33 shows a torus opened out into a flat sheet and divided into triangular simplices. Let $c_2 = \sigma_0 + ... + \sigma_{15}$. It can be shown that $\partial(c_2) = 0$, so c_2 is a cycle. Since there are no 3-simplices this is a non-bounding 2-cycle and $H_2(K) \cong \mathbb{Z}$. This corresponds to the void inside the torus. There are two principle groups of non-bounding 1-cycles. The first is the 1-cycles homologous to $c_{1,1} = \langle v_0, v_1 \rangle + \langle v_1, v_2 \rangle + \langle v_2, v_0 \rangle$, which is associated with the hole that you could poke your finger through. The second is the group of 1-cycles homologous to $c_{1,2} = \langle v_0, v_3 \rangle + \langle v_3, v_4 \rangle + \langle v_4, v_0 \rangle$ which is associated with the hole that you could tie a piece of string around. Thus $H_1(K) \cong \mathbb{Z}$. Since the torus is one connected piece, $H_0(K) \cong \mathbb{Z}$. The pth *Betti number* of a topological object is the maximum number of independent free subgroups of $H_p(K)$. Thus the Betti numbers of the torus are $B_0 = 1$, $B_1 = 2$, $B_2 = 1$, and $B_p = 0$ for $p > 2$.

The homology of the Möbius band

Figure 5.34 shows the Möbius band which can be made by taking a rectangle, twisting it, and joining the ends. This object has the counter-intuitive property of having a single side. $c_1 = \langle v_0, v_1 \rangle + \langle v_1, v_2 \rangle + \langle v_2, v_3 \rangle$ and $c_1' \langle v_0, v_1 \rangle + \langle v_1, v_2 \rangle + \langle v_2, v_3 \rangle$ are non-bounding cycles. They are homologous since $c_1 = c_1' + \partial(\sigma_0 + \sigma_1 + ... + \sigma_5)$. Thus $H_1(K) \cong \mathbb{Z}$ and, since the band has one piece, $H_0(K) \cong \mathbb{Z}$. Thus $B_0 = 1$, $B_1 = 1$, and $B_p = 0$ for $p \geq 2$.

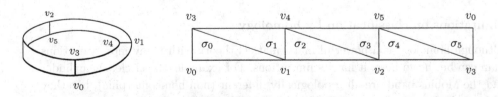

Fig. 5.34 A triangulation of the Möbius band

The homology of the Klein bottle

Figure 5.35(a) illustrates a topological object obtained by taking a rectangle, rolling it into a cylinder, and twisting the cylinder through itself to join the two ends. Like the Möbius band, this object also has just one side. Figure 5.35(b) shows a triangulation for the Klein bottle. A true Klein bottle cannot be made in 3-space because passing the cylinder through itself without cutting the surface (and thereby changing the topology) is impossible.

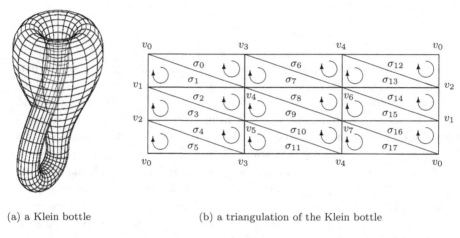

(a) a Klein bottle (b) a triangulation of the Klein bottle

Fig. 5.35 The Klein bottle

The Klein bottle has two 1-cycles, $a = \langle v_0, v_1 \rangle + \langle v_1, v_2 \rangle + \langle v_2, v_0 \rangle$ and $b = \langle v_0, v_3 \rangle + \langle v_3, v_4 \rangle + \langle v_4, v_0 \rangle$. It can be shown that $\partial(\sigma_0 + \sigma_1 + ... + \sigma_{17}) = 2a$. Thus, although a is not a bounding cycle, $2a$ is a bounding cycle. This is an example of *torsion* so the subgroup of $H_1(K)$ associated with a is not a free group but is equivalent to $\mathbb{Z}$ mod 2, written as $\mathbb{Z}_2$. The homological structure of the Klein bottle is written as $H_1(K) \cong \mathbb{Z} \oplus \mathbb{Z}_2$, with $H_p(K) = 0$ for $p > 1$.

The Klein bottle is a remarkable if counter-intuitive object. For example, the YouTube video (http://www.youtube.com/watch?v=sRTKSzAOBr4) shows how the Klein bottle can be made from two Möbius bands.

Limitations on classification by homology

Although homology is a powerful mathematical theory with many more variations than can be shown here, it has its limitations. For example, the circle, the cylinder and the Möbius band are all topologically different (non-homeomorphic), but they all have the homology groups $H_0(K) = 0$, $H_1(K) \cong \mathbb{Z}$, and $H_p(K) = 0$ for $p > 1$.

5.16 From cohomology in physics to q-connectivity in social science

The theory of Q-analysis developed by R. H. Atkin has its origins in his work to unify physics by what he calls "The Law of the Trivial Cocycle". His recent book [Atkin (2010)] gives a comprehensive summary of this theory which is based on the argument that the space-time backcloth of many physical phenomena is acyclic, *i.e.* without holes or non-bounding cycles so that $H_p(K) = 0$ for all p.

Cohomology and Atkin's Cocycle Law

Cohomology theory sets an essential context for hypernetwork theory, but it is not practical to give a full and detailed exposition. For simplicity the following development will omit many details which can be found, for example, in Section 2.6 of [Hilton and Wylie (1965)]. For example, torsion-free complexes will be assumed.

Let K be an oriented simplicial complex. A *p-cochain*, c^p, is a function ϕ defined on the oriented p-simplices of K with values in an abelian group G such that $\phi(c_1 + c_2) = \phi(c_1) + \phi(c_2)$. This can be written as $c^p(\sigma_{p,i}) = g_i$.

A mapping $\phi : A \to B$ between groups A and B is a *homomorphism* if $\phi(a+b) = \phi(a) + \phi(b)$. Every cochain c^p establishes a homomorphism $c^p : K_p \to G$ with

$$c^p(\sum_i m_i \sigma_{p,i}) = \sum_i m_i(c^p(\sigma_{p,i}))$$

and every such homomorphism determines a unique cochain.

Using *inner product* notation for a cochain d^p and a chain c_p, $d^p(c_p)$ will be written as (c_p, d^p). Let $\sigma^{p,j}$ be a cochain with $(\sigma_{p,i}, \sigma^{p,j}) = 1$ if $i = j$ and is otherwise zero. Every cochain c^p can then be written in the form $c^p = \sum m_i \sigma^{p,i}$.

The following definition of δ lies at the heart of Atkin's theory of physics. Given a $(p-1)$-cochain d^{p-1}, the p-cochain δd^{p-1} is given by:

$$(\sigma_p, \delta d^{p-1}) = (\partial \sigma_p, d^{p-1})$$

This establishes a relationship between functions defined on the p-simplices of K and functions on its $(p-1)$-simplices. In particular, if ϕ is a mapping from K to G, it means that $\phi \sigma^p$ can be evaluated by the values of ϕ on the boundary of σ.

Although this may seem strange, t corresponds to the everyday use of integrals. For example, let K be the simplicial complex formed from all intervals of $\mathbb{R}$, $\langle v_0, v_1 \rangle$ and their vertices $\langle v_i \rangle$. Let $\delta : \phi^0 \to \phi^1$ with $\phi^1 \langle v_0, v_1 \rangle \stackrel{\text{def}}{=} \int_{\langle v_0, v_1 \rangle} \phi^0 \, d\phi^0$. Then

$$\int_{\langle v_0, v_1 \rangle} \phi^0 \, d\phi^0 \stackrel{\text{def}}{=} (\langle v_0, v_1 \rangle, \phi^1) = (\langle v_0, v_1 \rangle, \delta \phi^0) = (\partial \langle v_0, v_1 \rangle, \phi^0)$$
$$= (\langle v_1 \rangle - \langle v_1 \rangle, \phi^0) = \phi^0 \langle v_1 \rangle - \phi^0 \langle v_0 \rangle$$

It can be shown that $\delta^2 = 0$ so that coboundaries, cocycles and cohomology can be defined in ways analogous to cycles and boundaries. The details can be found in the literature, *e.g.* [Argoston (1976)][Hilton and Wylie (1965)].

The interpretation of δ as a mapping between a function and its integral generalises to higher dimensions so that integrals of two-dimensional objects can be evaluated on their bounding contours, and integrals of three dimensional objects can be evaluated on their bounding surfaces. The definition of δ can be considered to provide a generalised Stokes' theorem [Hilton and Wylie (1965)].

In his book *Mathematical Physics*, [Atkin (2010)] reviews many phenomena in physics and shows how they can be expressed in the language of algebraic topology. He writes, "The development of traditional Physics has been based on the idea that the underlying space available for for carrying the measures of generating elements is *homologically trivial*, this being the *Euclidean space* E^3. That is to say that the homology groups are $H_p(K) \cong 0$ meaning that there are *no holes* in the space – which is naturally said to be acyclic" (p. 177).

Atkin makes a distinction between the "physicist's continuum" and the "mathematician's continuum" where the former is determined by a mesh of observable rational points and the latter determined by the real numbers. He writes "a 1-simplex is adequately defined by a pair $P_1 = (s_1, t_1)$ and $P_2 = (s_2, t_2)$ and does not require there to be any other points (of the assumed continuum) between them ... The acyclic nature of the backcloth, which was there before, is still there. Nor does it depend on whether the sets of points S or T are finite or infinite." ([Atkin (2010)], p. 178). In this context Atkin shows that many phenomena in physics are manifestations of his *Cocycle Law*, $\delta c = 0$.

The homology groups of relations

In 1951 C. H. Dowker showed that a relation R between two sets A and B has two associated simplicial complexes [Dowker (1952)]. These are the simplicial complexes introduced in the previous chapter as $K_A(B, R)$ and $K_B(A, R)$.

Remarkably, Dowker showed that the two simplicial complexes of a relation have the same homology groups. This is illustrated in Table 5.9 and Fig. 5.36.

	b_1	b_2	b_3	b_4	b_5	b
a_1	1	0	1	1	0	0
a_2	1	1	0	0	1	0
a_3	0	1	1	0	0	1

Table 5.9 The relation R between $A = \{a_1, a_2, a_3\}$ and $B = \{b_1, b_2, b_3, b_4, b_5, b_6\}$

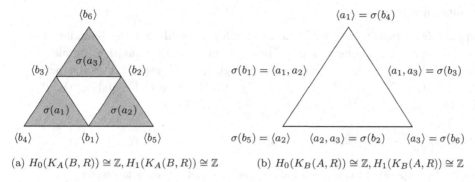

Fig. 5.36 The conjugate complexes of a relation have isomorphic homology

Figure 5.36 shows the conjugate complexes for the relation R. They are both simply connected with $H_0(K_A) \cong \mathbb{Z}$ and $H_0(K_B) \cong \mathbb{Z}$. They both have a 1-dimensional cycle around a 2-dimensional hole with $H_1(K_A) \cong \mathbb{Z}$ and $H_1(K_B) \cong \mathbb{Z}$. There are no p-cycles for $p > 1$ so $H_p(K_A) \cong 0$ and $H_p(K_B) \cong 0$. Thus the conjugate complexes K_A and K_B have the same homology structure.

From cohomology in physics to q-connectivity in social science

Dowker's insight that relations have associated complexes led Atkin to suggest the completely original idea of bridging the use of simplices and cochains to represent physical space and dynamics and the possibility of using simplices and cochains to represent relational structure and dynamics in social systems.

Atkin set out the argument in his paper "From cohomology in physics to q-connectivity in social science" [Atkin (1972)]: "This paper is a review of a personal search over many years to find a formulation for physical science which would not do too much violence to accepted theories and yet show a way for its extension and generalization into fields of social science. Since the latter seems to require a language which mathematicians would call "combinatorial", being concerned with finite sets, it became a search for a mathematical language which would describe certain key properties of the familiar continuum but which would carry these over when that continuum should be replaced by a finite set of points."

The details of Atkin's research programme can be found in his many papers, and in his recent book [Atkin (2010)]. It included, literally, moving from the established mathematics of algebraic topology to the creation of new combinatorial structures designed to represent social phenomena. These include the theory of Q-analysis and a different way of looking at the mappings defined on socially determined simplicial complexes, leading to the definition of a new "face operator" alongside the boundary operator, ∂. Q-analysis theory goes far beyond the listing of q-connected components with new ways of looking at multidimensional connectivity including a theory of "pseudo-homotopy", and the definition of simplex formation as structural events.

The face operator

In moving from physics to social science Atkin suggested that the boundary operator, ∂, could be augmented by a new operator, $\hat{f}$ that maps a simplex to a chain without the sign of the faces changing. Whereas ∂ has the action $\partial \langle v_0, v_1, ..., v_p \rangle = \sum_{i=0}^{p}(-1)^i \langle v_0, ..., \hat{v}_i, ..., v_p \rangle$ with the term $(-1)^i$ making the sign alternate, Atkin defined the *face operator* to have the form:

$$\mathtt{f} \langle v_0, ..., v_p \rangle \stackrel{\text{def}}{=} \sum_{i=0}^{p} \langle v_0, ..., \hat{v}_i, ..., v_p \rangle.$$

Thus the face operator maps a simplex to the chain of its unsigned faces.

A drawback of this operator is that it generates copies of the faces when applied more than once, e.g. $\mathtt{f}^2 \langle v_0, v_1, v_2, v_3 \rangle = \mathtt{f}(\langle v_1, v_2, v_3 \rangle + \langle v_0, v_2, v_3 \rangle + \langle v_0, v_1, v_3 \rangle + \langle v_0, v_1, v_2 \rangle) = (\langle v_2, v_3 \rangle + \langle v_1, v_3 \rangle + \langle v_1, v_2 \rangle) + (\langle v_2, v_3 \rangle + \langle v_0, v_3 \rangle + \langle v_0, v_2 \rangle) + (\langle v_1, v_3 \rangle + \langle v_0, v_3 \rangle + \langle v_0, v_1 \rangle) + (\langle v_1, v_2 \rangle + \langle v_0, v_2 \rangle + \langle v_0, v_1 \rangle) = 2(\langle v_2, v_3 \rangle + 2 \langle v_1, v_3 \rangle + 2 \langle v_1, v_2 \rangle + 2 \langle v_0, v_3 \rangle + 2 \langle v_0, v_2 \rangle + 2 \langle v_0, v_1 \rangle$. To correct this Atkin introduced the *exponential face operator* defined as $\hat{\mathtt{f}}^k = \frac{1}{k!} \mathtt{f}^k$ with

$$\hat{\mathtt{f}} \sigma_p \stackrel{\text{def}}{=} \mathtt{f} \langle v_0, ..., v_p \rangle = \sum_{i=0}^{p} \langle v_0, ..., \hat{v}_i, ..., v_p \rangle, \text{ and}$$

$$\hat{\mathtt{f}}^k \sigma_p \stackrel{\text{def}}{=} \frac{1}{k!} \mathtt{f}^k \sigma_p = \sum_{\sigma_{p-k} \lesssim \sigma_p} \sigma_{p-k}$$

so that $\hat{\mathtt{f}}^k(\sigma_p)$ is the sum of all the $(p-k)$-faces of σ_p, with each face occurring once only. $\hat{\mathtt{f}}$ will be called $\mathtt{f}$-*hat*. $\hat{\mathtt{f}}$ is defined on chains by the rule that

$$\hat{\mathtt{f}}^k \sum_{i=1}^{n} \alpha_i \sigma_i \stackrel{\text{def}}{=} \sum_{i=1}^{n} \alpha_i \hat{\mathtt{f}}^k \sigma_i$$

For $\sigma_p + \sigma'_p$ this has the property that all the k-faces of $\sigma_p \cap \sigma'_p$ appear twice in $\hat{\mathtt{f}}^k(\sigma_p + \sigma'_p)$ since $\hat{\mathtt{f}}^k(\sigma_p + \sigma'_p) = \left(\sum_{\sigma_{p-k} \lesssim \sigma_p \text{ and } \sigma_{p-k} \not\lesssim \sigma_p \cap \sigma'_p} \sigma_{p-k} + \sum_{\sigma_{p-k} \lesssim \sigma_p \cap \sigma'_p} \sigma_{p-k} \right) + \left(\sum_{\sigma_{p-k} \lesssim \sigma'_p \text{ and } \sigma_{p-k} \not\lesssim \sigma_p \cap \sigma'_p} \sigma_{p-k} + \sum_{\sigma_{p-k} \lesssim \sigma_p \cap \sigma'_p} \sigma_{p-k} \right) = \sum_{\sigma_{p-k} \lesssim \sigma_p \text{ or } \sigma_{p-k} \lesssim \sigma'_p \text{ but } \sigma_{p-k} \not\lesssim \sigma_p \cap \sigma'_p} \sigma_{p-k} + \sum_{\sigma_{p-k} \lesssim \sigma_p \cap \sigma'_p} 2 \sigma_{p-k}$.

To provide an alternative where each face appears once, even when simplices intersect, a new operator $\widehat{\mathtt{f}^k}$, called $\mathtt{f}$-*wide-hat*, is defined as:

$$\widehat{\mathtt{f}^k}(\sigma_p + \sigma'_p) \stackrel{\text{def}}{=} \sum_{\sigma_{p-k} \lesssim \sigma_p \text{ or } \sigma_{p-k} \lesssim \sigma'_p} \sigma_{p-k}$$

When $\sigma_p \cap \sigma'_p = \emptyset$, $\widehat{\mathtt{f}^k} = \hat{\mathtt{f}}^k$. Also $\widehat{\mathtt{f}^p}(\sigma_p + \sigma'_p) = \sum_{\langle v \rangle \lesssim \sigma_p \text{ or} \langle v \rangle \lesssim \sigma'_p} \langle v \rangle$ is the chain of vertices that belong to σ_p or σ'_p, with each vertex occurring once.

5.17 Shomotopy

In topology two curves on a surface are homotopic if one can be continuously transformed into the other. Atkin suggested a discrete analogue of this for simplicial complexes which he called "pseudo homotopy" or *shomotopy*.

Let K be a simplicial complex and let the notation $[\sigma, \sigma']$ represent a chain of q-connection between σ and σ'. The chains $[\sigma_r, \sigma_s]$ and $[\sigma_h, \sigma_k]$ are defined to be q-adjacent if and only if:

(a) σ_r is q-near σ_h.
(b) σ_s is q-near σ_k.
(c) the sequences σ_t and σ_m which define the chains $[\sigma_r, \sigma_s]$ and $[\sigma_h, \sigma_k]$ are separately ordered (denote the ordering $\leq$).
(d) for every σ_t in $[\sigma_r, \sigma_s]$ there exists a σ_m in $[\sigma_h, \sigma_k]$ such that σ_t is q-near σ_m, and if $\sigma_{t_1} \leq \sigma_{t_2}$, then $\sigma_{m_1} \leq \sigma_{m_2}$.
(e) for every σ_m in $[\sigma_h, \sigma_k]$ there exists a σ_t in $[\sigma_r, \sigma_s]$ such that σ_m is q-near σ_t, and if $\sigma_{m_1} \leq \sigma_{m_2}$, then $\sigma_{t_1} \leq \sigma_{t_2}$.

The idea behind this definition is that a chain can be "q-continuously" transformed into another. It can be illustrated by showing the chains as horizontal sequences in the q-graph as shown in Fig. 5.37(a). Here there is a simple q-nearness relationship between the σ_t and the σ_m and intuitively one chain can be transformed into the other in a way that respects q-connectivity, or q-continuously.

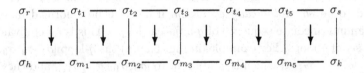

(a) $[\sigma_r, \sigma_s]$ can be transformed into $[\sigma_h, \sigma_k]$ q-continuously in the q-graph and they are q-adjacent

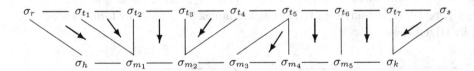

(b) $[\sigma_r, \sigma_s]$ can be q-continuously shrunk to $[\sigma_h, \sigma_k]$ in the q-graph and they are q-adjacent

Fig. 5.37 q-adjacency in the q-graph

Figure 5.37 shows the general case of different length chains. The transformation from the longer chain to the shorter chain is q-continuous, but in this case there must be triangles in the q-nearness graph. q-similarity is an equivalence relation on the set of q-chains and partitions them into equivalence classes.

Loops and q-holes

A chain of q-connection $[\dot{\sigma}_r, \sigma_s]$ is defined to be a q-loop when $\sigma_r = \sigma_s$. It is interesting to consider those loops that go round holes in a simplicial family. This is illustrated in Fig. 5.38(a) where the hole is a non-bounding p-dimensional cycle, similar to the hole in a torus that you could poke your finger through.

(a) a homological hole   (b) a q-hole

Fig. 5.38   Holes as cycles in the $\boldsymbol{q}$-graph

In contrast to this, Fig. 5.38(b) shows a p-dimensional cycle in the q-graph that is filled in since all the simplicies are $q-1$-near. This is not a hole in the homological sense of being a group of non-bounding cycles. Instead it is a q-dimensional depression like a crater.

In the same way that real holes can be represented by homologous groups of non-bounding cycles, q-holes can be represented by classes of q-similar loops. Two q-loops are said to be *shomotopically equivalent* if one can be deformed into the other through an intermediate sequence of q-adjacent loops. This is an equivalence relation on the set of q-loops. For example, in Fig. 5.39(a) the q-loop $[\sigma_1, \sigma_2, \sigma_3, \sigma_4]$ is q-adjacent to $[\sigma_5, \sigma_6, \sigma_7, \sigma_8]$ which is q-adjacent to $[\sigma_9, \sigma_{10}, \sigma_{11}, \sigma_{12}]$, and they are all shomotopically equivalent.

A class of q-loops that is shomotopically equivalent to a simplex is said to be *trivial*. This is illustrated in Fig. 5.39(b) where $[\sigma_1, \sigma_2, \sigma_3, \sigma_4]$ is q-adjacent to $[\sigma_5, \sigma_6, \sigma_7, \sigma_8]$ which is q-adjacent to the single simplex $[\sigma_9]$, and they form a shomotopically trivial class.

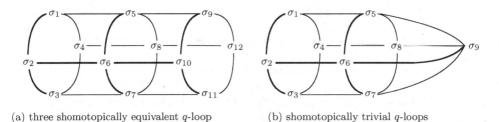

(a) three shomotopically equivalent q-loop   (b) shomotopically trivial q-loops

Fig. 5.39   Shomotopy loops

The shomotopy bottle of Q-analysis

To compute the shomotopy classes of a simplicial family involves finding the classes of shomotopically equivalent loops. Figure 5.37(b) suggests an approach based on shrinking cycles to shorter cycles. However this is problematic as illustrated by the *shomotopy bottle* construction in Fig. 5.40 where the lines between the simplices indicate q-nearness.

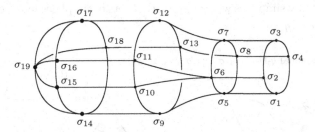

Fig. 5.40 A q-shomotopy bottle – links in the q-graph indicate q-nearness

Here $[\sigma_9, \sigma_{10}, \sigma_{11}, \sigma_{12}, \sigma_{13}]$ is q-similar to $[\sigma_{14}, \sigma_{15}, \sigma_{16}, \sigma_{17}, \sigma_{18}]$ which can be shrunk to $[\sigma_{19}]$ so these cycles are all shomotopically trivial. In contrast, $[\sigma_1, \sigma_2, \sigma_3, \sigma_4]$ is q-adjacent $[\sigma_5, \sigma_6, \sigma_7, \sigma_8]$, but this cannot be shrunk to $[\sigma_9, \sigma_{10}, \sigma_{11}, \sigma_{12}, \sigma_{13}]$. This suggests an approach based on shrinking the largest cycles, since $[\sigma_9, \sigma_{10}, \sigma_{12}, \sigma_{12}, \sigma_{13}]$ can be shrunk to both $[\sigma_1, \sigma_2, \sigma_3, \sigma_4]$ and $[\sigma_{19}]$. Since q-similarity is transitive, $[\sigma_1, \sigma_2, \sigma_3, \sigma_4]$ is q-shomotopic to $[\sigma_{19}]$ and hence shomotopically trivial.

q-holes and obstruction as cycles in the q-complex

In Fig. 5.40 there is something unsatisfactory about the q-loop $[\sigma_1, \sigma_2, \sigma_3, \sigma_4]$ being shomotopically trivial. q-holes are interesting because they can be interpreted as *obstruction* to flows across simplicial families. The q-shomotopy to $[\sigma_{19}]$ seems to be distant to q-loop $[\sigma_1, \sigma_2, \sigma_3, \sigma_4]$ where the loop could be creating a local obstruction to q-transmitted flows. This suggests a more local approach to defining q-holes as cycles in the q-graph. However, as Fig. 5.41 shows, cycles in the q-graph can be associated with "strong" holes (q-homology) and no hole at all (q-hubs).

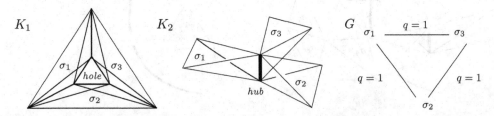

Fig. 5.41 K_1 has a "strong" hole, K_2 has a 1-hub, but both have the same q-graph, G

5.18 From the q-graph to the q-complex

It is clearly not satisfactory to have holes defined in ways that exclude the strong concept of hole of homology theory. This is due to the ambiguity in the q-graph between holes and hubs. This problem is easy to rectify by defining the q-complex of a simplicial family K to be the simplicial complex with simplices $\langle \sigma_1, \sigma_2, ...\rangle$ where $|\sigma_1 \cap \sigma_2...| \geq q$. This augments the edges of q-graph which denote two simplicies being q-near by simplices which denote that sets of simplices have common p-dimensional face, $p \geq q$.

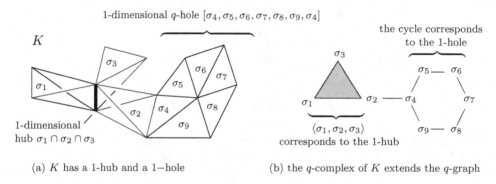

(a) K has a 1-hub and a 1−hole

(b) the q-complex of K extends the q-graph

Fig. 5.42 The q-complex disambiguates hubs and holes

To illustrate this, consider the simplicial family shown in Fig. 5.42(a). On the left are three tetrahedra σ_1, σ_2 and σ_3 where $|\sigma_1 \cap \sigma_2 \cap \sigma_3| = 1$, *i.e.* these simplices have a 1-dimensional hub. This is represented by a solid triangle in the q-complex (Fig. 5.42(b)). This explicitly solves the problem of ambiguity in the q-graph so that a *q-loop*, defined as a cycle in the q-complex corresponds to the intuitive notion of a hole, while a "filled in", or bounding, cycle does not. Furthermore, a "homological" hole characterised by a free non-bounding q-cycle in a complex always has associated q-holes (Fig. 5.43).

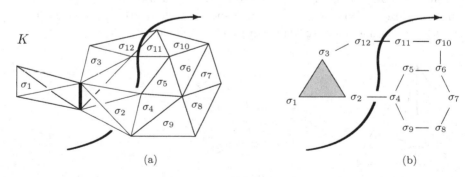

Fig. 5.43 Homological holes are always cycles in the q-complex

Shomotopy and the q-complex

Defining q-holes in terms of cycles in the q-complex gives a new perspective on Atkin's shomotopy. His definition of q-adjacency can be extended as follows. Let two chains $[\sigma_1, \sigma_2, ..., \sigma_m]$ and $[\sigma'_1, \sigma'_2, ..., \sigma'_m]$ be *strongly adjacent* if $|\sigma_i \cap \sigma_{i+1} \cap \sigma'_i \cap \sigma'_{i+1}| \geq q$ for $i = 1, ..., m-1$ (the generalisation to different length chains is straight forward but will not be developed here). This is illustrated in Fig. 5.44 where the cycle $[\sigma_i, \sigma_{i+1}, \sigma'_{i+1}, \sigma'_i]$ is filled in if $|\sigma_i \cap \sigma_{i+1} \cap \sigma'_{i+1} \cap \sigma'_i| \geq q$. This shows that the simplex $\langle \sigma_i, \sigma_{i+1}, \sigma'_i, \sigma'_{i+1} \rangle$ exists in the q-complex. Then $[\sigma_1, \sigma_2, \sigma_3, \sigma_4]$ is strongly q-similar to $[\sigma'_1, \sigma'_2, \sigma'_3, \sigma'_4]$ in Fig. 5.44(a) while none of the chains in Fig. 5.44(b) is q-similar to the others. Intuitively there is no hole in the q-complex of Fig. 5.44(a) while there are three in Fig. 5.44(b).

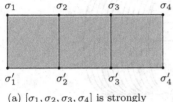
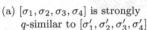

(a) $[\sigma_1, \sigma_2, \sigma_3, \sigma_4]$ is strongly q-similar to $[\sigma'_1, \sigma'_2, \sigma'_3, \sigma'_4]$

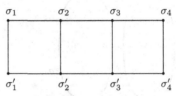

(b) none of the chains in this q-complex is q-similar

Fig. 5.44 Strongly q-similar chains

Let two q-loops be defined to be *strongly q-shomotopic* if there is an intermediate sequence of strongly q-adjacent q-loops between them. In other words the strong q-shomotopy relation on the set of q-loops is the transitive closure of the strong q-similarity relation.

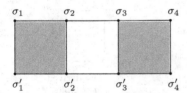
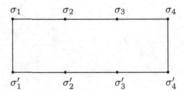

Fig. 5.45 Strongly q-similar chains

For example, in Fig. 5.45(a) there is one q-hole which can be characterised by any of the strongly shomotopic cycles $[\sigma_2, \sigma_3, \sigma'_3, \sigma'_2, \sigma_2]$, $[\sigma_1, \sigma_2, \sigma_3, \sigma'_3, \sigma'_2, \sigma'_1, \sigma_1]$, $[\sigma_2, \sigma_3, \sigma_4, \sigma'_4, \sigma'_3, \sigma'_2, \sigma_2]$, and $[\sigma_1, \sigma_2, \sigma_3, \sigma_4, \sigma'_4, \sigma'_3, \sigma'_2, \sigma'_1, \sigma_1]$. In comparison Fig. 5.45(b) shows a single cycle as representative of the q-hole.

Shomotopy as the homology of the q-complex

It is very satisfactory that strong shomotopic equivalence corresponds well to intuitive holes, and the q-complex suggests some interesting new possibilities.

Let a *minimum cycle* in the q complex be a q-loop $[\sigma_1, \sigma_2, ..., \sigma_m, \sigma_{m+1} = \sigma_1]$ such that, for $j > i$, σ_i is q-near σ_j if and only if $j = i+1$ for $i = 1, ..., m$.

Intuitively, strong shomotopy means that q-holes can be identified by cycles in the q-complex and that in simple cases each minimum cycle corresponds to a unique q-hole that is not q-adjacent to any other. Adding filled-in cycles (simplex) can create other cycles, but these are shomotopically equivalent to the minimum cycle.

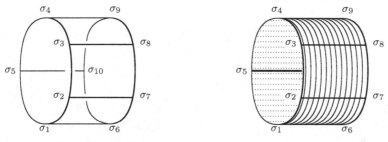

Fig. 5.46 q-holes in the q-complex

Figure 5.46(a) shows a q-complex with no simplices with dimension greater than one. Every simple cycle in this complex is associated with a distinct q-hole. In contrast Fig. 5.46(b) shows a q-complex with the filled in cycles $\langle \sigma_1, \sigma_2, \sigma_7, \sigma_6 \rangle$, $\langle \sigma_2, \sigma_3, \sigma_8, \sigma_7 \rangle$, $\langle \sigma_3, \sigma_4, \sigma_9, \sigma_8 \rangle$, $\langle \sigma_4, \sigma_5, \sigma_{10}, \sigma_9 \rangle$, and $\langle \sigma_5, \sigma_1, \sigma_6, \sigma_{10} \rangle$. In this case the homologous cycles $[\sigma_1, \sigma_2, \sigma_3, \sigma_4, \sigma_5, \sigma_1]$ and $[\sigma_6, \sigma_7, \sigma_8, \sigma_9, \sigma_{10}, \sigma_6]$ are q-similar and hence shomotopically equivalent.

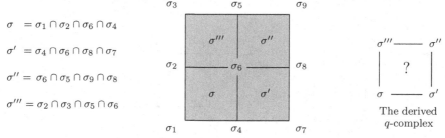

Fig. 5.47   Higher level shomotopy in the derived q-complex

Figure 5.47 shows four sets of q-near simplices with intersections σ, σ', σ', σ'. It s natural to ask if these *shomotopically derived* simplices have intersections. For example, is $\sigma \cap \sigma'$ non-empty? To go one step further, is $\sigma \cap \sigma' \cap \sigma'' \cap \sigma'''$ non-empty? If it is empty there is a derived hole in the derived q-complex, suggesting the possibility of a higher level shomotopy.

5.19 Designing the backcloth to carry the traffic

The natural sciences have progressed by inferring the dynamics of systems from data with successes in physics followed by chemistry and biology. In principle scientists studied these systems 'from the outside", unable to change their fundamental properties, *e.g.* humans cannot change gravity, cannot change the fundamental particles, cannot change molecular structures, and cannot change the fundamentals of life. Of course, in practice scientists try to change them all, with considerable success in chemistry and biology. These interventions are made not just for curiosity, but because they are useful. In other words, instead of studying systems as they are, disciplines such as engineering, medicine, and genetic modification create systems as they *ought* to be [Simon (1965)].

Most socio-technical systems are *artificial* – they are *designed*. This involves hypothesising new backcloth structures that will support desirable new and hypothetical traffic.

Recall, for example, the "magic roundabout" in Chapter 4. The original roundabout had a traditional design with associated maximum flows (capacity). As traffic demand increased this capacity was inadequate. In response the scientists at the UK Transport and Road Research Laboratory suggested a radical redesign to enable the system to carry more traffic. This design was based on the heuristic that disconnecting the paths across the roundabout would decrease interactions between vehicles and improve traffic flows. Generalising this, when designing systems the following heuristics apply:

Backcloth-traffic design heuristic 1: to increase interactions the backcloth should be as highly connected as possible.

Backcloth-traffic design heuristic 2: to decrease interactions the backcloth should be as disconnected as possibe

These hypotheses have practical methodological implications. First the *desired traffic* must be specified. Then the designer has to understand what kinds of backcloth structure can support the desired traffic This requires deciding which vertices are relevant and which relations are relevant, and how these vertices should be put together to support the desired traffic, *i.e.* how the system should be designed.

Almost any social system can be made more "complex". The art of the science of complex systems is to give guidelines on designing the backcloth so that it constrains traffic in desired ways. Although social systems cannot be "engineered" or "controlled" in mechanistic ways, they are certainly designed to make them *predisposed* to behave in desirable ways.

Design necessarily specifies the parts of systems and how they should related to each other in order to support desirable traffic. The relational structure of hypernetworks is an essential part of this.

Chapter 6

Hypernetworks

Hypernetworks are characterised by three major ideas: the first is that of "relational simplex" or *hypersimplex*; the second is that hypersimplices provide an unambiguous way of discriminating levels in multilevel systems; and the third is that these structures can support multilevel system backcloth and traffic dynamics.

Figure 6.1 shows how the common relational structures form a unified whole. On the top line, relations between pairs of things are given more structure. Graphs are given orientation to become directed graphs, or digraphs. Digraphs with numbers associated with their vertices and edges become networks. In contrast, vertically there is generalisation from binary relations between pairs of things to n-relations between any number of things. On the bottom line, hypergraphs edges are sets of vertices which become simplices when ordered. Hypernetworks, sets of hypersimplices, provide the last piece in this relational jigsaw.

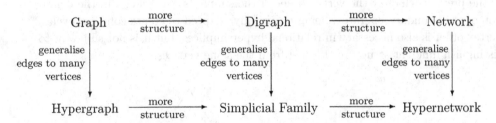

Fig. 6.1 Hypernetworks generalise all the common network structures

A *hypersimplex* is an ordered set of vertices with an explicit n-ary relation, *e.g.* let the hypersimplex $\langle \triangleleft, \triangleright, \odot; R \rangle$ represent the configuration $\triangleleft \odot \triangleright$, one of many possible relations on the objects $\triangleleft$, $\triangleright$, and $\odot$. R' is another, where $\langle \triangleleft, \triangleright, \odot; R' \rangle$ represents the configuration $\triangleright \odot \triangleleft$. *Hypernetworks* are sets of hypersimplices.

Hypernetworks do not compete with hypergraphs or networks – they naturally generalise both. For every hypergraph there is a unique hypernetwork, as there is for every network. In any situation the simplest structure can be used, but when needed, hypernetworks offer extra representational power.

6.1 Hypersimplices and hypernetworks

A *hypersimplex* is a simplex that carries its defining relation explicitly following the vertices, *e.g.* the blocks b_1, b_2 and b_3 in Fig. 6.2 are combined by the relation R to create the hypersimplex $\sigma = \langle b_1, b_2, b_3 ; R \rangle$, where the "arch" σ exists at a higher more aggregate level than its parts. To emphasise their relational nature, hypersimplices may also be called *relational hypersimplices*.

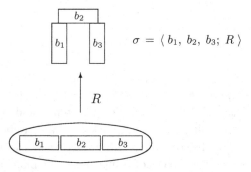

Fig. 6.2 Parts combined into a whole by R forming a *relational hypersimplex*

Figure 6.2 gives a diagrammatic way of showing R assembling the blocks into the arch. Implicitly in this diagram the blocks are ordered by their numbers.

Vertex Order in Relational Hypersimplices

In the previous chapter the vertex order of the simplices was crucial in the definition of the boundary operator, being beautifully adapted to create coherent cycles. Vertex order is also necessary in relational hypersimplices, but it is not sufficient to distinguish different structures formed from the same vertices.

Fig. 6.3 The relational simplices $\langle a, c, t; R_{\text{act}} \rangle$ and $\langle a, c, t; R_{\text{act}} \rangle$

Consider the words *act* and *cat* formed by 3-ary relations on the letters a, c and t. These could be discriminated by ordering the vertices appropriately, since $\langle a, c, t \rangle \neq \langle c, a, t \rangle$. For such structures the *order* of the vertices matters, and for hypersimplices in general $\langle ..., v_i, ..., v_j, ...; R \rangle \neq \langle ..., v_j, ..., v_i, ...; R \rangle$. However, although vertex ordering can usefully discriminate simplices such as $\langle a, c, t \rangle$ and $\langle c, a, t \rangle$, it is not sufficient to differentiate all n-ary relational structure.

Hypernetworks 153

Fig. 6.4 Twelve configurations of the elements $\{\Diamond, \heartsuit, \Box\}$

For example, Fig. 6.4 shows the three objects $\Diamond$, $\heartsuit$, and $\Box$ related in twelve ways. There are only six ways of ordering these objects as vertices, namely $\langle \Diamond, \heartsuit, \Box \rangle$, $\langle \Diamond, \Box, \heartsuit \rangle$, $\langle \heartsuit, \Diamond, \Box \rangle$, $\langle \heartsuit, \Box, \Diamond \rangle$, $\langle \Box, \Diamond, \heartsuit \rangle$, and $\langle \Box, \heartsuit, \Diamond \rangle$, and this is not enough to distinguish the twelve different relationships that define the configurations.

Since the vertices can be related in more ways than they can be ordered, to remove ambiguity it is necessary to make explicit the n-ary relation that defines any particular structure. For illustration, the definition of relational hypersimplices allows c_1 to be written as $\langle \Diamond, \heartsuit, \Box; R_1 \rangle$ which can be discriminated from, for example, $c_7 = \langle \Diamond, \heartsuit, \Box; R_7 \rangle$ as shown below:

$$\langle \Diamond, \heartsuit, \Box; R_1 \rangle = \begin{smallmatrix}\Box\heartsuit\\\Diamond\end{smallmatrix} \neq \begin{smallmatrix}\Diamond\\\Box\heartsuit\end{smallmatrix} = \langle \Diamond, \heartsuit, \Box; R_7 \rangle$$

Hypernetwork theory extends conventional graphs, networks and simplicial complexes to make explicit the n-ary relations. Relational hypersimplices allow structures to be discriminated when they have the same constituent parts. In general $\langle v_1, ..., v_n; R \rangle \neq \langle v'_2, ..., v'_n; R' \rangle$, even when $\langle v_1, ..., v_n \rangle = \langle v'_2, ..., v'_n \rangle$.

6.2 Examples

Example: The knight fork

Figure 6.5 shows three configurations of chess pieces. The configuration on the left, $\langle$rook, knight, king; $R_1\rangle$, is called a *knight fork* because the white knight threatens the black rook at the same time that it puts the black king in check. Unless black has a piece that can take it, the white knight can take the black rook because black must move the king out of check. The configuration in the centre, (b) $\langle$rook, knight, king; $R_2\rangle$, is not a knight fork, even through the knight puts the king in check. The configuration on the right is another knight fork, but it is clearly different to that on the left. Thus, the same three pieces are assembled by three different relations, R_1, R_2 and R_3 to form three different structures.

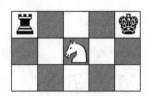 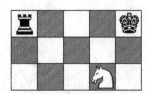 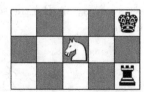

(a) $\langle$rook, knight, king; $R_1\rangle$ (b) $\langle$rook, knight, king; $R_2\rangle$ (c) $\langle$rook, knight, king; $R_3\rangle$

Fig. 6.5 The knight fork in chess

Example: Chemical Isomers

Chemical molecules are assemblies of atoms. For example propanol assembles three carbon atoms with eight hydrogen atoms and on oxygen atom, written as C_3H_8O or C_3H_7OH.

```
      H H H                          H                     H H       H
      | | |                          |                     | |       |
  H — C — C — C — O — H       H O H                 H — C — C — O — C — H
      | | |                   | | |                       | |       |
      H H H               H — C — C — C — H               H H       H
                              | | |
                              H H H
```

(a) n-propyl alcohol   (b) isopropyl alcohol   (c) methyl-ethyl-ether

Fig. 6.6 Chemical isomers as relational simplices

Figure 6.6 shows the atoms of propanol arranged in a variety of ways. The first two show the isomers n-propyl alcohol and isopropyl alcohol. The oxygen atom is attached to an end carbon in the first isomer and to the center carbon in the second, but the C-O-H hydroxyl group substructure is common to both. The rightmost isomer of C_3H_8O, methoxyethane, has the oxygen atom connected to two carbon atoms and there is no C-O-H substructure. This makes it an ether, methyl-ethyl-ether, rather than an alcohol. Thus the relational simplices of the isomers have the same vertices, but the assembly relations are different. n-propyl alcohol and isopropyl alcohol share the hydroxyl group substructure C-O-H and are similar, but methyl-ethyl-ether does not and has different properties. Thus

$\langle$ C, C, C, H, H, H, H, H, H, H, O ; $R_{n-\text{propylalcohol}}\rangle \quad \neq$

$\langle$ C, C, C, H, H, H, H, H, H, H, O ; $R_{\text{isopropylalcohol}}\rangle \quad \neq$

$\langle$ C, C, C, H, H, H, H, H, H, H, O ; $R_{\text{methyl}-\text{ethyl}-\text{ether}}\rangle$

Example: Organising boy-scouts on a hike

[Goldratt (1994)] suggests two ways of organising scouts on a hike (Fig. 6.7). Under the 6-ary relation R_1 they are ordered with the slowest at the back. Under R_2 the slowest is at the front. For $\langle s_1, s_2, s_3, s_4, s_5, s_6; R_1\rangle$ the dynamics can be troublesome with the slowest scout getting left further and further behind. For $\langle s_1, s_2, s_3, s_4, s_5, s_6; R_2\rangle$ the scouts all reman in a compact group behind the slowest at the front.

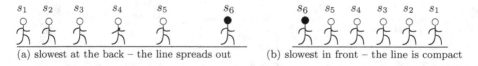

(a) slowest at the back – the line spreads out   (b) slowest in front – the line is compact

Fig. 6.7 $\langle s_1, s_2, s_3, s_4, s_5, s_5; R_1\rangle$ is troublesome; $\langle s_1, s_2, s_3, s_4, s_5, s_5; R_2\rangle$ works well

Example: IQ Questions

Figure 6.8 is based on an a question in an IQ test. To answer the question, one line of reasoning goes as follows: on the top row the circle (second object) is inside the triangle (first object); on the second row the square (first object) is inside the circle (second object); to follow the pattern on the last row the second object (a triangle) should be inside the first object (a square) and the answer is (B). Thus the structure behind this question is the relational hypersimplex $\langle x_1, x_2; R_{x_2 \text{ is inside } x_1}\rangle$. Although one might be tempted to answer (A), this is not the right pattern because its relation is $R_{x_1 \text{ is inside } x_2}$ rather than the correct relation $R_{x_2 \text{ is inside } x_1}$.

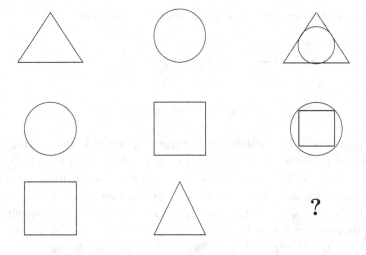

Which of the shapes below completes the sequence?

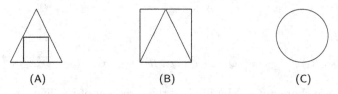

(A) (B) (C)

Fig. 6.8 An IQ test question

Another question in this IQ test asked, "A forest is to a tree as a tree is to a ?", giving the options orchard, plant, jungle and leaf. The relation $R_{x_2 \text{ is part of } x_1}$ applies to forest and tree, and which suggests the answer is given by relational hypersmplex simplex $\langle$ tree, leaf; $R_{x_2 \text{ is part of } x_1}\rangle$. If so the answer is "leaf".

Another question asked, "Car is to road as train is to", giving the options surface, locomotive, rails and wheels. Here the relational hypersimplex simplex is $\langle x_1, x_2; R_{\text{travels on}}\rangle$ and the answer is "rails".

Not all IQ questions are based on 1-dimensional hypersimplices and as simple as this, as illustrated by the 4-ary relation in Fig. 6.9.

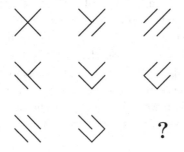

Which of the figures shown below completes the sequence?

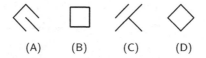

(A)   (B)   (C)   (D)

Fig. 6.9   A higher dimensional question

Let these configurations be numbered 1 to 9 from top left to bottom right, with configurations 1 to 3 in the top row, *etc.* The question is which of (A), (B), (C) or (D) is configuration number 9? Inspection shows that each of configuration is made up of four from eight diagonal lines. These lines are arranged in a square whose quadrants will be labeled Q_1, Q_2, Q_3 and Q_4. When the lines in the quadrants slope backward they will be denoted $\overleftarrow{Q_1}$, $\overleftarrow{Q_2}$, $\overleftarrow{Q_3}$ and $\overleftarrow{Q_4}$. When they slope forward they will be denoted $\overrightarrow{Q_1}$, $\overrightarrow{Q_2}$, $\overrightarrow{Q_3}$ and $\overrightarrow{Q_4}$. Thus the first three configurations in Fig. 6.9 are the relational simplices shown below.

$$\sigma_1 = \langle\, \overleftarrow{Q_1},\, \overrightarrow{Q_2}\, \overleftarrow{Q_3}\, \overrightarrow{Q_4};\, R_1 \rangle \qquad \sigma_2 = \langle\, \overleftarrow{Q_1},\, \overrightarrow{Q_2}\, \overrightarrow{Q_3}\, \overrightarrow{Q_4};\, R_2 \rangle \qquad \sigma_3 = \langle\, \overrightarrow{Q_1},\, \overrightarrow{Q_2}\, \overrightarrow{Q_3}\, \overrightarrow{Q_4};\, R_3 \rangle$$

The sequences are characterised by flipping the direction of the lines, *e.g.* σ_2 is obtained from σ_1 by flipping $\overleftarrow{Q_3}$ to become $\overrightarrow{Q_3}$. σ_3 is obtained from σ_2 by flipping $\overleftarrow{Q_1}$ to become $\overrightarrow{Q_1}$. Let ϕ be the *flipping operator* with $\phi\overrightarrow{Q_i} = \overleftarrow{Q_i}$ and $\phi\overleftarrow{Q_i} = \overrightarrow{Q_i}$. The top and middle rows of Fig. 6.9 have the progression

$$\langle x_1, x_2, x_3, x_4; R_i \rangle \longrightarrow \langle x_1, x_2, \phi x_3, x_4; R_{i+1} \rangle \longrightarrow \langle \phi x_1, x_2, \phi x_3, x_4; R_{i+2} \rangle.$$

Applied to the bottom row this gives

$$\sigma_7 = \langle\, \overleftarrow{Q_1},\, \overleftarrow{Q_2}\, \overleftarrow{Q_3}\, \overleftarrow{Q_4};\, R_7 \rangle \rightarrow \langle\, \overleftarrow{Q_1},\, \overleftarrow{Q_2}\, \overrightarrow{Q_3}\, \overleftarrow{Q_4};\, R_8 \rangle \rightarrow \langle\, \overrightarrow{Q_1},\, \overleftarrow{Q_2}\, \overrightarrow{Q_3}\, \overleftarrow{Q_4};\, R_9 \rangle = \sigma_9$$

and σ_9 is therefore configuration (D) in Fig. 6.9.

Example: The café wall illusion

Fig. 6.10 A row of black and white tiles

Figure 6.10 shows a row of black and white tiles in which the tops and bottoms of the tiles are parallel.

Figure 6.11 shows a remarkable illusion first observed by Richard Gregory. The story is that he was in a cafe which had black and white tiles on the wall arranged in the offset manner shown, and he sensed that alternate rows of tiles got narrower or wider from left to right. For example, as I look at row r_3 it gets narrower from left to right, while row r_4 gets wider from left to right.

Let $s_{i,j}$ be the jth tile from the left in the ith row, and $\sigma(r_i) = \langle s_{i,1}, s_{i,2}, s_{i,3}, s_{i,4}, s_{i,5}, s_{i,6}, s_{i,7}, s_{i,8}, ; R_1 \rangle$ represent the ith row of tiles. This structure has the property that its sides appear parallel.

Let $\sigma(\text{offset tiled wall}) = \langle r_1, r_2, r_3, r_4, r_5, r_6, r_7, r_8\, ; R_2 \rangle$ be the rows assembled into the configuration shown in Fig. 6.11(a). Then this structure has the emergent property of illusory bent lines which is not possessed by the rows when viewed in isolation.

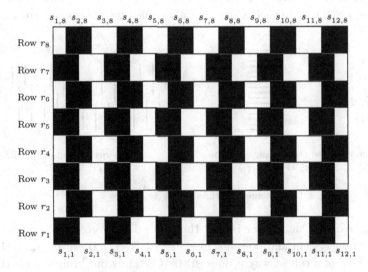

Fig. 6.11 The café wall illusion

Let $\sigma(\text{normal tiled wall}) = \langle r_1, r_2, r_3, r_4, r_5, r_6, r_7, r_8\, ; R_3 \rangle$ be the rows assembled into the configuration shown in Fig. 6.12. This structure assembles the rows in the usual way, and it does not have the emergent property of illusory bent lines.

158   Hypernetworks in the Science of Complex Systems

Fig. 6.12   The café wall illusion disappears for normal tilings

The sun illusion and virtual contours

Figure 6.13(a) shows the set of lines $\ell_1, ..., \ell_{16}$ arranged in a circle by the relation R_1. The resulting structure $\langle \ell_1, ..., \ell_{16}; r_1 \rangle$ has the emergent property that most people see a clear white disk at the centre of the lines, the so-called *sun illusion*. Figure 6.13(b) shows the same set of lines assembled under a different relation, $R)2$. Now there is no disk but a rectangle shape emerges. Figure 6.13(c) shows a set of twenty eight lines assembled in such a way that a so-called *virtual contour* emerges.

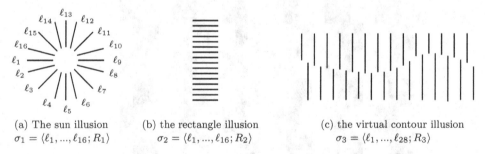

(a) The sun illusion
$\sigma_1 = \langle \ell_1, ..., \ell_{16}; R_1 \rangle$

(b) the rectangle illusion
$\sigma_2 = \langle \ell_1, ..., \ell_{16}; R_2 \rangle$

(c) the virtual contour illusion
$\sigma_3 = \langle \ell_1, ..., \ell_{28}; R_3 \rangle$

Fig. 6.13   Emergent features in line assemblies

Does the disk exist? One answer is that the white disk emerges as a quirk of our visual perception system and does not exist "objectively". However, since the disk can be observed in a robust way it does exists. Certainly our brains *create* the disk from the atomic observation of the lines, but the regularity and replicability of the process makes the white a disk a reality.

It would be a simple matter to implement R_1 as a computer program in a way that recognised the disk and made it explicit. Furthermore, the computer program could then store information about the disk in a descriptor simplex of σ_1.

Example: Combinatorial causes of road accidents

In a study of road accidents, drivers who had been involved in accidents were interviewed to find out the possible causes. The telephone interviews were unstructured and open-ended with the interviewer trying to elicit the causes from the interviewee rather than suggesting particular causes. For example, interviewees were asked the speed at which they were traveling, which would often elicit a response that the driver was or was not going too fast for the conditions. The interviewees were remarkably willing to discuss their accidents and remarkably honest about their own culpability. Some typical examples of the fifty seven reported accident hypersimplices are:

⟨mechanical failure, need to stop, lack anticipation, stress; R_1⟩
⟨carelessness, unexpected manoeuvre; R_8⟩
⟨change in road layout, poor signposting, bad visibility; R_{16}⟩
⟨speed, lack of concentration; R_{23}⟩
⟨inexperienced driver, car in wrong position; R_{31}⟩
⟨poor visibility, lack of caution, road wet; R_{23}⟩
⟨not paying attention, to near/too fast, brakes poor, unexpected manoeuvre; R_{51}⟩
⟨narrow road, speed R_{53}⟩

These combinations of causes were expressed in the soup of everyday language, and the words were aggregated into the following intermediate words at *Level N+1*:

D1–Stress	D2–carelessness	D3–Poor anticipation
D4–Too close	D5–Looking wrong way	D6–Alcohol
D7–Health/Tiredness	D8–Young male ego	D9–Inexperience
D10–Unfamiliarity with vehicle	D11–Cyclist blind	D12–In a hurry
D13–Unfamiliar with road	D14–Speed	D15–Mistaken priority
V1–Mechanical failure	R1–Difficult configuration	R2–Poor visibility
R3–Poor signposting	R4–Difficult surface	R6–Heavy traffic
A1–Unexpected event	A2–Slow vehicle in front	

A star-hub analysis of the data produced the following Galois pairs:

⟨D2–Carelessness, R1–Difficult configuration⟩	⟨2, 5, 9, 12, 35, 40, 42, 51, 57⟩
⟨D1–Stress, R1–Difficult configuration⟩	⟨1, 2, 20, 26, 34, 51, 52⟩
⟨D2–Carelessness, R2–Poor visibility⟩	⟨2, 3, 4, 35, 38, 40⟩
⟨D14–Speed, R1–Difficult configuration⟩	⟨10, 12, 22, 39, 43, 53⟩
⟨D1–Stress, R2–Poor visibility⟩	⟨2, 3, 11, 13, 26⟩
⟨R1–Difficult configuration, R2–Poor visibility⟩	⟨2, 26, 35, 40, 43⟩
⟨R2–Poor visibility, R4–Difficult road surface⟩	⟨11, 13, 26, 36, 38⟩
⟨R2–Poor visibility, A1–Unexpected event⟩	⟨11, 13, 16, 36, 54⟩
⟨R2–Poor visibility, R3–Poor signposting⟩	⟨2, 16, 26, 56⟩
⟨D1–Stress, D13–Unfamiliar with road⟩	⟨2, 3, 25, 52⟩
⟨D2–Carelessness, A1–Unexpected event⟩	⟨1, 9, 10, 41⟩
⟨R2–Poor visibility, R4–Difficult road surface⟩	⟨11, 13, 26⟩

⟨R2–Poor visibility, R4–Difficult road surface, A1–Unexpected event⟩ ⟨11, 13, 36⟩
⟨D2–Carelessnes, R1–Difficult configuration, R2–Poor visibility⟩ ⟨2, 35, 40⟩

Figure 6.14 shows a graphical summary of the stars and hubs, where the numbers show the number accidents associated with the hypersimplices.

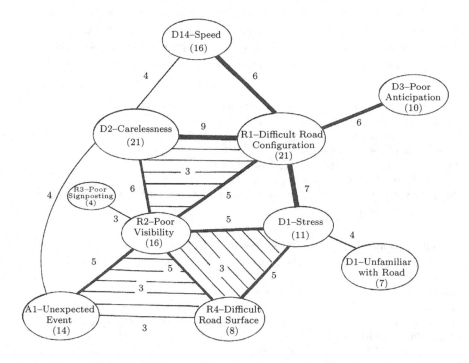

Fig. 6.14 Frequencies of occurences of accident factors

During this research it became clear that some accident factors played a greater role than others, and the interviewees were asked to rate the importance of the factors on a five-point scale, with 1 being less important and 5 being very important.

Then, for example, σ(Accident-2) = ⟨D1–Stress(5), D2–Careless(3), D13–Unfamiliar road(5), D15–Mistaken priority(5), R1–Difficult config(5), R2–Poor visibility(3), R3–Poor signposting(5)⟩, and σ(Accident-2) = ⟨D1–Stress(5), D2–Careless(4), D6–Alcohol(1), D7–Tired(5), D13–Unfamiliar road(3), D15–Speed(3) R2–Poor visibility(2)⟩.

Let $\mu(v_i)$ be the weighting given to accident factor v_i. The value on the whole simplex can be defined to be the *fuzzy conjunction* defined as $\mu\sigma = \min\{\mu(v_i) \mid v_i \lesssim \sigma\}$. Then for a fuzzy value of 3, σ(Accident-2) and σ(Accident-3) share the face ⟨D1-Stress, D2-Careless, D13-Unfamiliar road⟩, and they are 3-fuzzy 2-near.

Example: Abstracting features from noisy images

Figure 6.15 is based on a microscope image of a zebra fish. Advanced laser microscopy produces sequences of three-dimensional images of the organism *in vivo* through time, allowing scientists to study the cells as they divide and plot the development of the organism. This requires machine vision to detect the cells and match them through the sequence of images over a period of hours. The images are noisy and have variable contrast and illumination making it hard to segment them accurately into polygons containing the cells.

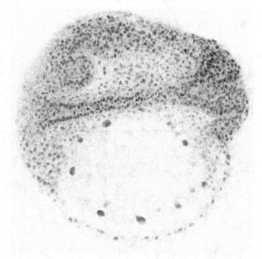

Fig. 6.15 An *in-vivo* microscope image of a zebra fish embryo

Figure 6.16 shows a relation R defined between neighbouring pixels of an image with pixel p R-related to pixel p' if p' is darker than p, and darker than any other pixel adjacent to p. The relation for the zebra fish image is shown in Fig. 6.17(b). As can be seen, the graph is made up of locally connected components where each has a local collector – the darkest pixel in the component. The transitive closure of R is an n-ary relation on the pixels and the hypersimplices are polygons. Rather remarkably, this simple approach segments the image to enclose the cell in a polygon made up of all the connected pixels, as shown in Fig. 6.17(c).

      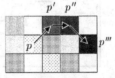

(a) the neighbours of p (b) p' is the darkest neighbour of p (c) a light-to-dark path

Fig. 6.16 The darkest neighbour relations

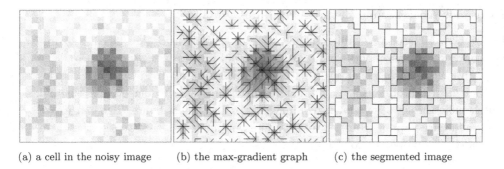

(a) a cell in the noisy image   (b) the max-gradient graph   (c) the segmented image

Fig. 6.17   Oriented links in image segmentation

Figure 6.18 shows a segmentation of a larger region based on a variant of this method. It should be noted that this segmentation is parameter free, *i.e.* the program has no greyscale or gradient thresholds, but works exclusively on the "darker than" relation. In this application the method of segmentation is remarkably robust.

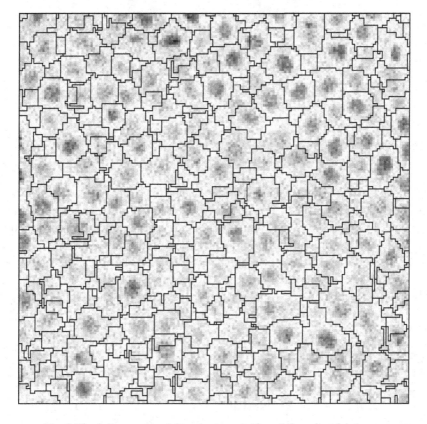

Fig. 6.18   A hypernetwork-based segmentation of the zebra fish image

6.3 The Fundamental Question of Hypernetworks

The nature of intersections and connectivity in hypernetworks is much more subtle than it is for simplicial complexes. The intersection of simplices is easy to define in simplicial families. For simplices $\sigma = \langle x_0, x_1, .., x_p \rangle$ and $\sigma' = \langle y_0, y_1, ..., y_{p'} \rangle$, $\sigma \cap \sigma' \overset{\text{def}}{=} \langle z_0, z_1, ..., z_q \rangle$ where $\{z_0, z_1, ..., z_q\} = \{x_0, x_1, .., x_p\} \cap \{y_0, y_1, ..., y_{p'}\}$. To begin extending this to hypersimplices let

$$\sigma \cap \sigma' \overset{\text{def}}{=} \langle x_0, x_1, ..., x_m; R \rangle \cap \langle y_0, y_1, ..., y_{p'}; R' \rangle \overset{\text{def}}{=} \langle z_0, z_1, ..., z_p; R'' \rangle$$

Question: What is $\sigma \cap \sigma'$ for hypersimplices σ and σ'?

This will be called the *Fundamental Question of Hypernetwork Theory*.

This question is central to the concept of connectivity in complex systems and the way that connected simplices constrain system dynamics, Whereas it seems reasonable that the vertex list for $\sigma \cap \sigma'$ should be those vertices common to both simplices, it is not clear how R'' should be defined as a combination of R and R'.

To illustrate the problem consider companies $c_1, ..., c_6$ and the hypersimplices $\langle c_1, c_2, c_3, c_4; R_1 \rangle$ and $\langle c_3, c_4, c_5, c_6; R_2 \rangle$ where R_1 is the "consortium-1" relation and R_2 is the "consortium-2" relation. Let $\langle c_3, c_4; R(R_1, R_2) \rangle$ be the intersection of the two consortia. The problem is how to interpret the relation $R(R_1, R_2)$. Neither R_1 nor R_2 are defined on $\langle c_1, c_2 \rangle$ since they are 4-ary relations. Perhaps one can consider the "restriction" of these 4-ary relations to $\langle c_1, c_2 \rangle$? This, and other ideas, will be explored to try to answer the Fundamental Question.

Intersections of hypersimplices

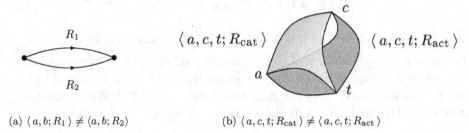

(a) $\langle a, b; R_1 \rangle \neq \langle a, b; R_2 \rangle$ (b) $\langle a, c, t; R_{\text{cat}} \rangle \neq \langle a, c, t; R_{\text{act}} \rangle$

Fig. 6.19   Relations links and relational hypersimplices

Figure 6.19(a) shows the relations R_1 and R_2 between the vertices a and b with the 1-simplices $\langle a, b; R_1 \rangle$ and $\langle a, b; R_2 \rangle$ touching at the vertices. By analogy, Fig. 6.19(b) shows two triangles touching at the vertices. Is this what is wanted in the definition of $\sigma \cap \sigma'$?

164 Hypernetworks in the Science of Complex Systems

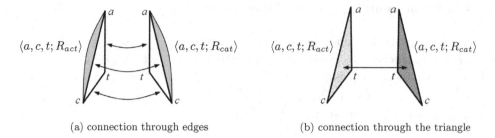

(a) connection through edges   (b) connection through the triangle

Fig. 6.20 Connectivity between relational hypersimplices

Why should the intersection of the simplices be just the vertices? The intersection $\langle a,c,t;R_{\text{act}}\rangle \cap \langle a,c,t;R_{\text{cat}}\rangle$ could just be the vertices $\langle a\rangle$, $\langle c\rangle$ and $\langle t\rangle$ as drawn in Fig. 6.19(b), but it could also be the edges $\langle a,c\rangle$, $\langle a,t\rangle$ and $\langle c,t\rangle$ as shown in Fig. 6.20(a), or the whole triangle as shown in Fig. 6.20(b).

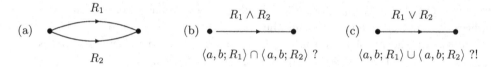

Fig. 6.21 Possible intersections of 1-dimensional hypersimplices

To try to understand how higher dimensional hypersimplices might intersect in hypernetworks, consider the more familiar case of networks. with the relationships R_1 and R_2 drawn in Fig. 6.21(a) as two separate oriented arcs. In this case the domain (vertex set) of the combination of the relations is the same as the domain of each relation, and it seem natural to write

$$\langle a,b;R_1\rangle \cap \langle a,b;R_2\rangle \stackrel{\text{def}?}{=} \langle a,b;R_1 \wedge R_2\rangle$$

where a is $R_1 \wedge R_2$ related to b if it is both R_1 and R_2 related to b.

In contrast to the simplicity of being connected at the vertices, hypersimplices present a new set of questions relating to connectivity. For example, Fig. 6.19(b) shows the simplices "cat" = $\langle a,c,t;R_{\text{cat}}\rangle$ and "act" = $\langle a,c,t;R_{\text{act}}\rangle$ connected at their vertices. This could give a partial solution to defining $\sigma \cap \sigma'$, namely:

$$\langle v_0,v_1,...,v_p;R_1\rangle \cap \langle v_0,v_1,...,v_p;R_2\rangle \stackrel{\text{def}?}{=} \langle v_0,v_1,...,v_p;R_1 \wedge R_2\rangle$$

Fig. 6.19(c) suggests this can be extended to a "union" of hypersimplices:

$$\langle v_0,v_1,...,v_p;R_1\rangle \cup \langle v_0,v_1,...,v_p;R_2\rangle \stackrel{\text{def}?}{=} \langle v_0,v_1,...,v_p;R_1 \vee R_2\rangle$$

Logical operations on hypersimplices

This partial solution to defining $\sigma \cap \sigma'$ is possible because this is a special case in which the domains of the relations are the same. In fact the "intersection" ("union") is more like a conjunction (disjunction). To make this clear the *conjunction operator* $\wedge$, *disjunction operator* $\vee$, and also a *negation operator* can be defined on hypersimplices as follows. Let $\sigma_1 = \langle x_1, x_2, ..., x_n; R_1 \rangle$ and $\sigma_2 = \langle x_1, x_2, ..., x_n; R_2 \rangle$. Then:

Conjunction: $\sigma_1 \wedge \sigma_2 = \langle x_1, ..., x_n; R_1 \rangle \wedge \langle x_1, ..., x_n; R_2 \rangle \stackrel{\text{def}}{=} \langle x_1, ..., x_n; R_1 \wedge R_2 \rangle$

Disjunction: $\sigma_1 \vee \sigma_2 = \langle x_1, ..., x_n; R_1 \rangle \vee \langle x_1, ..., x_n; R_2 \rangle \stackrel{\text{def}}{=} \langle x_1, ..., x_n; R_1 \vee R_2 \rangle$

Negation: $\neg \sigma_1 = \neg \langle x_1, ..., x_n; R_1 \rangle \stackrel{\text{def}}{=} \langle x_1, ..., x_n; \neg R_1 \rangle$

Care has to be taken when applying logical operators to hypersimplices. For example, Fig. 6.22 shows two relations on the lines ℓ_1, ℓ_2 and ℓ_3. Although they share the same vertices, the relations $R_F \vee$ and R_h cannot both hold at the same time, $R_F \vee$ implies $\neg R_h$ and R_h implies $\neg R_F \vee$. In this case $\langle \ell_1, \ell_2, \ell_3; R_F \rangle \wedge \langle \ell_1, \ell_2, \ell_3; R_h \rangle = \sigma_{-1}$ where σ_{-1} is the *null simplex*.

Fig. 6.22 $R_F \wedge R_h$ is the null relation since both relations cannot hold simultaneously

The faces of relational hypersimplices

The intersection of two hypersimplices can get even more complicated. For example, consider the words *sing* and *ring*, which could be written as $\langle s, i, n, g \rangle$ and $\langle r, i, n, g \rangle$. Then one might write $\langle s, i, n, g \rangle \cap \langle r, i, n, g \rangle = \langle i, n, g \rangle$. However what about the simplex $\langle s, i, n, g, i, n, g \rangle$ which has two $\langle i, n, g \rangle$ faces, as shown in Fig. 6.23. Are these two $\langle i, n, g \rangle$ faces the same?

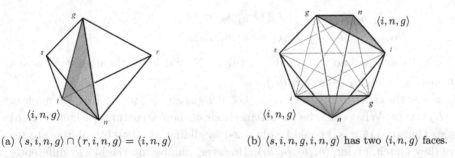

(a) $\langle s, i, n, g \rangle \cap \langle r, i, n, g \rangle = \langle i, n, g \rangle$ (b) $\langle s, i, n, g, i, n, g \rangle$ has two $\langle i, n, g \rangle$ faces.

Fig. 6.23 Repetition of faces is allowed in relational simplices

The hypersimplices $\langle i,n,g; R_{\text{PresentParticiple}}\rangle$ and $\langle i,n,g; R_{\text{VerbStemPart}}\rangle$, can discriminate these "ing" faces, but even if this were a solution there is another problem: it is not possible to decided if the letters form the present participle or are part of the verb stem without seeing *all* the letters of the word, *i.e.* the relations $R_{\text{PresentParticiple}}$ and $R_{\text{VerbStemPart}}$ act on the whole word $\langle s,i,n,g,i,n,g\rangle$, not just the face $\langle i,n,g\rangle$.

The ambiguity in the case is due to the symbols i, n and g between repeated, and taking their meaning from the ordering. To make the problem even more extreme consider a case in which *all* the vertices are the same.

Figure 6.24 shows a set of identical *blocks* $B = \{b_0, b_1, b_2, b_3, b_4, b_5, b_6, b_7, b_8, b_9\}$. Let three blocks $\langle v_1, v_2, v_3\rangle$ make an *arch* under the 3-ary relation R_{arch} if v_1 and v_3 are vertical blocks from B with a gap between them, and v_2 is a horizontal block for B and b_2 rests on b_1 and b_3. By this definition $\sigma(a_1)$ is an arch but $\sigma(a_0)$ is not.

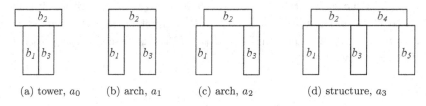

(a) tower, a_0 (b) arch, a_1 (c) arch, a_2 (d) structure, a_3

Fig. 6.24 Simple structures built from blocks

Under this definition $\sigma(a_1)$ and $\sigma(a_2)$ are both arches, but they are not identical. If it were desirable to discriminate them other definitions could be used. For example, "Let the vertices v_1, v_2, and v_3 be R_2 related if they are all blocks in B, and they are assembled as shown in Fig. 6.24(c) with v_1 corresponding to b_1, v_2 corresponding to b_2 and v_3 corresponding to b_3". The arch in Fig. 6.24(b) cannot be matched with that in Fig. 6.24(c) and is not an arch by this definition.

What about the "double arch" $\sigma(a_3) \stackrel{\text{def}}{=} \langle b_1, b_2, b_3, b_4, b_5; R_3\rangle$? The blocks b_1, b_2 and b_3 satisfy the definition of R_2 and can be assembled into the R_2 arch $\langle b_1, b_2, b_3; R_2\rangle$. The blocks b_3, b_4 and b_5 in $\sigma(a_3)$ also satisfy the definition of R_2 and they can be assembled into the R_2 arch $\langle b_3, b_4, b_5; R_2\rangle$. This suggests writing:

$$\langle b_1, b_2, b_3; R_2\rangle \lesssim \langle b_1, b_2, b_3, b_4, b_5; R_3\rangle.$$
$$\langle b_3, b_4, b_5; R_2\rangle \lesssim \langle b_1, b_2, b_3, b_4, b_5; R_3\rangle.$$

Thus $\sigma(a_3)$ contains two $\sigma(a_2)$ substructures. Perplexingly these share a vertex $\langle b_1, b_2, b_3; R_2\rangle \cap \langle b_3, b_4, b_5; R_2\rangle \stackrel{?}{=} \langle b_3\rangle$.

Consider the structures s_1 and s_2 in Figs. 6.25(a) and (b) which are both made of three R_2 arches. What could be the intersections of these structures? Figure 6.25(a) suggests that $\sigma(s_1) \cap \sigma(s_2)$ could be obtained by sliding $\sigma(s_1)$ on top of $\sigma(s_2)$ to see where they match, giving $\langle b_1, b_2, b_3, b_4\rangle$. However, numbering the blocks differently gives a different result as shown in Fig. 6.25(b).

Hypernetworks

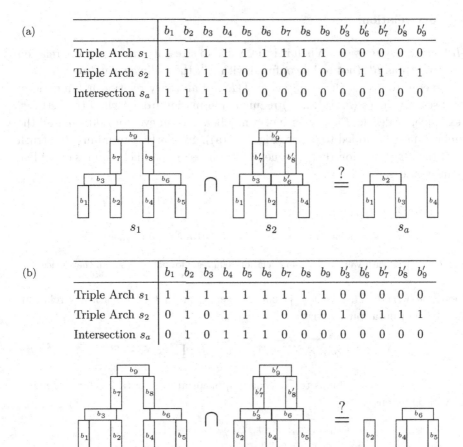

(a)

	b_1	b_2	b_3	b_4	b_5	b_6	b_7	b_8	b_9	b'_3	b'_6	b'_7	b'_8	b'_9
Triple Arch s_1	1	1	1	1	1	1	1	1	1	0	0	0	0	0
Triple Arch s_2	1	1	1	1	0	0	0	0	0	1	1	1	1	1
Intersection s_a	1	1	1	1	0	0	0	0	0	0	0	0	0	0

(b)

	b_1	b_2	b_3	b_4	b_5	b_6	b_7	b_8	b_9	b'_3	b'_6	b'_7	b'_8	b'_9
Triple Arch s_1	1	1	1	1	1	1	1	1	1	0	0	0	0	0
Triple Arch s_2	0	1	0	1	1	1	0	0	0	1	0	1	1	1
Intersection s_a	0	1	0	1	1	1	0	0	0	0	0	0	0	0

Fig. 6.25 $\sigma(s_1) \cap \sigma(s_2) = ?$

In this case the intersection of the structures is determined more by the arbitrary numbering of the identical blocks than by the relational structure.

The examples of $\langle a, c, t; R_{\text{act}} \rangle \cap \langle a, c, t; R_{\text{cat}} \rangle$, $\langle s, i, n, g, i, n, g \rangle \cap \langle r, i, n, g \rangle$ and $\sigma(s_1) \cap \sigma(s_2)$ show that the concept of intersections of structured sets of vertices is much more subtle than the intersections of sets or the intersections of sequences of abstract vertices. This is because, to some extent, the list of vertices is a *syntactic* construct while the relation is a *semantic* construct. They are of course complementary since the ordered list of vertices is the *domain* of the relation, and it is necessary for the vertices to be in the right order for the relation to be *well formed* in a logical sense. For example, $\langle d, o, g; R_{\text{cat}} \rangle$ is well formed, even though the simplex does not exist while $\langle \heartsuit, \square, \triangle; R_{\text{cat}} \rangle$ is not well formed because the relation R_{cat} does not know what is meant by the symbols $\heartsuit$, $\square$ and $\triangle$.

6.4 n-ary relations

Let R be an n-ary relation. When it is desirable to stress its *dimension* the relation will be written as R^n (not to be confused with $\mathbb{R}^n$, the n-dimensional reals).

In general, to decide the *truth value*, $\mathsf{T}(R)$, of an n-ary relation holding on an ordered set of vertices $\langle v_1, v_2, ..., v_n \rangle$ requires information from each of the vertices. For example, to decide if a sequence of n numbers has an even or odd sum, all the n numbers must be added together (Fig. 6.26(a)). Similarly determining the truth value of the n-ary relation that a sequence of n lines are parallel requires every line to be inspected Fig. 6.26(b).

$\langle 17, 8, 3, v_4; R_{sum\ is\ odd/even} \rangle$ $\qquad\qquad$ $\langle\ /\ ,\ /\ ,\ /\ ,\ /\ ,\ /\ ,\ /\ ,\ /\ ,\ /\ , v_9; R_{parallel} \rangle$

(a) a 4-ary relation $\qquad\qquad\qquad\qquad$ (b) a 9-ary relation

Fig. 6.26 The truth value of an n-ary relation requires information from all the vertices

Let $\mathbb{T}$ be a *truth set*, *e.g.* $\mathbb{T} = \{True, False, ...\}$ and let R be an n-ary relation. By an abuse of notation we write

$$R : X_1 \times X_2 \times ... \times X_n \to \mathbb{T}, \quad \text{or} \quad R : \prod_{i=1}^{n} X_i \to \mathbb{T}$$

where $R\langle x_1, x_2, ..., x_n \rangle$ belongs to $\mathbb{T}$. Where appropriate the notation for hypersimplices will be extended to

$$\langle\, x_1, ..., x_n;\ R,\ \tau,\ \Delta t\, \rangle$$

where $\tau = R\langle x_1, x_2, ..., x_n \rangle$ is the truth value of the relation in time "interval" Δt. This can be a single point in time, a set of points in time, a time interval, or a set of time intervals, *e.g.* $\Delta t \equiv$ "every Thursday". When $\tau =$ True, the expression $\langle\, x_1, ..., x_n;\ R,\ \tau,\ \Delta t\, \rangle$ means that the n-relation R holds on the simplex $\langle x_1, ..., x_n \rangle$ for all the times in Δt.

The abstract expression $\langle\, x_1, ..., x_n;\ R,\ \tau,\ \Delta t\, \rangle$ cannot be evaluated since none of the x_i is explicit, R is not explicit, and Δt is not explicit. The truth value τ cannot be evaluated until everything is made explicit by substituting these symbols with 'real things', when they are said to be *instantiated*, as shown in Fig. 6.27.

$\langle\, x_1, ..., x_4;\ R,\ \tau,\ \Delta t\, \rangle$ $\qquad\qquad$ $\langle\ |\ ,\ \overline{\ \ }\ ,\ |\ ,\ \_\_\ ;\ R_{\text{square}},\ \tau,\ \text{Today} \rangle$

(a) uninstantiated $\qquad\qquad\qquad\qquad$ (a') instantiated

$\langle\, x_1, ..., x_4;\ R,\ \tau,\ \Delta t\, \rangle$ $\qquad\qquad$ $\langle \text{Brazil, Russia, India, China};\ R_{\text{BRIC}},\ \tau,\ \text{Now} \rangle$

(b) uninstantiated $\qquad\qquad\qquad\qquad$ (b') instantiated

Fig. 6.27 Instantiating variables and relations to evaluate truth values

Figure 6.27(a') shows four lines and the relation of forming a square today. It is instantiated because all the necessary information is available to make a decision. Whether or not the lines actually form a square is another matter – this requires a computation associated with R_{square}.

Figure 6.27(b') shows the countries Brazil, Russia, India and China and the relation of them currently forming the so-called BRIC countries. They do, so here $\tau = \text{True}$.

To compute the truth value of an n-ary relation requires a number of conditions be satisfied. The first is that simplex $\langle x_1, ..., x_n \rangle$ is *well formed* for R. This means that each of the x_i is of the appropriate type to make R computable. For example, suppose that R is the "parallel line" relation. Then if any of the x_i is not a 'line' it will not carry the necessary information for the computation, and the simplex will not be well formed.

The distinction here is more practical than philosophical. If the instantiation of a vertex is not well formed the computational function will not be able to make sense of the data it receives. Strongly-typed computer languages may guarantee that instantiated simplices are well formed. However, consider objects represented by two data points, $(x_{i,1}, y_{i,1}), (x_{i,2}, y_{i,2})$. Let n objects can be defined to be parallel if $(y_{i,1} - y_{i,2})/(x_{i,1} - x_{i,2})$ is the same for them all. If all the objects are lines this will work well. However suppose one of the objects is a box with bottom left corner given by $(x_{i,1}, y_{i,1})$ and top right corner given by $(x_{i,2}, y_{i,2})$. Then $\tau = \text{True}$ in the following:

$$\langle /, /, /, /, /, /, /, \Box; R_{parallel}, \tau = \text{True} \rangle$$

This could be considered to be an error in the definition of "parallel" if, despite $\Box$ being well formed, it is a not formed as intended. On the other hand, this ambiguity could be considered to be very interesting, since it may enable new things to be created, new things to be discovered, and new things to happen.

The details of what it means for a particular simplex to be "well formed" for a particular n-ary relation will depend on the implementation of the computation. However it will always be the case that the instantiation of each vertex provides necessary information for the system to decide if that instantiation is appropriate and "allowable", before it attempts to compute the truth value of the n-ary relation.

Ticking the boxes for the vertex parts

The distinction between the computation that establishes that the n vertices of a potential hypersimplex are "allowable" for a well formed relation R^n and the computation that established that these vertices are related under R^n is important. Generally the instantiated vertices "bring their data with them" and the eligibility computation simply involves checking these data against simple criteria.

170 Hypernetworks in the Science of Complex Systems

Checking each instantiated vertex against the criteria is a box-ticking exercise. In principle the box for each vertex can be ticked independently of all the others. Thus this part of the computation of an n-ary relation has computational complexity linear in n.

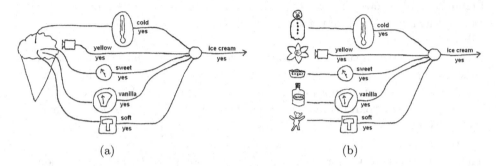

(a) (b)

Fig. 6.28 Sensors for observing a descriptor simplex

Figure 6.28 illustrates the difference between ticking the vertex boxes and computing the relational whole. On the left is an ice cream being sensed for coldness, yellowness, sweetness, vanilla flavour, and softness. All of the these are present simultaneously and the whole is recognised to be a vanilla ice cream. However, on the right these sensors are being individually "fooled". The temperature sensor is put in snow, the colour sensor is shown a daffodil, the sweetness sensor is given sugar, the vanilla sensor is given vanilla essence from a bottle, and the softness sensor touches a cuddly toy. Since all the elements are in place, the vanilla ice cream detector can be fooled into making a wrong decision – all the boxes are ticked but what is sensed is not ice-cream. Although the individual sensors are responding correctly as vertices, their responses are not *combined* together in the correct way to respond to ice creams and only ice creams.

This illustrates that the method of observation plays a crucial role in observing n-ary relations. In this case the observations are not completely independent because they are made simultaneously, which imposes a relation on the vertices. However even this is not sufficient. In Fig. 6.28(a) the sensors are "hard-wired" to look at the same object in an appropriate way. The notion of Gestalt is precisely that our brains are hard-wired to receive and combine signals from our senses.

Computation to establish the relational whole

In comparison to the linear complexity of ticking the boxes for the eligibility of the vertices, the computation of the relation R^n usually has higher complexity. In practice it is done by humans, requires special hardware, or done by standard computers with typical algorithms having computational complexity of order $n \log n$, n^2, or worse. In real world systems the great majority of relational decisions are made by people using a combination of process, intuition, theory and data.

Relational computation by humans

Human being have an astonishing ability to compute n-ary relations; intuitively with no apparent effort, by surveying the available information and reaching implicit conclusions, or by explicit procedures including individual and collective reasoning.

As individuals we are able to operate in completely new environments outside of previous experience. We structure the new environment on an *ad-hoc* basis and navigate by intuition. This applies to both physical environments such as an exotic city or social environments such as starting a new job. Generally situations have to be assessed in real time for immediate responses. We have an amazing ability to be aware of possibly relevant vertices in the environment, to be aware of possibly relevant n-ary relations on those vertices, and to instantly compute their truth values as necessary.

Previously examples were shown of IQ questions that involved abstracting n-ary relations and using them to answer questions. To further illustrate our fabulous cognitive ability to process n-ary relations, consider the following:

> Aoccdrnig to a rscheearch at Cmabrigde Uinervtisy, it deosn't mttaer in waht oredr the ltteers in a wrod are, the olny iprmoetnt tihng is taht the frist and lsat ltteer be at the rghit pclae. The rset can be a toatl mses and you can sitll raed it wouthit porbelm. Tihs is bcuseae the huamn mnid deos not raed ervey lteter by istlef, but the wrod as a wlohe.

The suggestion that only the first and last letters matter is an exaggeration. For example, let $\sigma_1 = \langle a, b, C, d, g, i, m, r, e; R_1 \rangle$ = "Cambridge". Then $\sigma_2 = \langle a, b, C, d, g, i, m, r, e; R_2 \rangle$ = "Cmabrigde" is relatively highly connected to "Cambridge" since it shares the 4-face "C$\star$bri$\star\star$e". In contrast $\sigma_3 = \langle a, b, C, d, g, i, m, r, e; R_3 \rangle$ = "Cridgambe" is just 1-near to "Cambridge". Actually the idea that words can be often be read when their internal letters are scrambled was in Graham Rawlinson's Ph.D. thesis at Nottingham University [Rawlinson (1976)], as unearthed by Cambridge researcher Matt Davies. Details of the text above and Rawlinson's work are on the websites http://www.mrc-cbu.cam.ac.uk/people/matt.davis/Cmabrigde/ and http://www.mrc-cbu.cam.ac.uk/people/matt.davis/cmabridge/rawlinson.html.

The main problem with using our inbuilt automatic n-ary relation processing ability is that, because everything is implicit, errors can be hard to detect and their effects can propagate. Our outstanding individual and social ability to compute relations is presumably a major evolutionary reason for our success, since we can respond to unknown and unforeseen dangers very well with great resilience as a species. Of course individuals and groups sometimes respond inappropriately and crash out of the system.

Generally n-ary relations are too complicated to be decided by intuition alone, and more explicit and systematic methods are required for their computation.

Relational computation by specialised hardware

The hypothetical ice-cream detector considered earlier is an example of specialised hardware to perform the computation of n-ary relations. As another example, consider a device designed to recognise symbols made of nine light sensors, s_1 to s_9 as illustrated in Fig. 6.29. The individual sensors are set such that they are activated when the light level entering them falls below a threshold, *i.e.* when a dark object is placed in front of it.

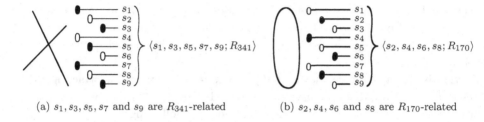

(a) s_1, s_3, s_5, s_7 and s_9 are R_{341}-related (b) s_2, s_4, s_6 and s_8 are R_{170}-related

Fig. 6.29 A hardware device to measure n-ary relations directly

There are $2^9 = 512$ possible combinations of sensor responses, and this number of possible 9-ary relations on the sensors. The relations are numbered such that R_1 means that sensor just s_1 is activated, R_2 means that just R_2 is activated, R_3 means that sensors s_1 and s_2 are activated, as so on until R_{511} means that all sensors are activated.

It is supposed that the sensors are wired together to form a device with 512 output wires. On receiving a stimulus, wire w_k is activated if the sensors corresponding to R_k are activated. For example, wire w_k could have its voltage set to 5v while all the other wires are set to 0v. This device is an *n-ary relation detector*.

Figure 6.29(a) shows the response to an X shape presented to the device, namely that relation R_{341} holds *simultaneously* on the vertices s_1, s_3, s_5, s_7 and s_9, so wire w_{341} is activated, In contrast, Fig. 6.29(b) shows the response to an O shape presented to the device, where the relation R_{170} holds on the vertices s_2, s_4, s_6 and s_8, so wire w_{170} is activated.

Figure 6.30 shows part of a face recognition system. Typically these systems work on the basis of n-ary relations between n identified points on the face such as the centre of the eyes and the mouth, as in the triangles illustrated here.

Fig. 6.30 Part of an automatic face recognition system

Relational computation by sequential computers

It is counterintuitive that, although computing an n-ary relation requires all vertices to be instantiated and all these data to be used, the actual computation can work through the vertices in a step-by-step way.

To illustrate this consider the n-ary relation of computing the parity of a binary number (odd if the number of bits set to 1 is odd and even if the number of bits set to 1 is even). One way to perform the computation is go through the bits counting the number of 1s. Let c be this count, initially set to the value of the first bit, which does not have to be considered again. Then, using mod 2 addition, add the value of the second bit to c. This second bit does not have to be considered again. Thus to compute the parity of the whole, each bit is being inspected in isolation to the others – which is the antithesis of the notion of Gestalt?

Let $\sigma = \langle b_1, b_2, b_3, b_4, b_5, b_6, b_7, b_8 \rangle$ be the sequence of digits in a one-byte binary number. Then the parity computation begins with the computation of $\phi\langle b_1, b_2\rangle = (\phi\langle b_1\rangle + \phi\langle b_1\rangle)$ mod 2. It continues with the computation of of $\phi\langle b_1, b_2, b_3\rangle = ((\langle \phi\langle b_1, b_2\rangle + \phi\langle b_2\rangle))$ mod 2, and so on until the calculation of $\phi\langle b_1, b_2, b_3, b_4, b_5, b_6, b_7\rangle + \phi\langle b_8\rangle)$ mod 2, which is either 1 or 0. Seen this way, the computation proceeds across the faces of σ, *constructing* the n-ary relation as the computation proceeds.

Heuristics for computing n-ary relations

Although aimed at large complex systems, hypernetwork theory can contribute to implicit decisionmaking by making explicit the questions: "what are the relevant n-ary relations and what are the elements?", or "what are the vertices and how are they assembled into a whole?", with related questions such as "are all the necessary vertices available?", "what is known about the vertices?", and "is it possible to decide if the n-ary holds with the information available?".

6.5 Hyperfaces for defining intersections of hypersimplices

One of the most difficult challenges in hypernetwork theory is understanding the nature of faces of hypersimplices. Removing any of the vertices of a hypersimplex can be compared to sticking a pin into a balloon – the n-ary relation structure necessarily collapses. Indeed this prompts the question whether it is possible for a hypersimplex to have "hyperfaces"?

Recall that for simplices, $\widehat{f^k}\sigma_p = \sum_{\sigma_{p-k}} \sigma_{p-k}$. Let this be extended to the hypersimplex $\sigma = \langle x_0, ..., x_{p-}; R^p \rangle$ as

$$\widehat{f^k}\langle x_0, ..., x_{p-1}; R^p\rangle \stackrel{\text{def}}{=} \sum_{\sigma_{p-k} \lesssim \sigma} \langle \sigma_{p-k}; R^{p-k}\rangle$$

with the special case $\widehat{f^p}\langle v_0, ..., v_{p-1}; R \rangle = \langle v_0; \widehat{f^p R^p} \rangle + \langle v_1; \widehat{f^p R^p} \rangle + ... \langle v_{p-1}; \widehat{f^p R^p} \rangle$, where $\widehat{f^p R^p}$ is a unary relation defined on isolated vertices. This suggests the possibility of moving from R^p with full information about the n-ary relation to $\widehat{f^p R^p}$ with almost no information about the n-ary relation.

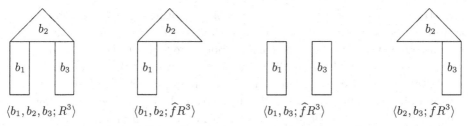

Fig. 6.31 The restriction of a 3-ary relation to 2-ary relations

Figure 6.31 shows the 3-ary relation R^3 *restricted* to the faces of $\langle b1, b_2, b_3 \rangle$ with whatever residual structure remains after the removal of one of the vertices.

Consider the newspaper headline hypersimplex

⟨ Rail, chiefs, urge, people, not, to, travel, in, south-west, and, steer, clear, of, floodwater, after, several, stranded, motorists, are, rescued; R^{20}⟩.

with twenty words as vertices, and the relation R^{20} that puts all these words in the right place. This has many hyperfaces, including

$\sigma_1 = \langle$Rail, chiefs, urge, people, not, to, travel; $\widehat{f^{13} R^{20}} \rangle$.
$\sigma_2 = \langle$Rail, chiefs, urge, people, to, steer, clear, of, floodwater; $\widehat{f^{11} R^{20}} \rangle$.
$\sigma_3 = \langle$after, several, stranded, motorists, are, rescued; $\widehat{f^{14} R^{20}} \rangle$.

Within the original sentence there are meaningful substructures as hypersimplices. Although these have an obvious connectivity through shared word, this alone does not capture well the meanings. σ_1 and σ_2 are 4-near but express subtly different things. σ_3 seems to say something new, namely that "motorists have been stranded and rescued" and motivates σ_1 and σ_2.

Defining $\sigma \cap \sigma'$.

Returning to the expression

$$\langle x_0, x_1, ..., x_{m-1}; R^m \rangle \cap \langle y_0, y_1, ..., y_{n-1}; R'^n \rangle = \langle z_0, z_1, ..., z_{q-1}; R''^q \rangle,$$

it seems that $\widehat{f^{m-q} R^m} \wedge \widehat{f^{n-q} R^n} = R''^q$. The previous section shows that the definition of $\sigma \cap \sigma'$ does not follow any simple rules, and can appear to be be completely arbitrary, sometimes just depending on the numbering the vertices or other serendipity choices. Partly this is due to the generality of allowing anything to be a vertex, and allowing any collection of vertices to form a simplex. It also reflects ability of the human mind to deal with many connectivities simultaneously. In any particular application it is important to know why a vertex in one simplex should be considered to be shared with another simplex. This depends on the application.

Definition: Let $\sigma_x = \langle x_0, x_1, ..., x_m \rangle$ and $\sigma_y = \langle y_0, y_1, ..., y_m \rangle$ be simplices. σ_x and σ_y are $\phi\psi$-q-connected by the simplex $\sigma_z = \langle z_0, z_1, ..., z_q \rangle$ if there exist mappings ϕ and ψ of the numbers $i = 0, 1, ..., q$ with $0 \leq \phi(i) \leq m$ and $0 \leq \psi(i) \leq n$. These mappings induce mappings of the vertices, also denoted ϕ and ψ with $\phi z_i = x_{\phi i}$ and $\psi z_i = y_{\psi i}$.

$$x_{\phi i} \xleftarrow{\phi} z_i \xrightarrow{\psi} y_{\psi i}$$

Figure 6.32(a) shows the general case that $X \neq Y \neq Z$. Figure 6.32(b) shows the usual case in which $X = Y = Z$ and the simplex σ_z is a face of σ_x with the same vertex ordering, and a face of σ_y with a different vertex ordering.

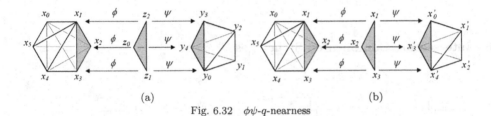

Fig. 6.32 $\phi\psi$-q-nearness

Figure 6.33 shows $\sigma(\text{singing}) = \langle v_0, v_1, v_2, v_3, v_4, v_5, v_6 \rangle$ and $\sigma(\text{bringing}) = \langle v'_0, v'_1, v'_2, v'_3, v'_4, , v'_5, v'_6, v'_7 \rangle$. The abstract simplex $\langle v''_0, v''_1, v''_2 \rangle = \langle i, n, g \rangle$ with $\phi v''_0 = v_4$ and $\psi v''_0 = v'_5$ establishes $v_4 \longleftrightarrow v'_5$. Similarly $v_5 \longleftrightarrow v'_6$ and $v_6 \longleftrightarrow v'_7$.

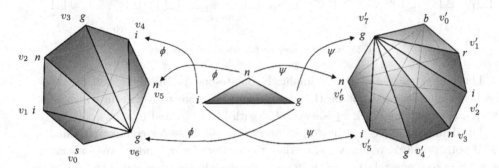

Fig. 6.33 $\langle s, i, n, g, i, n, g \rangle \xleftrightarrow{\phi\psi - 2near} \langle b, r, i, n, g, i, n, g \rangle$

This approach to defining connectivity allows the analyst to determine exactly how shared faces should be defined, and provides an unambiguous way of characterising the intersection of two relational simplices. Let σ and σ' be $\phi\psi$-connected by the simplex σ''. Then σ'' is said to be the $\phi\psi$-intersection of σ and σ', written $\sigma'' = \sigma \cap_{\phi\psi} \sigma'$. This definition is illustrated by $\langle s, i, n, g, i, n, g \rangle \cap_{\phi\psi} \langle b, r, i, n, g, i, n, g \rangle = \langle i, n, g \rangle$ in Fig. 6.33.

Establishing shared faces between simplices is just one part of the Fundamental Question. More interesting is the question of how the relations are combined in intersections. For example, Fig. 6.34 shows graphical primitives $\bigcirc, \circ, \circ, \smile$, and $\frown$ arranged into "smiley" and "frowny" faces, $\langle \bigcirc, \circ, \circ, \smile; R \rangle$ and $\langle \bigcirc, \circ, \circ, \frown; R \rangle$. The relation R that assembles these elements into faces is the same in both cases. The intersection of these structures is the head and two eyes assembled under a relation R' that is a "sub-relation" of R.

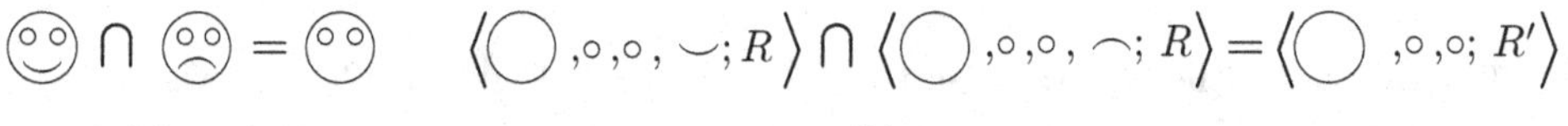

$$\bigodot \cap \bigodot = \bigodot \qquad \langle \bigcirc, \circ, \circ, \smile; R \rangle \cap \langle \bigcirc, \circ, \circ, \frown; R \rangle = \langle \bigcirc, \circ, \circ; R' \rangle$$

(a) (b)

Fig. 6.34 Intersecting structures can produce substructures

6.6 Intersecting Complicated Structures

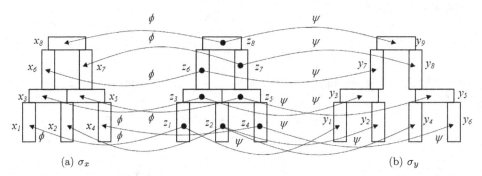

Fig. 6.35 ϕ and ψ may be one-to-many mappings

The mappings ϕ and ψ are implicitly one-to-one in the definition of $\phi\psi$-q-nearness, but this need not be the case. Figure 6.35 shows a case in which ψ is one-to-many and the block x_2 is associated with blocks y_2 and y_4.

The definition of $\psi\phi$ nearness open up new possibilities for connectivity analysis. If both ψ and ϕ are many-to-one then the vertices for σ_x and σ_y are associated by a many-to-many relation, with its associated Galois connection structure. Although this is an interesting possibility, it may reflect a lack of appropriate multilevel structure in which the analyst is attempting to do too much at a given level of representation. For example, what would it mean to write $\sigma(\text{Paris}) \cap \sigma(\text{London})$, or $\sigma(\text{water}) \cap \sigma(\text{electricity})$?

Chapter 7
Multilevel Systems

7.1 Systems of Systems of Systems

Let a *system* be defined to be a class of sets, a class of relations on those sets, and a class of mappings from the sets to number systems such as the integers, rationals and real numbers. Complex systems have many subsystems and can have ten or more discernible levels of description. They are made up of inextricably entangled multilevel social and physical subsystems with intra-level and inter-level bottom-up and top-down dynamics. They are *systems of systems*. In fact they are multilevel systems of systems of systems. Currently there is no coherent scientific way of combining the dynamics of the micro-, macro- and meso-levels. For many systems there are theories of system behaviour at different levels, but these theories are not coupled formally and in some cases the theories at different levels are based on contradictory premises. Although it is known that lower level dynamics impact bottom-up on higher level phenomena, and that higher level dynamics impact top-down on lower level phenomena, there is rarely any formal link or theory.

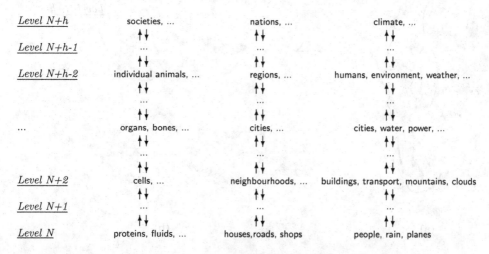

Fig. 7.1 Systems of systems of systems of systems ...

For example, biological systems have dynamics at the level of proteins, cells, organs, and bodies with much else in between. Subsystems include brains, the senses, digestion, and reproduction. Buildings and streets form local neighbourhoods which form cities, regions, and nations. Subsystems include the transportation system with bus, train, car, and pedestrian subsystems. The climate change system includes many physical subsystems such as the the oceans and the atmosphere and many social systems including industry, agriculture, transportation, housing, and all the other systems that sustain human activity.

For many systems there is partial knowledge about the subsystems, fixed at some high or low levels of representation, but the dynamics of the models are rarely integrated into a single whole. For example, microlevel theories of human behaviour in psychology are not well linked to the macroscopic phenomenon of disagreement on international climate change policy. Arguably, the need for a formalism to represent multilevel dynamics is this century's greatest obstacle to scientific progress and finding solutions to the enormous problems that lie ahead.

Figure 7.1 illustrates the difficulties in integrating systems of multilevel systems. Individual animals appear at *Level N+h-2* in the biological scheme but people appear at *Level N* in the climate change scheme, as does the built environment.

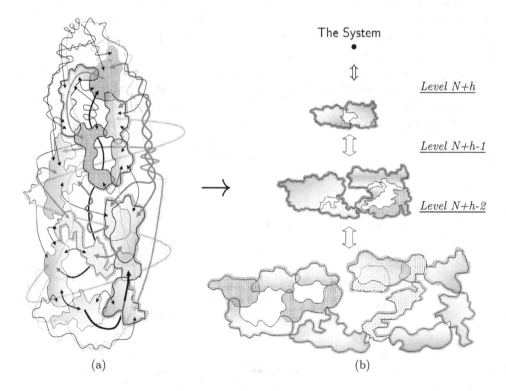

Fig. 7.2 Can highly entangled multilevel systems be resolved into well defined levels?

Can hierarchical schemes like those in Fig. 7.1 be integrated in a formal way? If so their levels will require relabelling so that constructs appear at consistent levels. It is an open question as to whether highly entangled multilevel systems, as in Fig. 7.2(a), can be resolved into well defined levels, as in Fig. 7.2(b).

7.2 The Intermediate Word Problem

When one first looks at a system the number of parts can be overwhelming. For example, look around you and take the "clipboard challenge". Suppose you were given a clipboard, paper and pen to list everything you can see, with a prize of £1 each – how much would you win? I am in my small study. I can see about fifty things on my desk alone, including books, papers, phones, pens, photographs, stapler, scissors, and so on. I have a computer of course, a webcam, a printer with cartridges, and plenty of cables. There are dozens of books and disks for my stereo. There are also lights, curtains, the door, windows, walls, radiator, floor, ceiling. I'd win a few thousand pounds, even for this small room. Within a short distance one could become a millionaire!

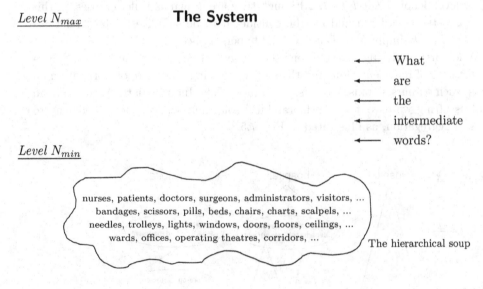

Fig. 7.3 The Intermediate Word Problem

Consider this challenge within a study of a hospital. You would observe many thousands of distinct objects which are shown inside an irregular shape in Fig. 7.3. This arbitrary and unstructured collection of things is called the *hierarchical soup*, "a prelogical primordial source containing the building blocks of all subsequent structures" [Gould *et al* (1984)].

180 Hypernetworks in the Science of Complex Systems

7.3 Relational Hypersimplices and Aggregation

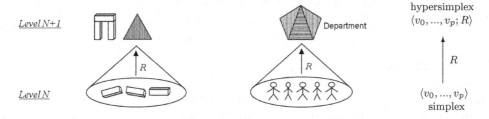

Fig. 7.4 Multilevel Aggregation

The goal of finding a well defined way of representing multilevel systems requires a way of distinguishing between levels. The notation *Level N+k* is used to indicate that levels are not absolute and if necessary *Level N + k* may be relabelled as *Level N + k'*. The notation $Level(x) = N + k$ will mean that x exists in the representation of the system at *Level N+k*.

Figure 7.4 shows the assembly of 'sets' of elements at *Level N* to structures at a higher level denoted *Level N+1*. Although they are drawn as Euler ellipses, in this book all sets are ordered and can be considered to be simplices. Thus, relations map *Level N+k* simplices to *Level N+k+1* hypersimplices.

When first analysing a system one sees 'The System', e.g. 'The Hospital', at a highest level of representation, and the soup containing a mixture of words from all levels, such as nurses, pills, trolleys, and offices. The Intermediate Word Problem involves lifting words out of the hierarchical soup and sorting them according to levels of aggregation, as illustrated in Fig. 7.5.

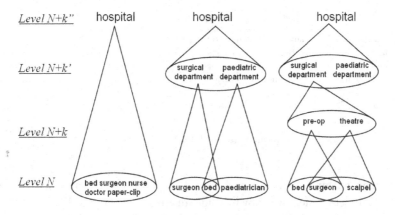

Fig. 7.5 Intermediate Words for a Hospital System

7.4 Structural aggregation in multilevel systems

The *mereological* axiom states that if x is a 'proper part' of y, then y is not a proper part of x, *i.e.* elements exist at a lower level of aggregation to the structures they form and there is an order relation on multilevel systems with some x existing at a lower level than some y, $Level(x) < Level(y)$.

The Axioms of Aggregation

Let $\sigma = \langle x_0, ..., x_n; R \rangle$ be a relational hypersimplex. Then:

(i) Mereology: $\quad Level(x_i) < Level(\sigma)$, for $i = 0, ..., n$.

(ii) Transitivity: $\quad \left. \begin{array}{l} Level(x) < Level(y) \\ Level(y) < Level(z) \end{array} \right\}$ imply $Level(x) < Level(z)$

Since $<$ is a mereological transitive order on the elements of a system, unambiguous *vertical* arrows can be drawn from lower level elements to higher level elements according to their part-whole relationships. This is used in the cone construction.

Naming mappings, intermediate words and hierarchical cones

In multilevel systems, hypersimplices are usually given *names* and treated as atomic objects at a higher level to their vertices. The *hypersimplex naming mapping* associates a string of characters with a hypersimplex, denoted $name(\sigma)$. The *inverse naming mapping* associates a hypersimplex with its name, $name^{-1}(name(\sigma)) = \sigma$.

Figure 7.6 shows the *hierarchical cone* construction for the relational hypersimplex simplex $\sigma = \langle x_0, x_1, ..., x_n; R \rangle$. The ordered set $\langle x_0, x_1, ..., x_n \rangle$, which is drawn as an ellipse, is said to be the *R-base* of σ, and σ is said to be the *apex* of its base.

On the left R maps its base to the hypersimplex $\sigma = \langle x_0, x_1, ..., x_n; R \rangle$. On the right R^{-1} maps $name(\sigma)$ to its base. For convenience, the base of the hypersimplex is defined to be the same as the base of its name. The name is an *intermediate word* and the pair $(name(\sigma), base(\sigma))$ is called a *hierarchical cone*.

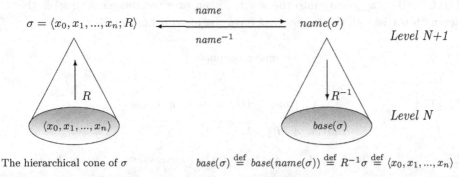

The hierarchical cone of $\sigma \qquad base(\sigma) \stackrel{\text{def}}{=} base(name(\sigma)) \stackrel{\text{def}}{=} R^{-1}\sigma \stackrel{\text{def}}{=} \langle x_0, x_1, ..., x_n \rangle$

Fig. 7.6 Intermediate words as the names of hypersimplices at the apex of hierarchical cones

182   *Hypernetworks in the Science of Complex Systems*

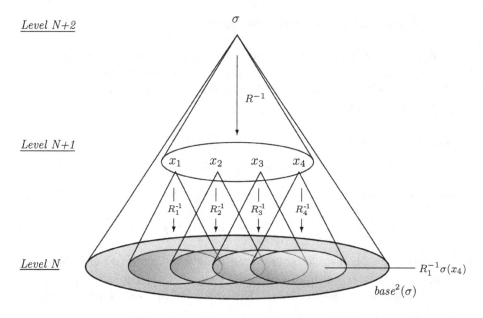

Fig. 7.7 $base^2\sigma = base(R_1^{-1}\sigma) \diamond base(R_2^{-1}\sigma) \diamond base(R_3^{-1}\sigma) \diamond base(R_4^{-1}\sigma)$

Figure 7.6 shows two levels of hierarchical cones. $R^{-1}(\sigma) = \langle x_1, x_2, x_3, x_4 \rangle$. There are four cones between *Levels* N and $N+1$ with bases $R^{-1}(x_i)$ for $i = 1, ..., 4$, and $base^2\sigma \stackrel{\text{def}}{=} base(R_1^{-1}\sigma) \diamond base(R_2^{-1}\sigma) \diamond base(R_3^{-1}\sigma) \diamond base(R_4^{-1}\sigma)$. In general

$$R^{(N+1)^{-1}} R^{N+2^{-1}}: \sigma \to base^2(\sigma) = R^{N+1^{-1}}\langle x_1, ..., x_n \rangle$$
$$\stackrel{\text{def}}{=} base(R_1^{-1}\sigma(x_1)) \diamond ... \diamond base(R_p^{-1}\sigma(x_n)).$$

which is the ordered set (simplex) of all the vertices that aggregate into σ.

This is illustrated by the example of a sentence in Fig. 7.8 where the *Level* $N+1$ words aggregate into $\sigma(\text{sentence\_example}) = \langle$ this, is, a, sentence; $R_1^{N+1}\rangle$ and the *Level N* letters aggregate into the words. Then $base^2(\text{sentence\_example})$ is the simplex with vertices all the letters that form the intermediate words.

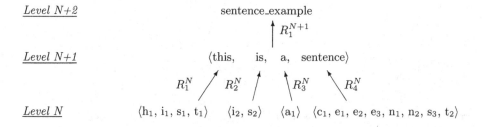

Fig. 7.8 $base^2(\text{sentence\_example}) = \langle$ a$_1$, c$_1$, e$_1$, e$_2$, e$_3$, h$_1$, i$_1$, i$_2$, n$_1$, n$_2$, s$_1$, s$_2$, s$_3$, t$_1$ t$_2\rangle$

7.5 AND aggregations and OR aggregations in multilevel systems

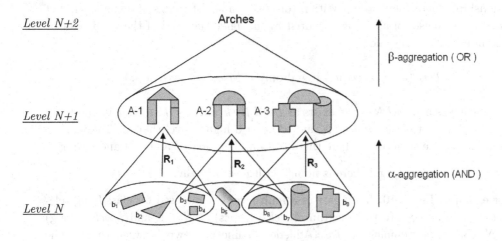

Fig. 7.9 Alpha-Beta AND-OR Aggregation

Aggregating parts to make structures is one major way of moving from a lower to a higher level of representation. Fig. 7.9 contrasts this a different way which collects things together at higher levels.

The relation R_1 is a 4-ary relation requiring all the blocks

$$b_1 \quad \text{AND} \quad b_2 \quad \text{AND} \quad b_3 \quad \text{AND} \quad b_4$$

for the relation to be tested. Thus, $R_1 : b_1, b_2, b_3, b_4 \rightarrow \langle b_1, b_2, b_3, b_4; R_1 \rangle$ and it is called an AND-aggregation or an α-aggregation.

In contrast to the AND-aggregation, the arches A-1, A-2 and A-2 are brought together as a set of Arches. To decide if a particular arch is a member of Arches it is sufficient to test if it of type

$$\text{A-1} \quad \text{OR} \quad \text{A-2} \quad \text{OR} \quad \text{A-2}.$$

Thus the aggregation of the arches into the set Arches is called an OR-aggregation or a β-aggregation.

Generally multilevel hierarchical schemes involve both kinds of aggregation. For example, in Fig. 7.8 the letters were α-aggregated into the words and the words were α aggregated into the sentence. This contrasts with Fig. 7.9 where the blocks α-aggregate into the arches and the arches β-aggregate into the set of arches.

As another example, gin, tonic, ice, and lemon α-aggregate into a gin-and-tonic, while gin-and-tonic β-aggregates into the set of cocktails.

7.6 Taxonomic aggregation in multilevel systems

Let a taxonomy be defined to be a set of words reflecting a supertype-subtype relationship, where the words partition (or cover) a set of things at some level. Let $base(w)$ be the set of all things denoted by w. Then the level of the words can be discriminated by the rule:

$$Level(w) < Level(w') \text{ if and only if } base(w) \subset base(w').$$

If w exists at *Level N+k*, w' exists at *Level N+k'* and $k < k'$ then w is said to be at a lower level than w' and w' is said to be at a higher level than w. Taxonomies form lattice-like structures where there is a link between w' and w if and only if

$$w < w', \text{ and there is no } w'' \text{ with } w < w'' < w'.$$

For example, Fig. 7.10 shows part of a multilevel scheme used for classifying television programmes. In this diagram $Level(\text{Athletics}) = N + 1$, and $Level(\text{running}) = N$. Since $base(\text{running}) \subset base(\text{Athletics})$ a line is drawn between them. The purpose of this scheme is to allow the broadcaster to know how many minutes of various types of programme were broadcast.

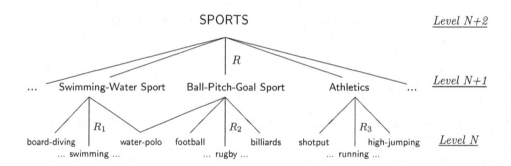

Fig. 7.10 Part of a possible SPORTS scheme

Classical taxonomies are *partitional* with $Level(w) = Level(w')$ and $w \neq w'$ implying $base(w) \cap base(w') = \emptyset$. Then the associated diagram is a tree. In Fig. 7.10 the R-base for SPORTS is ⟨Swimming-Water Sports, Ball-Pitch-Goal Sports, Athletics, ...⟩. The R_1-base for Swimming-Water Sport is ⟨board-diving, swimming, water-polo, ...⟩. The R_2 base for Ball-Pitch-Goal Sport is ⟨water polo, football, rugby, billiards,...⟩. Here water polo belongs to both Swimming-Water Sport and Ball-Pitch-Goal Sport with a line from each to water polo. Hence the diagram is not a tree-like, it is lattice-like.

Figure 7.10 is based on a scheme used to record statistics by a Swedish broadcaster. The usefulness of a scheme like this is that the traffic broadcasting time can be aggregated or disaggregated to greater level of generality or detail.

Sets as list hypersimplices

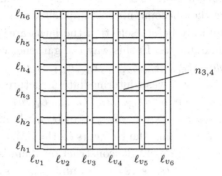

Fig. 7.11 The trellis list hypersimplex $\langle \ell_{v_1}, ... \ell_{v_6}, \ell_{h_1}, ..., \ell_{h_6}, n_{1,1}, ..., n_{6,6}; R_{\text{list}}\rangle$

Classical sets are not ordered and this makes them unsuitable for representing the vertices of hypersimplices. In general sets of vertices must be ordered before they can be used. For example, consider a bag of 30 mm x 2 mm nails, $\mathcal{N}$, and a bundle of 40 mm x 20 mm x 1000 mm wooden laths $\mathcal{L}$ used by a gardener to make a trellis for climbing roses, as shown in Fig. 7.11.

To make the trellis the gardener selects an arbitrary lath from the bundle as the bottom horizontal lath and places it in position on the ground, implicitly naming it ℓ_{h_1}. The other horizontal laths are similarly arbitrarily selected and named.

Then a lath is arbitrarily selected to become the leftmost vertical lath, implicitly naming it ℓ_{v_1}. The other vertical laths are similarly arbitrarily selected and named. Then an arbitrary nail, implicitly named $n_{1,1}$, is selected from the bag to fix the lowest horizontal lath, ℓ_{h_1} to the leftmost vertical lath, ℓ_{v_1}. Other nails, implicitly named $n_{i,j}$, are arbitrarily selected from the bag to fix lath ℓ_{h_i} to lath ℓ_{v_j}.

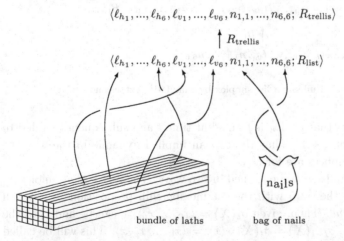

Fig. 7.12 Building a list hypersimplex by arbitrarily selecting elements

Before the nails are removed from the bag they are all 'the same', *i.e.* although they are all different nails they are *equivalent* for the purpose in hand. After selection for their given purpose they become unique, implicitly or explicitly identified by a name such as $n_{i,j}$. Similarly before selection the laths are all the same, but after selection they all become unique. In this way the gardener constructs what will be called a *list hypersimplex*, $\langle \ell_{v_1}, ... \ell_{v_6}, \ell_{h_1}, ..., \ell_{h_6}, n_{1,1}, ..., n_{6,6}; R_{\text{list}} \rangle$. This is a sort of "weak" hypersimplex, being a weak intermediate structure before the structuring relation R_{trellis} is applied to make $\langle \ell_{v_1}, ... \ell_{v_6}, \ell_{h_1}, ..., \ell_{h_6}, n_{1,1}, ..., n_{6,6}; R_{\text{trellis}} \rangle$.

Any classical finite set $S = \{s_1, s_2, ..., s_p\}$ can be made into a list hypersimplex by deleting repetitions and listing the elements in the order they "come out of the bag", to form $\langle s_0, s_1, s_2, ...s_p; R_{\text{list}} \rangle$. By construction there is a one-one correspondence $\langle s_0, s_1, s_2, ...s_p; R_{\text{list}} \rangle \leftrightarrow \langle s_0, s_1, s_2, ...s_p \rangle$, and $\langle s_0, s_1, s_2, ...s_p \rangle$ will be called a *list simplex*. The use of list simplices suppresses the symbol R_{list}, which is implied by context, and helps to make a distinction between uncommitted lists of vertices and structured lists of vertices as hypersimplices.

Building list simplices by cumulative extension

In Fig. 7.13 the list simplex $\langle \ell_{h_1}, ..., \ell_{h_6}, \ell_{v_1}, ..., \ell_{v_6}, n_{1,1}, ..., n_{6,6} \rangle$ is assembled under R_{trellis} to make $\sigma(\text{trellis}_i)$. If the gardener were making more than one trellis, this could be added to the set of trellises. Of course the 'set' of trellises is another list simplex, so the newly made $\sigma(\text{trellis}_i)$ can be added to it as a new vertex. This will be called the *cumulative extension* of the list of trellises. Not only does it extend the list, but it adds another extensionally defined element.

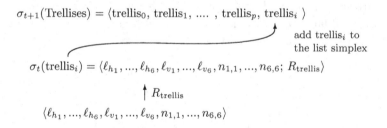

Fig. 7.13 Building a list simplex by cumulative extension

Figure 7.13 suggests that trellis_i is formed at time t and subsequently added to the list simplex at time $t+1$. Thus there is an implied dynamic in the way the simplices and hypersimplices are formed.

Extensional list simplices can be formed in two ways. The first is that an object, x, can be observed in the soup with the set membership proposition $p_X(x)$ well formed and $p(x) = True$. Then, given $\sigma_t(X) = \langle x_0, ..., x_p \rangle$, x can be added to the current extension of X, $\sigma_{t+1}(X) = \sigma_t(X) \diamond \langle x \rangle = \langle x_0, ..., x_p, x \rangle$. This will be called *set augmentation by cumulative intension* (Fig. 7.14(left)).

Multilevel Systems

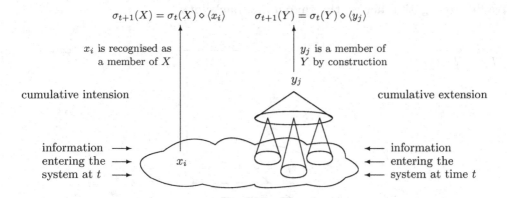

Fig. 7.14 Cumulative augmentation of list simplices

In contrast sets can be augmented *constructively* by cumulative extension, with the new element being *built* from lower level parts or information, as illustrated on the right of Fig. 7.14. In this case the set Y is constructed extensionally.

Cumulative intension is based on the name of the set and its intensional membership property. Anything that satisfies the intensional definition is added to the set hypersimplex. Cumulative extension is based on the name of the set and rules for building elements of the set from lower level elements or information. Anything that can be built is by definition a member of the list simplex.

Multilevel descriptor simplex taxonomies

Figure 7.15 shows three taxonomic cones of words used to describe movies. Between them are two *descriptor hypersimplices* at *Level N+1* and *Level N+2*, $\sigma^{N+2} = \langle \text{Relationships}, \text{Conflict}, \text{Crime}; R_{\text{movie}}^{N+2} \rangle$ and $\sigma^{N+1} = \langle \text{ Love, War, Murder}; R_{\text{movie}}^{N+1} \rangle$. These combinations of words could describe movies such as 'Casablanca' or 'Murder on the Orient Express', although more words would give a more complete description. Descriptor hypersimplices, as compound 'words', form multilevel OR-aggregating taxonomies with descriptors bound together by AND-aggregations.

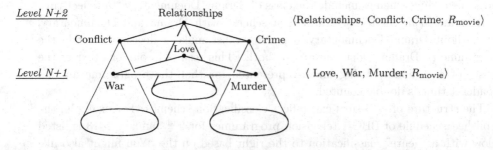

Fig. 7.15 Multilevel descriptor simplices

Traffic aggregation across the taxonomic backcloth

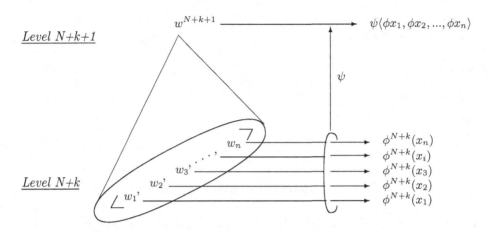

Fig. 7.16 Traffic aggregating over a hierarchical cone

Figure 7.16 shows a set of naming words at *Level N+k* aggregating into a word at *Level N+k+1*. ϕ^{N+k} is a mapping assigning numbers to the w_i. ψ is a *hierarchical traffic aggregation mapping* that takes the values of the traffic, ϕ, at *Level N+k* to *Level N+k+1*. For partitional taxonomies it is common to have $\psi : \phi\langle w_1, ..., w_n\rangle \rightarrow \sum_{i=1}^{n} \phi w_i$.

Example: television broadcast traffic

Most broadcasting organisations collect data on Drama and Documentaries they broadcast. Much drama is fictional while documentaries are factual. However, some programmes are both, *e.g. The Gathering Storm* made by the BBC on the life of Winston and Clementine Churchill set in the nineteen thirties is critically acclaimed for its historical accuracy and in this respect it is factual. However, even biographies and other historical records do not give every sentence or action and to present the factual information in an acceptable dramatic form requires some degree of invention. Programmes like this fall between Drama and Documentary and some coding schemes include the class of 'Drama Documentary' to reflect this.

If coding *The Gathering Storm* in a scheme with Drama and Documentary, but without Drama Documentary the coder must make a decision to code the programme as Drama, Documentary, or both. This results in one or other of the classes of programmes being under-represented, or that the programme and its broadcast time is double counted.

The structure of web sites can reflect classification schemes. For example, the published schedule of BBC-1 television programmes for 2nd January 2013 is listed below with a "genre" classification to the right based on the "You might also like ..." links given on the BBC website for each programme.

06:00	Breakfast: news, sport, business, weather	News & Weather
	News & Weather; Format: Magazine & Review	
09:15	Wanted Down Under: family documentary	Factual
	Factual>Families and Relationships; Format: Reality	
10:00	Homes Under the Hammer: property auction documentary	Factual
	Factual>Homes & Gardens>Homes; Factual>Money	
11:00	Saints and Scroungers: benefit–fraud documentary	Factual
	Factual>Crime & Justice; Factual>Life Stories; Factual>Money	
11:45	Cowboy Trap: house extension by a builder–from–hell	Factual
	Factual>Consumer; Factual>Homes & Gardens>Homes	
12:15	Bargain Hunt: antique buying documentary	Factual
	Factual>Antiques; Format: Games & Quizzes	
13:00	BBC News at One: national and international news & weather	News & Weather
	News & Weather; Format: Bulletins	
13:30	BBC London News: news sport and weather from London	News & Weather
	News & Weather; Format: Bulletins; Magazines & Review	
13:45	Doctors: medical soap opera	Drama
	Drama>Medical, Drama>Soaps	
14:15	The Two Ronnies Sketchbook: classic comedy	Comedy
	Comedy>Sketch	
14:45	Escape to the Country: house search documentary	Factual
	Factual>Homes & Gardens	
15:45	Perfection: Quiz game	Entertainment
	Entertainment; Format: Games & Quizzes	
16:30	Flog It!: Antique and bric–a–brac valuation	Factual
	Factual>Antiques	
17:15	Pointless: Quiz game	Entertainment
	Entertainment; Format: Games & Quizzes	
18:00	BBC News at Six: national and international news & weather	News & Weather
	News & Weather; Format: Bulletins	
18:30	BBC London News: news, sport and weather from London	News & Weather
	News & Weather; Format: Bulletins; Magazines & Review	
19:00	Celebrity Mastermind: Quiz game	Entertainment
	Entertainment; Format: Games & Quizzes	
19:30	Wallace and Gromit: comedy animation	Children's
	Children's>Entertainment & Comedy; Format: animation	
20:00	Holby City: medical drama	Drama
	Drama>Medical	
21:00	Africa–Kalahari: wild life documentary	Factual
	Factual>Science & Nature>Nature & Environment	
22:00	BBC News at Ten: national and international news	News & Weather
	News & Weather; Format: Bulletins	
22:25	BBC London News: news, sport and weather	News & Weather
	News & Weather; Format: Bulletins, Magazines & reviews	
22:35	A Question of Sport: light hearted sports quiz	Entertainment
	Entertainment>Sport; Format Games & Quizzes	
23:05	Match of the Day Live: Football match	Sport (ends 24:10)
	Sport>Events; Format: Performances & Events	

Table 7.1 Coded TV programmes. (Source: [BBC (2013)])

This coding results in the following statistics:

Factual	6 hours 30 minutes (36%)	Sport	65 minutes (6%)
News & Weather	6 hours 5 minutes (33%)	Children's	30 minutes (3%)
Entertainment	2 hours 0 minutes (11%)	Comedy	30 minutes (3%)
Drama	1 hour 30 minutes (8%)		

Statistics such as these are used in management and for policy purposes with the classes reflecting internal organisation structure, *e.g.* departments with autonomous budgets responsible for producing, news, weather, documentary, drama, comedy, and sports programmes. The broadcast times are a measure of the output of those departments. Also, in some countries the media are regulated and the statistics are used to ensure that no category dominates. For example, the 2006 Royal Charter for the BBC states that its mission to "Inform, Educate and Entertain", as established by Lord Reith in the nineteen twenties [BBC (2006)].

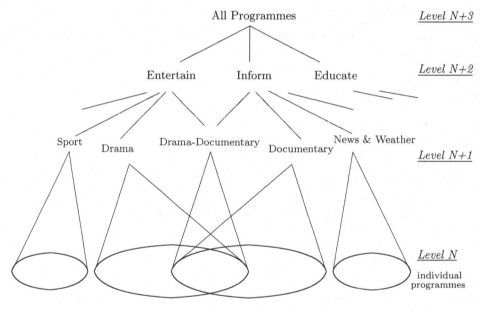

Fig. 7.17 Aggregating statistics over a multilevel scheme

Figure 7.17 shows Drama-Documentary at *Level N+1* in a scheme with individual programmes at *Level N*. Inform, Educate and Entertain are at *Level N+2* and at *Level N+3* is the class of All Programmes.

For the multilevel statistics to be coherent it is necessary to know for any *Level N+k* class c those programmes defined to be in $base^k(c)$. Let $\phi^{N+k}(c)$ be the broadcast time for programmes of type c, and let $\phi^{N+k}(p)$ be the duration of programme p. Then when $base$(Drama-Documentary) is non-empty, ϕ^{N+3}(All Programmes) $> \phi^{N+2}$(Inform) $+ \phi^{N+2}$(Educate) $+ \phi^{N+2}$(Entertain). To overcome this, it is a simple matter to calculate ϕ^{N+3}(All Programmes) as $\sum_{p \in base^3 (\text{AllProgrammes})} \phi(p)$.

7.7 Emergence in Aggregation and Disaggregation Dynamics

Emergence is ubiquitous in dynamic systems. Some emergence is useful, some is undesirable, and some is irrelevant. Some emergence is entirely predictable and some is entirely unpredictable. Emergence is particularly common when things are assembled to form new wholes. For example, in Fig. 7.18(a) combining a line and a circles gives two semicircles as completely new emergent objects. Similarly, in Fig. 7.18(b) a square emerges from the combination of two triangles, neither of which is a square.

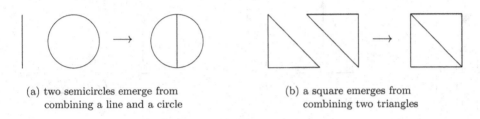

(a) two semicircles emerge from combining a line and a circle

(b) a square emerges from combining two triangles

Fig. 7.18 Emergence in wholes assembled from parts

Figure 7.19 illustrates the common possibility that a combination of objects (the rectangles r_1 and r_2) forms a new object (the cross σ) which has emergent features (the intersections of the lines) that enable a process of *deconstruction* in which new objects emerge (the short lines) which can be combined in new ways to create new objects (the square $\sigma' = \langle x_2, x_5, x_6, x_9; R' \rangle$). Here there is emergence at two levels, and possibly more.

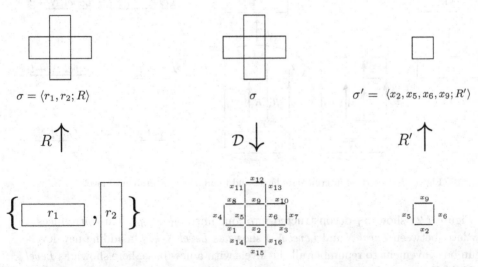

Fig. 7.19 Construction, deconstruction with emergent parts reconstructed to an emergent whole

192   Hypernetworks in the Science of Complex Systems

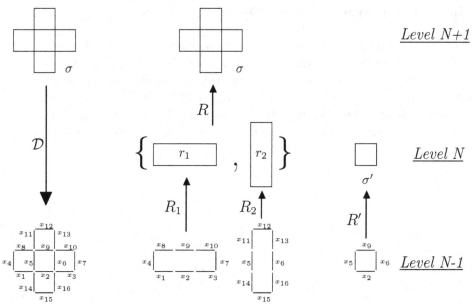

Fig. 7.20   Emergent hierarchical levels from construction and deconstruction

Figure 7.20 resolves the processes in Fig. 7.19 into different hierarchical levels. If the rectangle r_1 and r_2 are said to exist in the hierarchy at *Level N* the cross σ formed from them can be said to exist at *Level N+1*. The emergent new parts, $x_1, ..., x_{16}$ exist at a lower level to the rectangle because the relations R_1 and R_2 assemble subsets of them into R_1 and R_2. Let this level be called *Level N-1*.

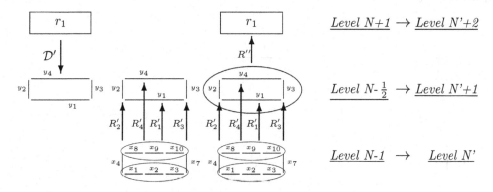

Fig. 7.21   Inserting intermediate levels and renumbering hierarchical levels

Figure 7.21 shows r_1 deconstructed into four lines, $y_1, ..., y_4$. This introduces a new level between *Level N* and *Level N-1* shows as *Level N-$\frac{1}{2}$*. On adding new levels it can be convenient to renumber all the levels with a new base, here shown as *Level N*, *Level N+1*, and *Level N+2*.

7.8 Combinatorial Explosion in Downward Emergence

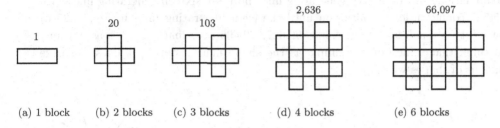

Fig. 7.22 Combinatorial explosion in downward emergence

It is astonishing how bottom-up aggregation and top-down disaggregation create astronomic numbers of possible higher level structures. This is illustrated in Fig. 7.22 where adding just one block to the original structure creates by downward disaggregation more lower level objects, namely the squares. However, increasing the number of these by even modest amounts creates a huge combinatorial explosion of possibilities at higher levels.

Putting together two blocks at *Level N* in Fig. 7.22 creates five squares, and these can be assembled in twenty configurations, as shown in Fig. 7.23.

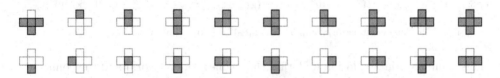

Fig. 7.23 The 20 contiguous configurations of squares generated by 2 blocks

Adding just one more block (Fig. 7.22(c)) creates 103 contiguous configurations, Adding another block (Fig. 7.22(d)) generates 2636 contiguous configurations, and adding just one more block results in 66,097 configurations being generated. It is surprising how the addition of one or two more objects can generate quite modest numbers of lower level emergent objects by top-down disaggregation, but how these can create combinatorially enormous numbers of configurations bottom-up at higher levels.

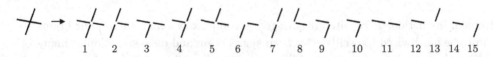

Fig. 7.24 Two crossed lines create four sub-lines with $2^4 - 1 = 15$ combinations

As Fig. 7.24 shows, just two generating lines create 4 sub-lines, and these can be combined in fifteen ($2^4 - 1$) ways to produce new configurations. Add one more generating line as in Fig. 7.25(c) and there are nine sub-lines with $2^9 - 1 = 511$ combinations. Add a fourth generating line and there sixteen generating lines with 65,535 configurations, and so on until the eight generating lines in Fig. 7.25(h) have fifty two sub-lines with 4,503,599,627,370,495 combinations. It is astonishing to think that something as simple as the shape made from eight lines can have billions of shapes within it.

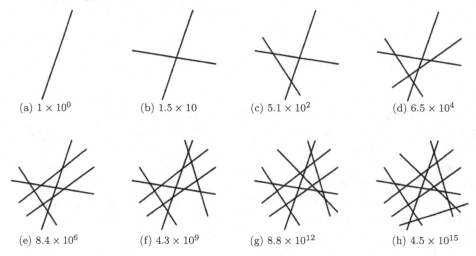

(a) 1×10^0 (b) 1.5×10 (c) 5.1×10^2 (d) 6.5×10^4

(e) 8.4×10^6 (f) 4.3×10^9 (g) 8.8×10^{12} (h) 4.5×10^{15}

Fig. 7.25 The combinatorial explosion created by adding lines to a drawing.

This combinatorial explosion underlies the richness of our world and the countless opportunities it offers. Just a handful of atomic particles form the few hundred elements of the periodic table form very large numbers of distinct molecules from which everything else in our universe is formed. This combinatorial explosion also exists in social structures, with just a few hundred or thousand people able to develop and sustain distinctive societies and cultures. Also, it exists at more abstract levels where atomic concepts are combined to form huge numbers of immensely complicated theories. Indeed, many things can be said with just a few hundred words, while almost everything can be said with a few tens of thousands.

7.9 Defining hierarchical levels by disaggregation

Although 'reductionism' is often disparaged in complex systems science, our brains seem to be hard-wired to oscillate between aggregation and disaggregation at many levels, with constructs emerging rapidly and being adopted or abandoned according to their usefulness. Whereas bottom-up construction provides a way to discriminate parts and wholes at different levels, top-down deconstruction is also a useful approach to establishing hierarchical levels.

Deconstruction: a piece of cake

Figure 7.26 shows a cake at *Level N* with a slice s_1 taken out. Let the slice s_1 and what is left of the cake, c_2 exist at *Level N–1*. Thus the cake is deconstructed by $\mathfrak{D}_1$ to give the deconstruction simplex $\mathfrak{D}_1(c_1) = \langle c_2, s_1; \mathfrak{D}_1 \rangle$. Suppose that s_1 is further deconstructed by $\mathfrak{D}_2$ to slices s_2 and s_3 at *Level N–2*. Slice s_2 is shown with a star at *Level N–2*.

However, instead of deconstructing s_2 by $\mathfrak{D}_2$ it could have been deconstructed by $\mathfrak{D}_3$ to give slices s_4 and s_5 at *Level N–2*. Then slice s_5 could be deconstructed by $\mathfrak{D}_4$ to give s_6 and s'_2 at *Level N–3*.

Suppose s_2 and s'_2 are 'the same'. In theory one could set up this experiment so that s_2 and s'_2 would be exactly the same whichever deconstruction path were chosen. How can it be that *exactly the same piece of cake can exist at Level N–2 and Level N–3*?

The answer to this question is that levels in multilevel systems are *constructed*. If construction route $\mathfrak{D}_2 \circ \mathfrak{D}_1$ is chosen the counter-factual $\mathfrak{D}_4 \circ \mathfrak{D}_3 \circ \mathfrak{D}_1$ cannot be chosen, and *vice-versa*. The construction of hierarchical schemes is *path dependent*.

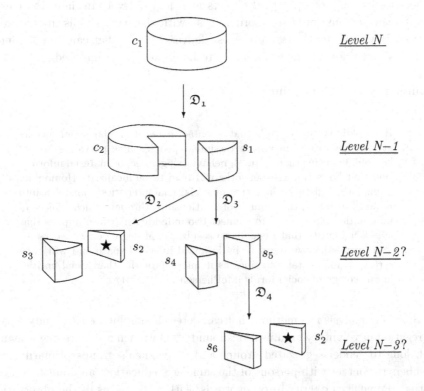

Fig. 7.26 Does the slice of cake s_2 exist at *Level N–2* or *Level N–3*?

7.10 Clustering and Hierarchical Set Definition

Clustering is a fundamental human activity that enables us to make sense of new or changing environments. Clustering provides an experimental way of moving from the multilevel soup into a better organised hierarchy of description. Clusters which are not useful are abandoned while useful clusters are developed to become structures in our representations.

We assemble clusters of things on a wide variety of criteria with two important principles being *similarity* and *complementarity*. For example, we cluster flying insects together because they all fly and, possibly, because some of them sting. In contrast, grocers may cluster things together on the same shelf because are complementary ingredients such as flour, baking powder, and sugar. Similarly the handyman may cluster heterogeneous things such as nails, screws, wood, glue, drills, and saws because in combination they can make new things.

A common reason for clustering is that things are *similar* or have some common property. Clustering elements is one of the fundamental processes in the science of multilevel systems since it is an or-aggregation, taking lower level elements to a 'collection' or 'bag' of *equivalent* objects at a higher level. Initially the clustered objects can be considered to unordered, as with the bag of nails in the previous section. However, the clustering process delivers objects that can be built into list simplices, ready for relational structure to be discovered or imposed.

Homophily and Clustering

> Homophily is the principle that a contact between similar people occurs at a higher rate than among dissimilar people. The pervasive fact of homophily means that cultural, behavioral, genetic, or material information that flows through networks will tend to be localized. Homophily implies that distance in terms of social characteristics translates into network distance, the number of relationships through which a piece of information must travel to connect two individuals. It also implies that any social entity that depends to a substantial degree on networks for its transmission will tend to be localized in social space and will obey certain fundamental dynamics as it interacts with other social entities in an ecology of social forms [McPherson *et al* (2001)].

Star-hub analysis is a method that can detect homophily, *e.g.* a study of gender segregation in voluntary organisations found: "(1) women are more likely than men to belong to gender-segregated groups and (2) women's groups primarily restrict members to contact with persons of the same age, education, and marital and work status" [Popielarz (1999)]. In other words, within the terms of the observations a significant number of women w_i in the study belonged to the star-hub pair

$$\langle w_1, w_2, ..., w_n \rangle \leftrightarrow \langle \text{age, education, martial status, work status} \rangle.$$

7.11 Multidimensional descriptor spaces

A common approach to clustering and classification is to make measurements of the elements and map them into a multidimensional space. For example, Fig. 7.27 shows sixteen items of fruit mapped to data points in two dimensional space. The horizontal axis is the longest length of the fruit as it lies on a surface and the vertical axis is the height. As can be seen strawberries (s) which have low values on both dimensions form a bottom-left cluster, and bananas (b) which are relatively long and thin form a bottom-right cluster. The oranges (o) are relatively large on both dimensions and form a centre-left cluster at the top. The pears (p) lying on their sides are relatively long and relatively wide, and form a cluster between the oranges and the bananas. These data illustrate ambiguity at the class boundaries which is a common problem. For example, the lowest data point under the word pears is close to the left-most data point above the word banana.

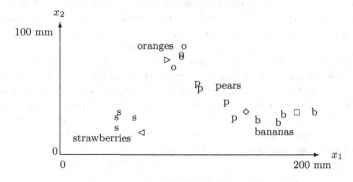

Fig. 7.27 Clustering fruit by size

Clustering can be useful in identifying class characteristics in training data and using these to classify new items. For example, if all the points s, o, p and b are training data, one could be confident that the point shown as ◁ is associated with a strawberry, that shown as ▷ is associated with an orange, and that shown as □ is associated with a banana. However, the data point shown as ◇ could be associated with either a long thin pear or a short fat banana.

7.12 Multilevel Descriptor Simplices

Let A be a set of objects and let d_j be a *descriptor* that maps each a in A to a member of a scale, S_i, $d_j : a \rightarrow d_j(a) \in S_j$. Given a set of such descriptors, we write $d(a) = \langle d_1(a), d_2(a), ..., d_n(a) \rangle$, which is defined to be a *descriptor simplex* for s. Note that the scale can be any set with any properties, including those in Steven's classification. Let $S = S_1 \times S_2 \times ... \times S_n$ be called *descriptor space*. For example, the descriptor space in Fig. 7.27 is $\mathbb{N} \times \mathbb{N}$ where $\mathbb{N}$ is the set of integers.

7.13 Personal Constructs, Triadic Sorting and Clustering

In "A Theory of Personality: the Psychology of Personal Constructs", Kelly wrote:

> Man looks at the world through transparent patterns or templates which he creates and then attempts to fit over the realities of which the world is composed. The fit is not always very good. Yet without such patterns the world appears to be such an undifferentiated homogeneity that man is unable to make any sense out of it. Even a poor fit is more helpful to him than nothing at all.
>
> Let us give the name *constructs* to these patterns that are tentatively tried on for size. They are ways of construing the world ...
>
> In general man seeks to improve his constructs by increasing his repertory, by altering them to provide better fits, and by subsuming them with superordinate constructs or systems ... Frequently his personal investment in the larger systems, or his personal dependence on it, is so great that he will forego the adoption of a more precise construct in the substructure. It may take a major act of psychotherapy or experience to get him to adjust his construction system to the point where the new and more precise constructs can be incorporated ...
>
> We assume that all of our present interpretations of the universe are subject to revision or replacement ... We take the stand that there are always some alternative constructions available to choose among in dealing with the world. No one needs to paint himself into a corner; no one needs to be completely hemmed in by circumstances; no one needs to be the victim of his biography. We call this philosophical position *constructive alternativism*. [Kelly (1963)]

The Intermediate Word Problem involves seeking words or phrases that express constructs in a well organised manner. In the first instance any such vocabulary will be *personal* in the sense of Kelly. In a social context it is necessary to *negotiate* constructs and the words that characterise them, trying to ensure that for all involved the intensions and extensions are the same. If this is not the case, we literally don't know what each other is talking about.

Kelly developed his theory as a fundamental postulate and various corollaries. The *Dichotomy Corollary* states that: "*A person's construct system is composed of a finite number of dichotomous constructs* ... The person's choice of an aspect determines both what shall be considered similar and what shall be considered contrasting. The same aspect, or the same abstraction, determines both. If we choose an aspect in which A and B are similar, but in contrast to C, it is important to note that it is the same aspect of all three, A, B, *and* C, that forms the basis of the construct. It is not that there is one aspect of A and B that makes them similar to each other an aspect that makes them contrasting to C". This is illustrated by A and B being men and C being a woman, where the dichotomous construct is man-woman. This is also called a *bipolar* construct with "man" and "woman" the *poles*, and could be written $\langle \text{man}, \text{woman}; R_{\text{construct}} \rangle$.

Kelly effectively asserts that a construct is a Gestalt formed from its two nondivisible poles, and that the poles cannot exist by themselves, as for example ⟨man⟩ and ⟨woman⟩. Certainly the construct ⟨man, beast; $R_{\text{construct}}$⟩ is different to that of ⟨man, woman; $R_{\text{construct}}$⟩, although here "man" is ambiguous and could mean "man as an individual male human being" or "humankind".

However, the *monopolar* construct ⟨man⟩ is not meaningless in the context of the *logical bipolar* construct ⟨man, not-man; $R_{\text{construct}}$⟩. Thus we can say that "Socrates is a man" without reference to either women or beasts, meaning that Socrates belongs to the set $Men \stackrel{\text{def}}{=} \{\, x \mid x \text{ is a man}\,\}$.

The *range of convenience* of a construct is the set of things to which it can be applied. For example, ships are outside the range of convenience of the construct ⟨man, woman; $R_{\text{construct}}$⟩. If the range of convenience of this construct is defined to be the set of all adult human beings then it could be argued that $Men \stackrel{?}{=} \{\, x \mid x \text{ is not a woman}\,\}$. However, it is less clear for the ⟨man, beast; $R_{\text{construct}}$⟩, $Men \stackrel{?}{=} \{\, x \mid x \text{ is not a beast}\,\}$, and such a formulation degrades the value of the construct. In as much as ⟨man, beast; $R_{\text{construct}}$⟩ might be one of *my* personal constructs I would use it to mean "x is a man *and* x is a beast" or "x is a man *and* x is not a beast". The former could apply to Herod and the latter to Ghandi.

In a clinical context a therapist talks to a client about their problems and, assuming people have different ways of construing the world, in this situation the most important constructs are those of the client rather than those of the therapist or the textbook. The therapist therefore has the problem of eliciting the client's constructs so that their problems can be articulated. In the context of knowledge engineering, the most important constructs are those of the people who will use a system rather than the software engineers who will implement it, and there is the problem of eliciting the constructs employed by users in expressing their knowledge. In both of these cases the method of *triadic sorting* can be used to elicit constructs.

The method of triadic sorting involves presenting a person with three things and asking them to sort them into a pair with some property that is different to a property possessed by the other. For example, suppose a person were presented with a metal knife, a metal fork and wooden spoon they might pair the knife and the fork saying that these are made of metal while the spoon is made of wood. Thus the construct ⟨metal, wood; $R_{\text{construct}}$⟩ has been elicited.

In some applications an elicited construct ⟨x, y⟩ is applied to other things with a *rating scale* such as $\{1, 2, 3, 4, 5\}$ where 1 and 5 respectively mean "at the extreme x or y ends of the construct", 2 and 4 mean respectively "more x than y" or "more y than x" and 3 means that "x and y apply equally". For example, on the ⟨metal, wood⟩ construct a nail would be coded as 1, a plank would be coded as 5, and the hammer used to drive the nail into the plank would be coded as 3. These numbers are then used to cluster the objects into more or less similar things.

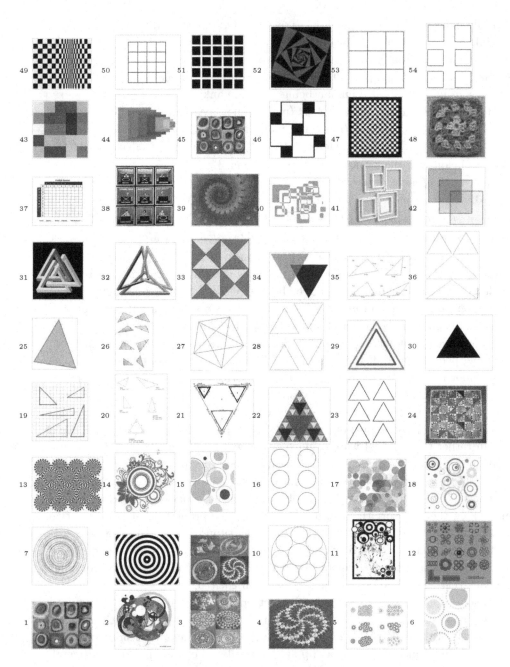

Fig. 7.28   Squares, triangles and circles images

The 54 images in Fig. 7.28 were obtained from an internet search using the terms "squares", "triangles" and "circles". Each is unique yet some share properties, *e.g.* images 10, 16, 19, 21, 23, 27, 28, 29, 35, 36, 50, 53 and 54 are line drawings.

For the method of triadic sorting triples were formed from the first, fourth and seventh rows of images as (1, 19, 37), (2, 20, 38), (3, 21, 39), (4, 22, 40), (5, 22, 41) and (6, 23, 42). Another six triples were formed for the second, fifth and eight rows of image, starting with (7, 25, 43) and ending with (12, 30, 48). A third set of six triples was formed from the third, sixth and ninth rows. These were used to elicit my personal constructs. For example, I sorted the triple (1, 19, 37) as follows: (1, 37)(19) line drawing, (1, 19)(37) table, (1)(19, 37) shape array, and (1)(19, 37) spiral shapes, (1, 37)(19) triangular shapes. The results of my triadic sorting exercise are given below. The pair-singleton groupings are not shown but just the triples generating the constructs:

(1, 19, 37) line drawing, table, shape array, spiral shapes, triangular shapes

(2, 20, 38) entangled, light, quiz

(3, 21, 39) shape array, spiral shape, whirlpool, enclosed shapes

(4, 22, 40) beads, fractal, plan-like, triangles, Catherine wheel

(5, 23, 41) small circles, groups of small shapes, triangles, array, enclosed squares, 3-D

(6, 24, 42) overlapping shapes, fractal, floating circles, concentric circles, squares

(7, 25, 43) concentric circles, large circles, triangle, shaded, square tiles, irregular, grid

(8, 26, 44) movement, whirlpool, concentric circles, triangles, array, overlapping, squares

(9, 27, 45) 2×2 shapes, hexagram, line drawing, array, concentric circles, 3×4 shapes

(10, 28, 46) dial, line drawing, triangles, head-to-toe, offset, black-and-white, cross shape

(11, 29, 47) bubbles, checkerboard, inside shape

(12, 30, 48) shape array, circle-shape, black, triangle, single shape, crochet, mat, square, rounded corners

(13, 31, 49) cogwheels, movement, triangles, intersecting, puzzle, topological, checkerboard, distorted

(14, 32, 50) 4×4-grid, grid, triangles, swirly shapes, leaves, tendrils, 3-D, straight lines, geometric

(15, 33, 51) bubbles, floating circles, triangular pattern, 5×5 grid, irregular, regular

(16, 34, 52) 2×3 grid, circles, down-pointing, triangles, overlapping, rose, rotate, dark, light

(17, 35, 53) disks, covered with shapes, line drawing, triangles, square grid, 3×3 grid

(18, 36, 54) concentric circles, covered with shapes, triangles, line drawing, 2×3 grid, square grid, simple

My constructs are all bipolar logical constructs so that, for example, the construct ⟨triangle, not-triangle⟩ can be applied independently to ⟨spiral shape, not spiral shape⟩. I would not use a "mixed" construct such as ⟨triangle, spiral shape⟩ because I feel that this creates an unnecessary coupling between something having the "triangle" property and it having the "spiral" property. This reflects a "convergent" way of construing the world, seeking to separate out different things as far as possible and avoiding ambiguity during analysis. It is different from a "divergent" way of construing the world in which "separable" constructs are deliberately conjoined to creating interesting if ambiguous pre-analytic connections. To some the convergent approach seems dull and uncreative, while to others the divergent approach seems to obfuscate, making things more complicated than necessary; no doubt the best approach is a combination of the two.

The constructs elicited in this exercise are listed below:

2 × 2 shapes	2 × 3 grid	3 × 3 grid
3 × 4 shapes	3-D	4 × 4-grid
5 × 5 grid	array	beads
black	black-and-white	bubbles
Catherine wheel	checkerboard	circle-shape
circles	cogwheels	concentric circles
covered with shapes	crochet	cross shape
dark	dial	disks
distorted	down-pointing	enclosed shapes
enclosed squares	entangled	floating circles
fractal	geometric	grid
groups of small shapes	head-to-toe	hexagram
inside shape	intersecting	irregular
large circles	leaves	light
line drawing	mat	movement
offset	overlapping	overlapping shapes
plan-like	puzzle	quiz
regular	rose	rotate
rounded corners	shaded	shape array
simple	single shape	small circles
spiral shapes	square tiles	square
square grid	squares	straight lines
swirly shapes	tendrils	topological
triangle	triangles	triangular pattern
triangular shapes	table	whirlpool

The Q-analysis relation between these constructs and the shapes in Fig. 7.28 is shown in Fig. 7.29. As can be seen the shapes cluster together according to the visual features elicited.

The image components $\{9, 1, 45, 3\}$ and $\{43, 51, 24, 38, 33\}$ at the bottom of Fig. 7.29 are grids of other designs. Above them is a large component of line drawings, $\{50, 53, 54, 36, 28, 36, 23, 21, 19, 16\}$.

At the top of Fig. 7.29 is the eccentric image 49, which is a deformed checkerboard grid very different from the other designs. Under this, image 52 is also relatively eccentric.

The component $\{7, 8\}$ is made of two circular designs, which the component $\{4, 39\}$ is dominated by spiral designs. Under this the component $\{18, 15, 11, 6, 17\}$ is dominated by repeated circles in all the images.

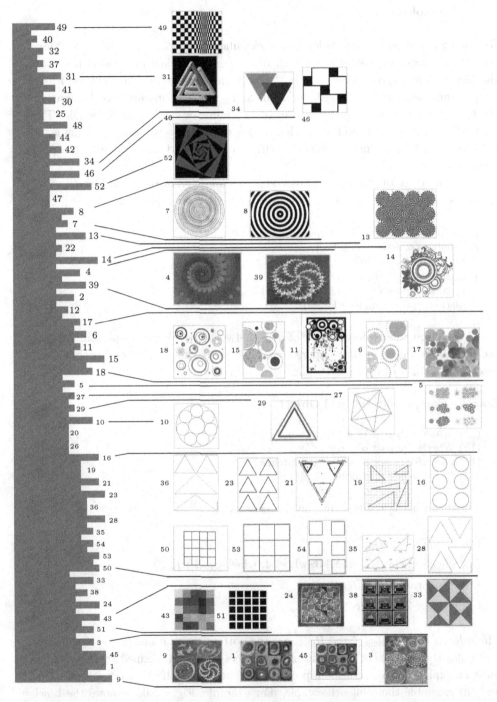

Fig. 7.29 Q-analysis of the image–descriptor relation

7.14 Mereology

Reasoning about parts and wholes goes back millennia to Plato and Aristotle. More recently the term *mereology* was coined in 1927 by Stanisław Leśniewski based on the Greek word $\mu\acute{\epsilon}\rho o\varsigma$ meaning "part". The term *meronymy* shares similar roots, $\mu\acute{\epsilon}\rho o\varsigma$ again meaning "part" with $o\nu o\mu\alpha$ meaning name. A *meronym* is the name or symbol for a part of a thing and a *holonym* is a name or symbol for the whole. The term holonymy, comes from the Greek $o\lambda o\nu$ for whole and again $o\nu o\mu\alpha$ for name. A holonymic relation connects terms denoting the whole and a term denoting a part or member of a whole.

The paper "A taxomony of part-whole relations" [Winston et al (1987)] gives six types of meronymic relations and investigates their transitivity:

1. component integral object (pedal-bike),
2. member-collection (ship-fleet),
3. portion-mass (slice-pie),
4. stuff-object (steel-car),
5. feature-activity (paying-shopping), and
6. place-area (Everglades-Florida).

This section will explore the use of hypernetwork formalism to represent some of the concepts covered in that paper, including various anomalous and pathological cases. This will be done by examining some of the examples.

Class 1: Component–Integral Object

(1a) A handle is part of a cup
(1b) Wheels are part of a car
(1c) The refrigerator is part of the kitchen
(1d) Chapters are part of books

These can be represented by the expressions

$\langle ..., \text{handle}, ... \; ; R_{\text{cup}} \rangle$

$\langle ..., \text{wheel-1, wheel-2, wheel-3, wheel-4, wheel-5}, ... \; ; R_{\text{car}} \rangle$

$\langle ..., \text{refrigerator}, ... \; ; R_{\text{kitchen}} \rangle$

$\langle ..., \text{chapter-2, chapter-2}, ... \; ; R_{\text{book}} \rangle$

where R_{cup}, R_{car}, R_{kitchen}, and R_{book} are the bottom-up relations that assemble the parts into the whole. For the relational simplices to be well defined these relations must be explicit with operational procedures that (i) identify the component parts and (ii) assemble them into the whole. For example, Fig. 7.30(a) shows the bowl and handle assembled to form a cup.

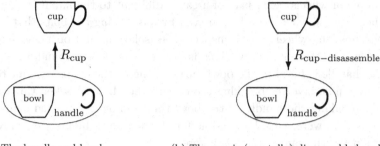

(a) The handle and bowl R_cup-assemble into a cup

(b) The cup is (mentally) disassembled under $R_{\text{cup}-\text{disassemble}}$ into a handle and a bowl

Fig. 7.30 The handle is part of the cup

However there is another common possibility, where an observer sees an object as "a cup" and mentally disaggregates it into parts. As argued previously humans cannot help doing this. When we see anything we automatically see (construct) its component parts. Often we give the components names, either the name of an intensionally defined class which matches the component, or the name of something similar that will mean this object in this context. For example, I don't know the names for the part of a cup without its handle, so I have used the word "bowl". This may or may not be the correct usage, but in this context it is clear to me, and hopefully the reader, what is meant by this word.

Thus the "being part of" component-integral object relation can either be tested bottom-up by checking that the parts do indeed assemble into the whole under a prescribed relation, or it can asserted top-down that within this whole there is identifiable substructure.

The use of language can be very individual. For me an object must have a handle to be called a cup (Fig. 7.31(a)). Usually I call a tall ceramic vessel with straight sides a mug, whether or not it has a handle (Figs 7.31(b) & (c)), and I call a vessel with curved sides and no handle a "beaker", especially if it is made of a non-ceramic material such as plastic ((Fig. 7.31(d)). If these objects were made of glass I would call them left to right (a) a glass cup, (b) a glass mug, (c) a glass, and (d) a glass. No doubt other native English speakers would use different names for some of the objects, and sometimes I might use the terms differently. Words are used differently around the UK, and my use of these terms reflects this.

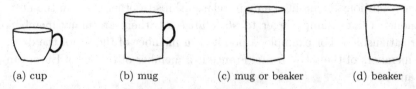

(a) cup (b) mug (c) mug or beaker (d) beaker

Fig. 7.31 Use of words can be individualistic and inconsistent

The example of wheels being part of a car is different to the handle being part of the cup because the wheels can be removed by unscrewing the bolts that attach them to the hubs, the wheel can be inspected in isolation, and on being screwed back the car will function normally. This is not the case for the handle of a cup. Generally the handle is fused to the bowl during firing as the cup is manufactured. Removing the handle involves breaking the ceramic bond. Even when this can be done without breaking the handle or the bowl the components created cannot be reassembled as they were, *e.g.* it's possible for the handle to be glued back to the bowl, but it's never the same as it was before.

While it makes sense to say that a particular refrigerator is part of a particular kitchen, there are refrigerators that are not parts of kitchen, and there are kitchens that have no refrigerator. Assembling the components into a kitchen is an α-aggregation. It is possible to form the set of all kitchens that have refrigerators, where this is a β-aggregation.

Some chapters are part of specific books and, no doubt, there exist books which are not divided up into chapters. It is not clear that "being a chapter" has a meaning outside of being a chapter in a specific book, so the chapter has meaning only in the context of other chapters and them being assembled into a book.

In all these cases the α-aggregation of specific parts into specific wholes can be made well defined, and in all cases the wholes can be gathered together by beta-aggregations according, for example, whether they have handles, have wheels, have refrigerators, or have chapters.

Class 2: Member-Collection

(2a) A tree is part of a forest
(2b) A juror is part of jury
(2c) This ship is part of a fleet

In principle, member-collection means a β-aggregation. However these example actually refer to α-aggregations. For example, a forest is a collection of trees with relational structure such as proximity and density. An arbitrary collection of trees is not a forest.

Again a jury is not a collective noun, it is a structure. It would be possible to find twelve jurors in a law court which collectively were not part of a particular jury, *e.g.* they might be members of twelve different juries.

Similar considerations apply to ships in fleets. To be part of a fleet a ship has to satisfy other conditions, including being in relationships to other ships in the fleet.

Even though these example refer to *structured sets* there are many member-collection relationships. For example, a poodle is a member of the set of the dogs, apples are members of the set of fruits, France is a member of the set of European countries, and so on.

Class 3: Portion-Mass

(3a) This slice is part of a pie
(3b) A yard is part of a mile
(3c) This hunk is part of my clay
(3d) She asked me for part of my orange
(3e) She asked me for some of my orange
(3f) The engine is part of the car

A pie is alpha-aggregated from ingredients at *Level N* by processes of mixing and cooking. Given a pie at *Level N+1* there are no slices *a priori*. Slices can be constructed *a posteriori* from a given pie by a "slicing" process. The creation of slices destroys the pie. Thus it makes sense to say that this slice at time t was sliced from the pie that existed at $t' < t$. More interesting is to ask what is the level of the slice with respect to the *Level N* ingredients and the *Level N+1* pie?

Yards and miles are units of measurement and take their meaning from the measurement process. A road (*Level N*) of length one mile can be deconstructed into a set of 1,760 pieces (*Level N-1*), each measuring one yard. These pieces of road at *Level N-1* can be alpha-assembled into the original road at *Level N*.

The terms hunk and clay imply that once there was a single hunk of clay that was disassembled into two or more portions, and "this hunk" is one of them.

Similarly, $\sigma(\text{orange}) \stackrel{R_{\text{disassemble}}}{\longrightarrow} \langle \text{peel, segment-1, segment-2, ...} \rangle$ shows that parts of oranges can be obtained by a disassembly processes, creating sets of new things from the whole. If "portion" is interpreted as "part of a disaggregation", then the peel and segments are portions of the original whole.

The engine being part of the car reflects either an assembly of components into a relational simplex, or a mental or practical dissassembly as discussed above. How do you know it's a portion of the car? Because I disassembled the car and this was one of the resulting objects.

Class 4: Stuff-Object

(4a) A martini is partly alcohol
(4b) The bike is partly steel
(4c) Water is partly hydrogen
(4d) The lens is made of glass

The martini is a relational hypersimplex $\langle \text{gin, martini, ice, olive}; R_{\text{make a martini}} \rangle$ at *Level N+1*. Both the gin and the martini at *Level N* can be mentally deconstructed into lower level parts including alcohol, water, and other things at *Level N-1*, with the conclusion that they both contain alcohol. One might reason "gin and martini contain alcohol *implies* $\langle \text{gin, martini, ice, olive}; R_{\text{make a martini}} \rangle$ contains alcohol because the process $R_{\text{make a martini}}$ does not destroy the alcohol in the gin and martini". Alternatively one could perform a chemical test to see if the assembly $\langle \text{gin, martini, ice, olive}; R_{\text{make a martini}} \rangle$ contains alcohol.

The bike being partly steel translates into "the bike is assembled from components, some of which are made from steel". Water being partly hydrogen translates into, "a water molecule is assembled from three atoms, two of which are hydrogen".

"The lens is made of glass" reflects the assembly process of the lens from its component parts, grains of sand, and the assembly process that melts and fuses them, shapes the resulting object as a lens, and polishes it, {sand grains} → ⟨lens⟩. "The lens is partly glass" suggests that it can be deconstructed to parts that include glass and other things.

Class 5: Feature-Activity

(5a) Paying is part of shopping
(5b) Bidding is part of playing bridge
(5c) Ovulation is part of the menstrual cycle
(5d) Dating is part of adolescence

"Shopping" is a collective noun for many different kind of shopping. For example, a trip to the supermarket can be described as

$\sigma_1 = $ ⟨drive car to supermarket, park the car, get out, walk to supermarket,
  get trolley, walk around aisles, select grocery items, put selected items in trolley,
  queue at checkout, paying, put items in bags, put bags in trolley,
  walk trolley back to car, put bags in car, return trolley to trolley station,
  walk back to car, get in car, drive home, get out of car,
  transfer bags to kitchen, put items where they are stored⟩.

while another shopping trip could be described as

$\sigma_2 = $ ⟨drive car to shopping mall, park the car, get out, walk to shopping mall,
  walk to fashion shop x, inspect clothes on racks, select garments, try items on,
  take selected items to checkout, paying, pick up bag with selected items,
  walk to fashion shop y, inspect shoes on stands, select items, try items on,
  take selected items to checkout, paying, pick up bag with selected items, ... ⟩

There are many variations on these shopping trips, and many other kinds of shopping trip to buy different things. For example a distinction might be made between "supermarket shopping" and "local food shopping" where for me the latter is different to the former, because I can walk to the local supermarket in my village while I must drive to the much larger supermarkets in the local towns.

Similarly a "fashion shopping" trip could begin with a taxi ride or taking the bus and end up being driven home by a friend or relative. Also there are different kinds of "fashion shopping". Apart from its supermarket, my village has a number of shops selling women's clothes, and until recently had a shoe shop. Thus "local clothes shopping" is possible in my village, as it is in parts of large cities.

At a lower level of representation, "paying" can be represented as sequences of relational simplices as events, $e.g.$

σ_1(paying) = ⟨take money from purse, hand over money to payee, receive change,
  receive receipt, put change in purse, put receipt in shopping; $R_{\text{paying}-1}$⟩

Multilevel Systems

$\sigma_2(\text{paying}) = \langle$take credit card from wallet, hand over credit card to payee,
enter code into machine, receive back credit card, recieve receipt,
put card back in wallet, put receipt in shopping bag; $R_{\text{paying}-2}\rangle$

$\sigma_3(\text{paying}) = \langle$take cheque book from pocket, write out cheque,
hand cheque to payee, receive receipt, return cheque book to pocket,
put receipt in shopping bag; $R_{\text{paying}-3}\rangle$

Again there are many variants of each of these types of paying. For example, when paying cash with a large denomination note one might be told that there is little small change in the till and be asked if one has something smaller, followed by looking through pockets for coins. As another example a credit card might be refused for some reason and have to be replaced by another, cash or a cheque.

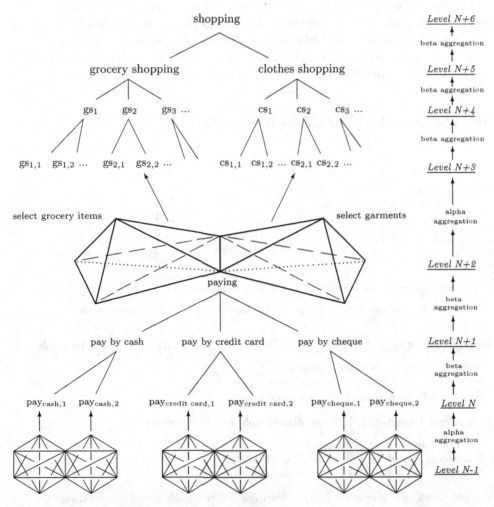

Fig. 7.32   Paying is part of shopping

It is surprising how complicated a simple idea like shopping and paying can be. In normal life one hardly gives these things a second thought, clustering together many slightly different shopping trips into "going to the (local) Co-op", "going supermarket shopping", "going to the mall", and so on. Behind all this are complicated alpha- and beta- aggregations as illustrated in Fig. 7.32. It is surprising that something apparently as simple as shopping can require eight levels of description.

Class 6: Place – area

(6a) The Everglades are in Florida
(6b) An oasis is part of a desert
(6c) The baseline is part of a tennis court

In spatial aggregation disjoint regions of space are drawn on maps, giving procedures to decide if one is inside the other. The Everglades are related to Florida by map inclusion, and many other relations too. By the same approach, it is possible to say if a particular oasis is in a particular desert. The case for the tennis court baseline is more complicated because baselines α-aggregate into tennis courts.

7.15 The intransitivity of meronymic relations

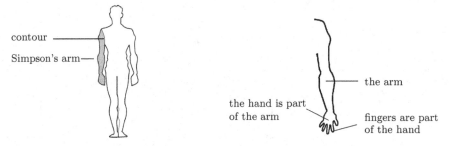

(a) Simpson's arm defined by a contour   (b) Fingers are part of the hand is part of the arm

Fig. 7.33 Simpson's finger is part of Simpson's arm is part of Simpson's body

Although at first sight "being a part of" is a transitive relation, as in the example

Simpson's finger is part of Simpson's hand
Simpson's hand is part of Simpson's body
Simpson's finger is part of Simpson's body

this apparent transitivity does not always hold as in the example

Simpson's arm is part of Simpson
Simpson is part of the Philosophy Department
Simpson's arm is part of the Philosophy Department

As noted previously this construction involves the composition of two relations for the arm which is a different relation to belonging to the Philosophy Department.

7.16 Hierarchical traffic aggregation

Figure 7.34 shows traffic aggregating over the backcloth from the micro-level to the whole system at the macro-level. At the micro- and meso-levels the dynamics are constrained by connectivity, while at the macro-level of the system there is just one vertex – the system itself.

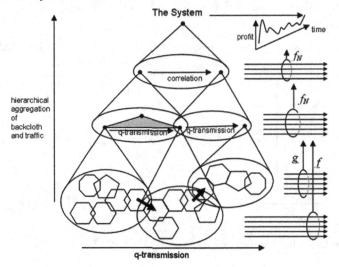

Fig. 7.34 Multilevel traffic aggregating across the multilevel backcloth

The objective here is to make hierarchical traffic aggregation well defined across backcloth levels *N+k* and *N+k+1*. Although the subscripts and superscripts may seem complicated, they are justified by the precision and clarity they give.

Given an assembly relation, $R^{N+k} : \langle v_1^{N+k}, ..., v_p^{N+k} \rangle \to \langle v_1^{N+k}, ..., v_p^{N+k}; R^{N+k} \rangle$ let $\tilde{R}^{N+K+1}$ be the *disassembly* relation

$$\tilde{R}^{N+k+1} : \langle v_1^{N+k}, ..., v_p^{N+k}; R^{N+k} \rangle \to \langle v_1^{N+k}, ..., v_p^{N+k} \rangle.$$

R^{N+k} maps an ordered set of vertices at *Level N+k* to a hypersimplex at *Level N+k+1*, and $\tilde{R}^{N+k+1}$ maps a hypersimplex at *Level N+k+1* to the ordered set of its vertices at *Level N+k*.

Let ψ^{N+k} be a mapping on the vertices v_i^{N+k}, $\psi^{N+k} : v_i^{N+k} \to \mathbb{R}$, where each $\phi^{N+k} v_i^{N+k}$ is a number. Let $\langle \phi^{N+k} v_1^{N+k}, ..., \phi^{N+k} v_p^{N+k} \rangle$ be the ordered set of these values, and let

$$\phi^{N+k} : \langle v_1^{N+k}, ..., v_p^{N+k} \rangle \to \langle \phi^{N+k} v_1^{N+k}, ..., \phi^{N+k} v_p^{N+k} \rangle$$

Let $\psi_{R^{N+k}}^{N+k}$ be a mapping associated with R^{N+k},

$$\psi_{R^{N+k}}^{N+k} : \langle \phi^{N+k} v_1^{N+k}, ..., \phi^{N+k} v_p^{N+k} \rangle \to \mathbb{R}$$

212   *Hypernetworks in the Science of Complex Systems*

Then $\psi_{R^{N+k}}^{N+k}$ is a *hierarchical traffic aggregation mapping*

$$\psi_{R^{N+k}}^{N+k} : \phi^{N+k} \to \phi^{N+k+1}$$

via

$$\phi^{N+k+1} = \psi_{R^{N+k}}^{N+k} \phi^{N+k} \tilde{R}^{N+k+1}$$

By construction, Fig. 7.35 is commutative with $\phi^{N+k+1} R^{N+k} = \phi^{N+k} \psi_{R^{N+k}}^{N+k}$.

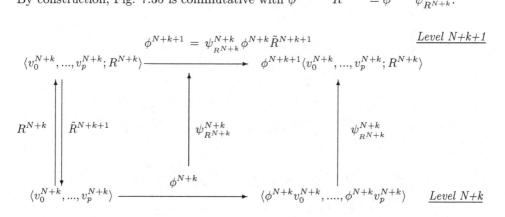

Fig. 7.35 The hierarchical aggregation mapping $\psi_{R^{N+k}}^{N+k} : \phi^{N+k} \to \phi^{N+k+1}$

Self-similar backcloth-traffic aggregation across hierarchical levels

Let $\langle v_0^{N+k}, ..., v_p^{N+k}; R^{N+k} \rangle$ become a vertex at *Level N+k+1*, with $v_i^{N+k+1} \stackrel{\text{def}}{=} \langle v_0^{N+k}, ..., v_p^{N+k}; R^{N+k} \rangle$. Then Fig. 7.35 can be redrawn as Fig. 7.36 so that the vertex mapping ϕ^{N+k} at *Level N+k* is transformed by the hierarchical aggregation mapping ψ^{N+k} into the vertex mapping ϕ^{N+k+1} at *Level N+k+1*.

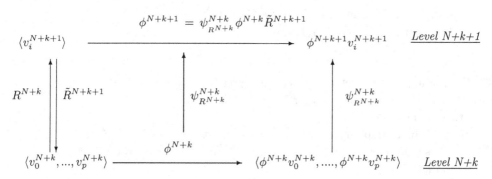

Fig. 7.36 $\psi_{R^{N+k}}^{N+k}$ transforms a vertex mapping at *Level N+k* to a vertex mapping at *Level N+k+1*

This achieves the goal of this section and a major objective of this book: the backcloth-traffic architecture shown here is self-similar between levels (Fig. 7.37).

Multilevel Systems

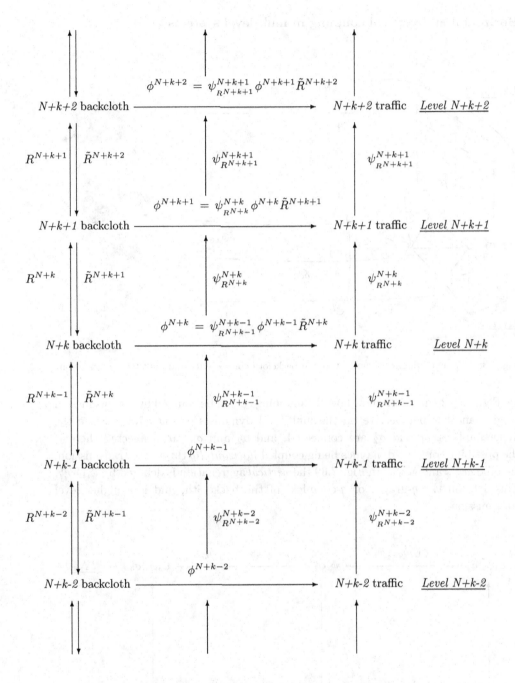

Fig. 7.37 Self-similar backcloth-traffic aggregation across hierarchical levels

Horizontal and vertical coupling in multilevel systems

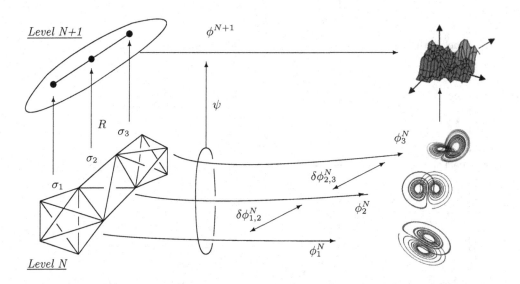

Fig. 7.38   System dynamics coupled through backcloth connectivity and multilevel aggregation

Figures 7.38 and 7.39 show how the mappings ϕ can be coupled by q-transmission to give another perspective on the multilevel dynamics of Fig. 7.37. Since the hypersimplices σ_1 and σ_2 are connected, and σ_2 and σ_3 are connected there is the possibility of their dynamics being coupled *horizontally* through q-transmission, denoted $\delta\phi_{1,2}^N$ and $\delta\phi_{2,3}^N$. In Fig. 7.39 the *vertical* aggregation is denoted by $\psi_{ij}\delta\phi_{ij}^N$. This acts on the q-graph, or q-complex, of the backcloth, and is a higher level transmission.

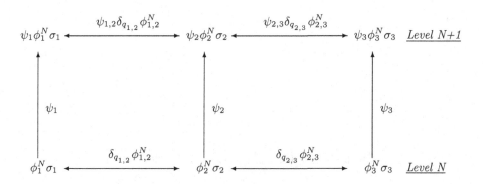

Fig. 7.39   Horizontally and vertically coupled dynamics

7.17 The Grand Challenge of multilevel systems

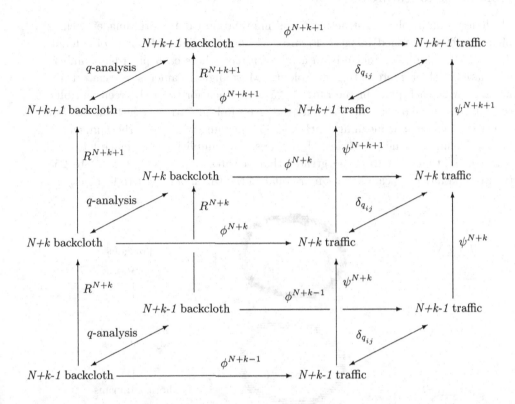

Fig. 7.40 A generalised data structure for hypernetworks

Combining Fig. 7.37 with 7.39 gives the diagram shown in Fig. 7.40. This diagram can be interpreted as a generalised data structure for multilevel hypernetworks.

Let a *multilevel backcloth-traffic system* be defined to be a system with

(i) three or more levels of vertices and hypersimplices
(ii) at least one level with hypersimplices renamed as vertices
(iii) time-dependent mappings ϕ^{N+k} defined on the vertices and hypersimplices

The <u>Grand Challenge for multilevel systems</u> is to find mappings ψ^{N+k} between the $\phi^{N+\tilde{k}}$ that unify the dynamics at each level with

$$\psi^{N+k+1} = \psi^{N+k} \phi^{N+k} \tilde{R}^{N+k+1}$$

7.18 Example: Machine vision

From pixels to features

Machine vision involves abstracting useful information from digital images such as photographs or scanned images. Figure 7.41 shows a digitized image of a hand-drawn circle. A digital image is an array of *pixels*, where each pixel has numbers associated with it representing its colour. Most digital cameras associate three numbers with each pixel in the range 0–255, one each for the red, green and blue components of the pixel. For simplicity the pixels in Fig. 7.41 have just one number called the *greyscale*, 0 meaning black and 255 meaning white and the numbers in between being intermediate levels of greyness. For simplicity this image has been *binarised* so that any greyscale greater than a threshold $T = 160$ is set to 255 (white) and any greyscale less than or equal to the threshold is set to 0 (black).

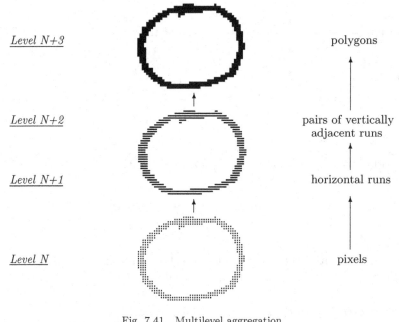

Fig. 7.41 Multilevel aggregation

The machine vision task is to find the object and recognise it. Initially all the machine knows is the x and y coordinates of each pixel and the greyscales. Although you instantly see the circle, the computer has to start from the atomic pixel data and construct the shape, and then it has to recognise it, *i.e.* classify it.

Level N to Level N+1

Let the individual pixels exist at *Level N*. A *horizontal run* of pixels of *length* ℓ is a hypersimplex of contiguous pixels $\langle p_{x,y}, p_{x+1,y}, ..., p_{x+\ell-1,y}; R_{x,y,\ell} \rangle$. A computer can form horizontal runs of black pixels by scanning the image from left to right until it encounters a black pixel and continuing until it finds a white pixel. It *constructs* a *dark run* hypersimplex. This is an "AND-aggregation" since all the pixels are necessary to make the run. Take any pixel away and it is not possible to decide if the remaining pixels form a contiguous run of dark pixels.

The construction of the dark horizontal runs augments the representation of the image with new objects at *Level N+1*. Every time the machine finds a dark run it has to allocate memory to store information on the set of pixels that make up the run. It also needs to store the address of that area of memory, and to do this it creates a *list simplex*, and the address of each run is added to it. The operation of adding a run to a list simplex at time t can be written as

$$\mathcal{A} : \langle r_0, r_1, \ldots, r_n; R_{\text{list of runs},t} \rangle + \langle r_{n+1} \rangle \to \langle r_0, r_1, \ldots, r_n; R_{\text{list of runs},t} \rangle \diamond \langle r_{n+1} \rangle$$
$$= \langle r_0, r_1, \ldots, r_n, r_{n+1}; R_{\text{list of runs},t+1} \rangle.$$

Note that adding to the list is part of the dynamics of the computation. If the run is created at time t it is appended to list simplex at time $t+1$, one "tick" later.

In hypernetwork theory lists are used instead sets. A list has no explicit relation on it, but since lists are *constructed* the elements are related by the process, here $R_{\text{list of runs},t}$, that creates the elements and adds them to the particular list at time t,. In computational terms it is usually not the element that is added to the list but a *pointer* to the element. The pointer is the numerical address of the start of the memory allocated to the element. A vertex of a list simplex of runs at *Level N+1* is a number that "points down" to the instantiation of the run as a list of pixel addresses at *Level N*.

Level N+1 to Level N+2

Given the list simplex of runs at *Level N+1* the machine can go through it to establish a vertical contiguity relation between pairs of runs. Let run r be all the pixels between (x_1, y) and (x_2, y) and run r' be all the pixels between (x'_1, y') and (x'_2, y'). Then runs r and r' can be defined to be *vertically adjacent* if (i) $y = y' + 1$ or $y = y' - 1$, and (ii) $x_2 \geq x'_1$ and $x'_2 \geq x_1$.

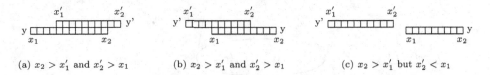

(a) $x_2 > x'_1$ and $x'_2 > x_1$ (b) $x_2 > x'_1$ and $x'_2 > x_1$ (c) $x_2 > x'_1$ but $x'_2 < x_1$

Fig. 7.42   Runs are vertically adjacent if $y' = y + 1$ or $y = y' + 1$, and $x_2 \geq x'_1$ and $x'_2 \geq x_1$

The hypersimplex $\sigma(e_i) = \langle r_1, r_2; R_{\text{vertically adjacent}} \rangle$ is special because it can be considered to be an edge in a graph or a network. Also for each run, $\langle r_i \rangle$, a list simplex can be made of its incident edges $\langle e_{ij}, e_{ik}, ..., e_{in} \rangle$ as shown in Fig. 7.43.

In this case the relation is symmetric, but more generally two list simplices can be made, one for in the in-edges and one for the out-edges. These list simplices form a data structure for the run-adjacency graph and this facilitate the computation of the connected graph components, which are used to make polygons at *Level N+3*.

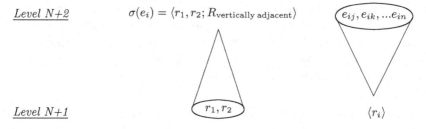

Fig. 7.43 Pairs of adjacent runs form a graph or network

Level N+2 to Level N+3

Let V^{N+1} be the set of horizontal runs of dark pixels, and let E^{N+2} be the set of network edges made of vertically adjacent pairs of runs. (V^{N+1}, E^{N+2}) is a network. It will be called the *run adjacency network*. The connected components in this network correspond to the polygons. The components partition the sets of edges and runs into subnetworks.

The polygon can be considered to be a list of pairs of *Level N+2* vertically adjacent runs. In practice it is also useful to have the list of *Level N+1* runs as shown in Fig. 7.44 with the natural bipartite relation between the lists.

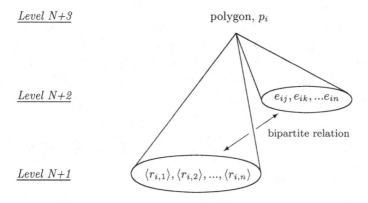

Fig. 7.44 Polygons are defined by components of the vertically connected run-pair network

Figure 7.45 shows the aggregations from pixels to polygons, including the bipartite network between runs and run-pairs used to find the polygons.

Multilevel Systems

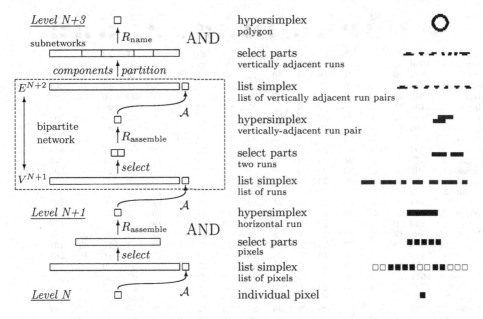

Fig. 7.45 Multilevel aggregation from pixels to polygons

Recognising polygons

Figure 7.46 shows a set of polygonal objects, p_0–p_{59} that the machine is required to recognise as "circles" or "rectangles". After the polygons have been constructed as

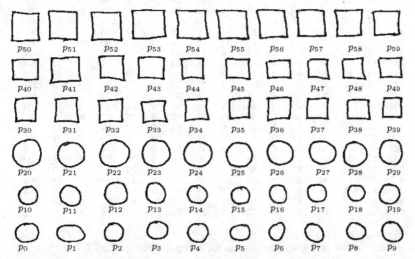

Fig. 7.46 Two kinds of polygon to be discriminated, circles and rectangles

	c_0	c_1	c_2	c_3	c_4	c_5	c_6	c_7	c_8	c_9	c_{10}	c_{11}	c_{12}	c_{13}	c_{14}	c_{15}
Polygon 0	1335	20	22	48	20	0	30	22	23	29	0	20	47	23	20	203
Polygon 1	1396	24	16	66	26	0	21	16	20	19	0	24	60	20	26	204
Polygon 2	928	20	22	29	16	0	30	22	17	31	0	20	38	17	16	198
Polygon 3	1007	26	14	30	25	0	28	14	14	27	0	26	31	14	25	201
Polygon 4	1301	27	22	33	27	0	27	22	16	33	0	27	39	16	27	228
Polygon 5	1136	21	21	38	22	0	28	21	20	30	0	21	38	20	22	190
Polygon 6	1090	18	19	42	22	0	25	19	20	28	0	18	37	20	22	204
Polygon 7	866	14	17	44	16	0	26	17	18	27	0	14	41	18	16	178
Polygon 8	842	15	23	21	17	0	28	23	23	30	0	15	19	23	17	162
Polygon 9	907	19	17	35	21	0	23	17	17	25	0	19	33	17	21	189
Polygon 10	1275	24	18	36	26	0	33	18	17	36	0	24	35	17	26	221
Polygon 11	1061	20	19	34	22	0	30	19	21	30	0	20	30	21	22	209
Polygon 12	1548	22	21	43	24	0	38	21	22	39	0	22	40	22	24	229
Polygon 13	1246	24	18	31	25	0	36	18	18	37	0	24	30	18	25	213
Polygon 14	1104	23	19	36	27	0	25	19	20	28	0	23	31	20	27	189
Polygon 15	1235	20	23	36	19	0	31	23	22	31	0	20	38	22	19	221
Polygon 16	899	19	14	30	20	0	35	14	13	37	0	19	30	13	20	167
Polygon 17	941	17	16	37	22	0	29	16	18	32	0	17	30	18	22	189
Polygon 18	943	19	14	37	18	0	33	14	15	31	0	19	37	15	18	191
Polygon 19	777	17	16	31	16	0	26	16	16	25	0	17	32	16	16	167
Polygon 20	2551	27	29	53	29	0	45	29	29	47	0	27	51	29	29	270
Polygon 21	2077	27	29	50	31	0	35	33	28	40	4	27	47	32	31	259
Polygon 22	2537	28	31	54	29	0	39	31	33	38	0	28	51	33	29	279
Polygon 23	2032	26	30	46	27	0	34	30	27	38	0	26	48	27	27	277
Polygon 24	2322	29	30	47	26	0	44	30	29	42	0	29	51	29	26	287
Polygon 25	2264	28	27	55	29	0	37	27	28	37	0	28	53	28	29	280
Polygon 26	1902	27	24	46	27	0	37	24	24	37	0	27	46	24	27	272
Polygon 27	1890	28	22	51	29	0	34	22	21	36	0	28	51	21	29	268
Polygon 28	1788	25	25	38	23	0	41	25	26	38	0	25	39	26	23	259
Polygon 29	1656	23	21	45	24	0	44	21	22	44	0	23	43	22	24	244
Polygon 30	1745	12	10	80	14	0	62	10	12	62	0	12	76	12	14	309
Polygon 31	1803	10	14	72	13	0	72	14	11	78	0	10	72	11	13	307
Polygon 32	1813	11	15	73	13	0	69	15	15	71	0	11	71	15	13	294
Polygon 33	1532	12	15	60	8	0	73	15	13	71	0	12	66	13	8	310
Polygon 34	1633	11	14	72	15	0	64	13	13	69	0	10	69	12	14	291
Polygon 35	1722	11	17	70	11	0	65	17	22	60	0	11	65	22	11	296
Polygon 36	1349	8	12	69	12	0	60	12	14	62	0	8	63	14	12	283
Polygon 37	1298	14	11	70	13	0	54	11	13	51	0	14	69	13	13	276
Polygon 38	1284	7	12	64	6	0	62	12	12	61	0	7	65	12	6	282
Polygon 39	1099	9	10	62	8	0	58	9	10	57	0	8	63	9	7	272
Polygon 40	2500	13	12	96	10	0	81	12	18	72	0	13	93	18	10	391
Polygon 41	1945	17	7	76	15	0	76	7	9	72	0	17	76	9	15	309
Polygon 42	1693	7	7	88	9	0	68	7	9	68	0	7	84	9	9	300
Polygon 43	1785	17	14	73	9	0	69	14	10	65	0	17	85	10	9	307
Polygon 44	1467	11	8	71	12	0	65	8	11	63	0	11	67	11	12	290
Polygon 45	1535	12	9	75	10	0	62	9	5	64	0	12	81	5	10	304
Polygon 46	1485	19	10	64	14	1	60	9	9	56	0	19	70	8	14	274
Polygon 47	1251	13	8	64	13	1	56	7	7	57	0	13	65	6	13	271
Polygon 48	1287	15	15	64	13	0	51	14	8	56	0	14	73	7	12	258
Polygon 49	1444	12	9	76	16	0	56	8	9	60	0	11	72	8	15	262
Polygon 50	2599	15	18	87	20	0	80	18	16	87	0	15	84	16	20	347
Polygon 51	2613	12	10	91	17	0	94	10	14	95	0	12	82	14	17	399
Polygon 52	2666	17	12	97	17	0	86	12	13	85	0	17	96	13	17	386
Polygon 53	3032	11	9	114	12	0	88	9	12	86	0	11	110	12	12	415
Polygon 54	2950	18	11	100	19	0	84	11	10	86	0	18	100	10	19	392
Polygon 55	3002	22	9	97	19	0	86	9	8	84	0	22	101	8	19	400
Polygon 56	2759	24	13	82	18	0	82	13	8	81	0	24	93	8	18	368
Polygon 57	2384	14	7	91	14	0	81	7	8	80	0	14	90	8	14	368
Polygon 58	2226	15	13	84	12	0	81	13	16	75	0	15	84	16	12	359
Polygon 59	2133	12	15	84	10	0	72	15	10	75	0	12	91	10	10	342

Table 7.1   Pixel configuration counts for the polygons in Fig. 7.46

Multilevel Systems

objects at *Level N+2* the task is to classify them, in this case discriminating the squares from the circles. There are many ways this could be done, but for simplicity the 2×2 pixel method will be used. The configuration counts for each polygon are given in Table 7.1.

There are many ways these data can be used to discriminate the shapes. For example configurations the sum of c_7, c_{11}, c_{13}, and c_{14} is generally higher for circles than squares, while the sum of c_3, c_6, c_9 and c_{12} is higher for squares than circles. These sums can be used to separate the shapes as shown in Fig. 7.47, discriminating circles from squares. In this way the aggregations go from pixels to features.

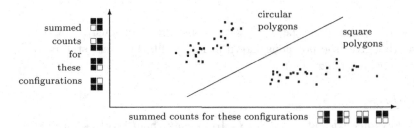

Fig. 7.47 Separating the polygons using the configuration counts

Multilevel face recognition

Figure 7.48 shows "smiley" and "frowny" faces. The next task for the vision system is to build a representation for the faces and to classify them as smiley or frowny.

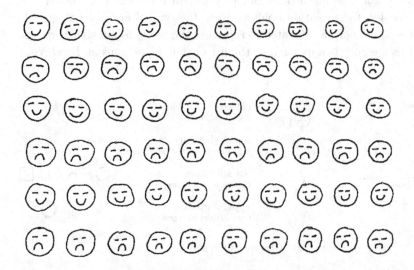

Fig. 7.48 Smiley and frowny faces to be recognised

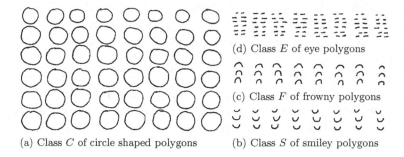

(d) Class E of eye polygons

(c) Class F of frowny polygons

(a) Class C of circle shaped polygons

(b) Class S of smiley polygons

Fig. 7.49   Polygons abstracted from smiley and frowny faces

The aggregation to polygons at *Level N+3* was considered previously. For the faces in Fig. 7.48 this gives the polygons shown in Fig. 7.49. The next step is to recognise the polygons and make the faces.

Level N+3 to Level N+4

The unclassified polygons exist at *Level N+3*. To classify them involves forming new structures, *i.e.* the polygon with its class symbol. For example, if S is the symbol representing "smile" and $\smile$ is a polygon, then $\langle S, \smile; R_S^{N+3}\rangle$ represents the polygon classified as a smile. Since it is a structure made from the smile symbol and the polygon at *Level N+3*, the hypersimplex $\langle S, \smile; R_S^{N+3}\rangle$ exists at *Level N+4*.

Figure 7.50 shows two list simplices identified by their pointers $\blacktriangleright_{PC}^{N+3}$ and $\blacktriangleright_{poly}^{N+3}$. One element is selected from each of these lists, say a pointer to the smile symbol and a pointer to a polygon that has a smile shape. This pair is presented to the relation R_S^{N+3} which decides if the relation holds between the symbol and the polygon. In this case it does so memory is allocated to represent the newly found structure, and a pointer to this memory is appended to the list of smiley polygons at *Level N+4*.

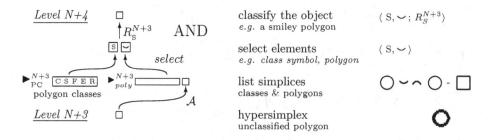

Fig. 7.50   Classifying the polygons

In order to compute whether or not the relation R_S^{N+3} holds the system requires emergent information about the polygon. This information does not exist in the

Multilevel Systems 223

pointers. Thus the system has to go down to the pointer to the polygon, which is a list of pointers to the run pairs, which is a list of two pointers to vertically adjacent runs each of which is a list of points to the actual pixels at *Level N*. At this level each pixel has three numbers associated with with it, x, y, and greyscale g. From these numbers the numbers x_{min}, x_{max}, y_{min} and y_{max} can be calculated, defining the bounding rectangle for the polygon. This can be used, for example, to calculate the pixel configuration counts in Table 8.1.

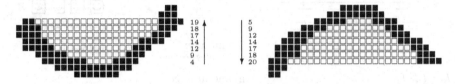

(a) horizontal gaps increasing upwards   (b) horizontal gaps increasing downwards

Fig. 7.51   Discriminating smile polygons from frown polygons

Neither the pixel configurations nor their dimensions can discriminate the smile polygons from the frowns because they are of similar size and reflections of each other about a horizontal axis. However, the gaps between the edges of the polygons are different. For example, in Fig. 7.51 the gap at the top between the black pixels is 19 white pixels, while the gap at the bottom is four white pixels. For the smile these gaps decrease top to bottom. In contrast, the frown has a gap of 20 white pixels at the bottom and a gap of four white pixels at the top. For smiles the gaps increase in size from the bottom to the top. This provides a robust way for R_S^{N+3} and R_F^{N+3} to discriminate the smiles from the frowns.

<u>Level N+4 to Level N+5</u>

Figure 7.52 shows circular and rectangular faces faces with smiles and frowns. It is assumed that the polygons have been found and classified by the methods of the previous sections. The next step is to assemble the polygons into faces.

Fig. 7.52   Circular and rectangular face to be recognised

To assemble the faces the system needs to collect the polygons together and to test that the polygons fit together to make faces, as shown in Fig. 7.53. There are four aggregation relations, R_{SC}^{N+4}, R_{FC}^{N+4}, R_{SR}^{N+4}, and R_{FR}^{N+4} making, respectively, Smiley Circular, Frowny Circular, Smiley Rectangular and Frown Rectangular faces at *Level N+5*. The symbols $\blacktriangleright_{SC}^{N+4}$, $\blacktriangleright_{FC}^{N+4}$, $\blacktriangleright_{SR}^{N+4}$ and $\blacktriangleright_{FR}^{N+4}$ are pointers to the lists of pointers to the various types of face.

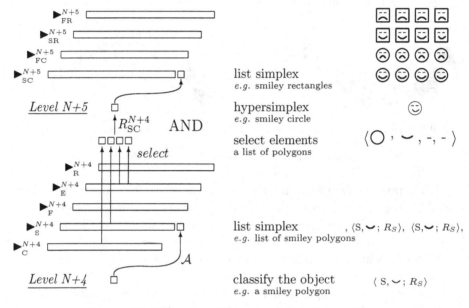

Fig. 7.53   Assembling polygons into faces

Level N+5 to Level N+6

The next aggregation is particularly interesting because it illustrates the idea of and "OR-aggregation" in contrast to the "AND-aggregations" seen so far.

Figure 7.54 shows an aggregation of circular and rectangular smiley faces, effectively forming the union of the lists of circular smiley faces and rectangular smiley faces, $\sigma(\blacktriangleright_S^{N+6}) \stackrel{\text{def}}{=} \langle \blacktriangleright_{SC}^{N+5}, \blacktriangleright_{SR}^{N+5}; R_{OR} \rangle$, as shown in Fig. 7.55.

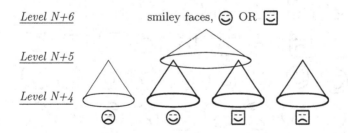

Fig. 7.54   Three out of twenty four ways of aggregating the lists of faces

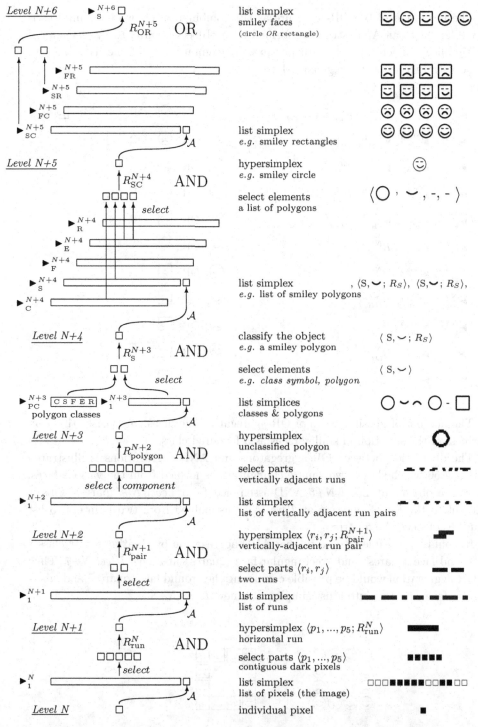

Fig. 7.55 Multilevel aggregation: from pixels to smiley faces

The formation of the OR-structure as the combination of two list simplices is very different to an AND-structure formed from objects pointed to by elements of list simplices. There are $2^4 = 16$ possible OR-combinations of the lists of faces. Those with two or more are given below:

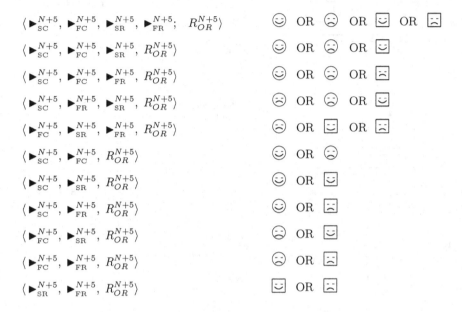

The number of possible ways of OR-aggregating the faces emphasises the arbitrariness of the selection of smiley faces as a favoured class.

The distinction between OR-aggregatons and AND-aggregations is illustrated in Fig. 7.56 which shows two square configurations formed from the various faces. These are objects at *Level N+6* AND-assembled from *Level N+5* parts. This is different to the list of pointers at *Level N+6* assembled from two pointers to list simplices of *Level N+5* parts.

It would be possible to do a further AND-aggregation by creating a new classes of "round face squares" and "rectangular face squares" at a new *Level N+7*. Then an OR-aggregation would be possible combining the "round face squares" and "rectangular face squares" into a list simplex at a new *Level N+8*.

Fig. 7.56 Square configurations created at *Level N+6* by AND-aggregations

Features as emergent properties

Where do features such as "circular" and "square" come from? Some of the polygons in Fig. 7.46 are relatively small, e.g. p_5 and p_{44}, while some are relatively big, e.g. p_{20} and p_{53}. Why not use the features "big" and "small"?

In their study of television programmes, [Gould et al (1984)] suggested *The Principle of Usefulness* which says "something is useful if it is useful". Although this is a deliberate tautology it provides a possible answer to the question of which of many possible emergent features should be included in the representation of a system. Emergent features will be selected because they are *useful*. This then shifts the question to "useful for what?", and this depends on the purpose of the system. Here the purpose is to recognise things.

How can it be known if any particular feature will be useful? In general it cannot be known *a priori*. It has to be tried and its usefulness determined. Where do the features come from to be tried? This is one of the fundamental questions of complex systems science. In general they come from "us". In machine vision, and much else, we act as gods, deciding which features will be included and which will not. The process that makes a candidate feature come into our heads seems to be based on serendipity, as does the decision on whether to give it a try. New constructs are constantly emerging in the social world and these are the building blocks of our vernacular models of the world.

Fig. 7.57 Vision systems to recognise car number plates cannot adapt to read cheques

In machine vision a system designed for one application usually fails for another, e.g. a system to recognise car number plates will fail to read bank cheques. Engineers creating these systems use features that they think will be useful and find useful features for the purpose in hand by trial and error. Number plates are generally "rectangular" while amounts of money written on cheques usually have "decimal points" and "commas" as shown in Fig. 7.57. Features like these are "hard wired" into vision systems and because they do not have the capability of generating new features, they are unable to adapt to new recognition tasks. Developing a system that can generate its own constructs is an outstanding challenge for hypernetwork research.

7.19 Example: Robot Football

Robot communities form a class of systems that is more complex than machines such as clocks and aeroplanes but less complex than human systems because they can be observed "from the outside". Communities of interacting robots can be set going and observed without the activities of the observer affecting their behaviour. Although humans often change their behaviour when they know they are being observed, robot systems can be created that do not. In particular teams of soccer-playing robots can be observed from the outside.

In 1997 Minoru Asada, Hiroake Kitano and others [Asada *et al* (1997)] suggested that a team of humanoid robots should beat the world champion soccer players as a challenge for machine intelligence superseding that of chess. The international RoboCup competition they inaugurated has various leagues including wheeled robots but the ultimate challenge is to have human-like robots compete with humans on equal terms.

Like the knight fork in chess, football has recognisable "set piece" positions. For example, Fig. 7.58(a) shows the *defenders dilemma* where black robot b_1 can attack white robot w_1 directly and risk an easy pass to w_2 who has an easy shot at goal, or it can try to move between w_1 and w_2 making it easier for white to pass by unimpeded on its left and shoot at goal. Assuming the pass is made to w_2 the goal keeper b_0 has the dilemma of attacking w_2 or defending the black team's goal, G_b.

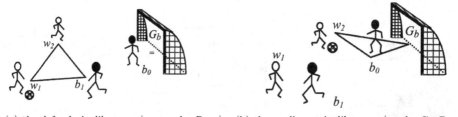

(a) the defender's dilemma, $\langle w_1, w_2, b_1; R_{DD}\rangle$ (b) the goalkeeper's dilemma, $\langle w_2, b_0, G_b; R_{GD}\rangle$

Fig. 7.58 Commonly occurring positions in football

In the same way that the path that leads to checkmate may be as interesting as the checkmate itself, in soccer the path of events that leads to scoring a goal can be as exciting as the goal itself. Although it does not guarantee scoring a goal, creating tactical positions such as those in Fig. 7.58 can be an *intermediate* strategic goal. Just like chess there can be positional play where, in the absence of an obviously advantageous tactical play, the strategic intermediate goal is to establish a "good position" which is predisposed to a good outcome [Johnson & Iravani (2007)].

Figure 7.59 shows the same three players configured differently by the 3-ary relations R_1 and R_2. Most of the interesting structures in robot football involve sequences of positions and for spectators, part of the pleasure is to watch the choreography as the sequence progresses. A *move* in robot football can be defined to

(a) $\langle w_1, w_2, b_1; R_1 \rangle$ (b) $\langle w_1, w_2, b_1; R_2 \rangle$

Fig. 7.59 The same set of players configured differently

be an assembly of structured players through time, *i.e.* a move is a *Level N+2* hypersimplex in a hypernetwork that combines both spatial and temporal structure as illustrated in Fig. 7.60.

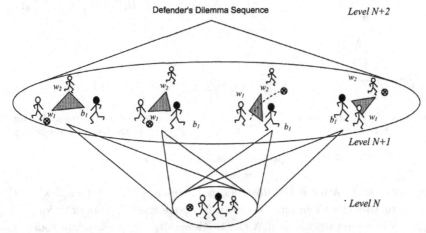

Fig. 7.60 A move such as the defender's dilemma sequence as a *Level N+2* hypersimplex

A move can involve a fixed set of players but in general some players will cease to be important as time progresses, as illustrated by b_1 being active in Fig. 7.61(a), less active in Fig. 7.61(b) and playing no part in Fig. 7.61(c). All of the hypersimplices are connected by the shared face $\langle w_1, w_2 \rangle$. An analysis of a robot football game using these structures is given in [Johnson & Iravani (2007)].

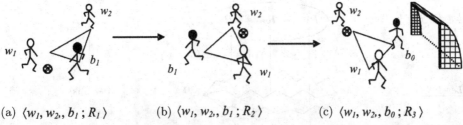

(a) $\langle w_1, w_2, b_1; R_1 \rangle$ (b) $\langle w_1, w_2, b_1; R_2 \rangle$ (c) $\langle w_1, w_2, b_0; R_3 \rangle$

Fig. 7.61 A move evolving through a sequence of connected hypersimplices

7.20 Multilevel Land-Use Transportation Systems

The goal in this section is to devise an *integrated* multilevel representation for roads and land uses at every level from the micro-microlevel of an individual room or a small piece of road, through the mesolevels of cities and regions to the macrolevels of entire countries, continents and the whole world.

Multilevel links and routes

This begins with the observation that a *route* between A and B is a structure formed from a related sequence of *links*, e.g. $\sigma(R_i) = \langle L_1, L_2, ..., L_6; \lambda \rangle$ as illustrated in Fig. 7.62. Here the relation is represented by the symbol λ so that R can be used to represent routes.

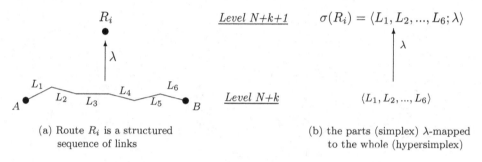

(a) Route R_i is a structured sequence of links

(b) the parts (simplex) λ-mapped to the whole (hypersimplex)

Fig. 7.62 λ maps the $N+k$ links to the $N+k+1$ route

Since the relation λ maps the unstructured sequence of links (simplex) to the route as a structure (hypersimplex) it moves from one level of representation, $N+k$, to a higher level of representation, $N+k+1$. Generally there are many routes between origin A and destination B and these can be grouped together under an *OR*-aggregation to form a new object that will be defined to be a *higher level link* at Level $N+k+2$, as shown in Fig. 7.63.

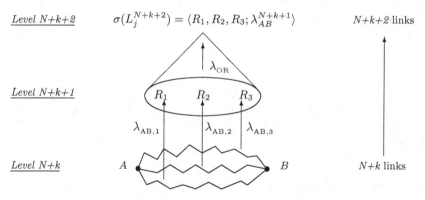

Fig. 7.63 Aggregation from $N+k$ links through routes to $N+k+1$ links

Multilevel Systems

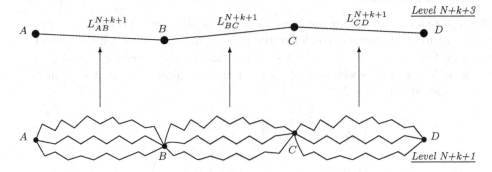

Fig. 7.64 Twenty seven routes at $N+k+1$ aggregate into one at $N+k+3$

Figure 7.65 shows three routes between A and B formed from *Level N+k* links, three routes between B and C and three more between C and D. Suppose the $N+k$ links are road segments. Each of these three sets of routes is shown to aggregate into a more abstract *Level N+k+1* link. These higher level links form a *Level N+k+3* route between A and D.

In this example the *Level N+k+2* links each represent three routes on the ground so that the route $\langle L_{AB}^{N+k+1}, L_{BC}^{N+k+1}, L_{CD}^{N+k+1}; \lambda_{AD}^{N+k+1} \rangle$ represents $3 \times 3 \times 3 = 27$ routes on the ground. A benign combinatorial explosion occurs higher levels.

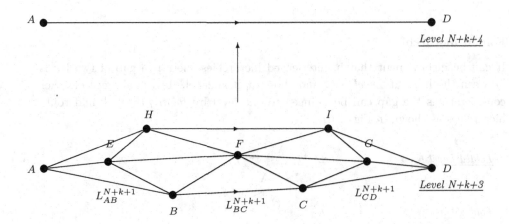

Fig. 7.65 The $N+k+4$ link represents more than 30,000 routes on the ground

Figure 7.65 shows how this $N+k+3$ route aggregates with others to form an $N+k+4$ link. At *Level N+k+3* there are 6 routes with four links, seventeen with five links, eighteen with six links and six with seven links. If each these $N+2$ links were made up of just three $N+k+1$ routes there would be $6 \times 3^4 + 17 \times 3^5 + 18 \times 3^6 + 6 \times 3^7$, more than 30,000, routes on the ground associated with the link at *Level N+k+4*.

232 Hypernetworks in the Science of Complex Systems

Interleaved hierarchies

This approach to representing road systems is potentially powerful since the higher level structures can represent combinatorially many links and routes on the ground. It is formed of two *interleaved hierarchies* of links and routes: structured sets of $N+k$ links form $N+k+1$ routes while sets of $N+k+1$ routes aggregate to form $N+k+2$ links *etc.*, as illustrated in Fig. 7.66.

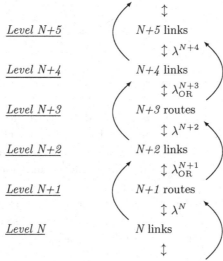

Fig. 7.66 The interleaved hierarchies of links and routes

Folded hierarchies

It can be inconvenient that in interleaved hierarchies there is a gap of two levels between the links at *Level N+k* and those at next level. In the absence of other considerations the gap can be reduced to one level by *folding* the link and route hierarchies as shown in Fig. 7.67.

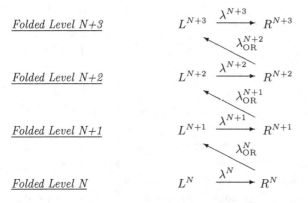

Fig. 7.67 The folded hierarchy of links and routes

Traffic on the multilevel backcloth

For road traffic systems the backcloth is the infrastructure of roads, junctions, plots of land, and land use activities such as "residence", "retail", and "industry". The traffic is, literally, the traffic of vehicles represented by various functions such as density of vehicles, speeds, and travel times. For the links these could be

$f(L_i^{N+k})$ is defined to be the flow of vehicles through a link L_i^{N+k}

$t(L_i^{N+k})$ is defined to be the travel time on link L_i^{N+k}

$\rho(L_i^{N+k})$ is defined to be the density of vehicles on L_i^{N+k}

If flow is defined to be the number of vehicles passing a point in unit time, the density around the observation point constrains the flow. In general as the density increases the speed decreases. This has little effect when densities are low but is significant after the density reaches the maximum flow threshold, $\rho_{\text{max flow}}$. Flows are unstable in this region where there can be shock waves as speeds reduce rapidly in a transition from low density to high density, as shown in Fig. 7.68.

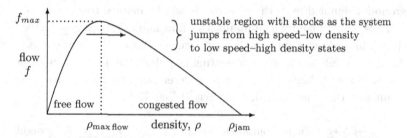

Fig. 7.68 The flow-density relation for road traffic

For the routes the traffic mappings include:

$f(R_i^{N+k+1})$ is the flow of vehicles across route R_i^{N+k+1}

$t(R_i^{N+k+1})$ is the travel time of vehicles across route R_i^{N+k+1}

The flow across a link is the sum of the flows on all the routes that traverse that link. The travel time on a route is the sum of the travel times on all the links that the route traverses. This can be expressed formally as follows.

Let $\sigma_R = \sigma(R_j^{N+k+1}) = \langle L_{j,1}^{N+k}, ..., L_{j,n_j}^{N+k}; R_{\text{AB}} \rangle$. Then

$$t(R_j^{N+1}) = t(\sigma(R_j^{N+k+1})) = \sum_{\langle L_i^{N+k} \rangle \lesssim \sigma_R} t(L_i^{N+k})$$

Let $\sigma_L = \sigma(L_i^{N+k}) = \langle R_{i,1}^{N+k-1}, ..., R_{i,n_i}^{N+k-1}; R_{\text{AB}} \rangle$. Then

$$f(L_i^{N+k}) = f(\sigma(L_i^{N+k})) = \sum_{\langle R_i^{N+k-1} \rangle \lesssim \sigma_L} f(R_i^{N+k-1})$$

In principle the flow of traffic is related to the mean speed and density. In practice the relationships are more complicated because the definitions of speeds and flows depend on how measurements are taken.

The *space mean speed*, $\overline{s}_m$ of traffic is the mean speed of the n vehicles on a link. $\overline{s}_{sms} = n/\sum_i(1/s_i) = n/\sum_i t_i$ where s_i the speed of vehicle i and t_i is the time it takes vehicle i to traverse the link. The *time mean speed* is the average speed of the n vehicles passing a point in unit time, $\overline{s}_{tms} = \sum s_i/n$.

Flows are measured as the number of vehicles passing an observation point and there can be a difference between the number of vehicles entering a link and the number of vehicles with a consequent change in density on the link. [Johnson (1981)] shows how the relationships above can be modified to represent the non-equilibrium dynamics on the links and routes.

The land use backcloth

The next step is to integrate the multilevel link-route representation with a multi-level zoning scheme providing a coherent representation of land use and transportation at all levels. The folded hierarchy of roads between junctions and routes on the ground assembled from those segments will be defined to exist at *Level N*. At *Level N-1* are plots of land with buildings such as houses, shops and factories. If required these can be divided into smaller areas such as individual rooms and parking spaces. At higher levels are zones representing neighbourhoods which aggregate into towns, and so on with higher level zones representing regions, countries, continents and ultimately the whole world, as shown in Fig. 7.69.

Level N+2	neighbourhoods		*Level N+8*	world
	↕			↕
Level N+1	bocks		*Level N+7*	continents
	↕			↕
Level N	streets		*Level N+6*	countries
	↕			↕
Level N-1	plots		*Level N+5*	regions
	↕			↕
Level N-2	houses, gardens, ...		*Level N+4*	counties/cities
	↕			↕
Level N-3	rooms, parking spaces, ...		*Level N+3*	towns

Fig. 7.69 Multilevel zones at all scales

This multilevel zoning scheme connects with the multilevel road traffic backcloth directly at *Level N* since some plots of land have the land uses "road" or "junction". By defining hierarchical sets of nodes on the boundaries of the hierarchical zones, the more abstract higher level links and routes can be integrated with the multilevel land-use scheme.

Multilevel Systems

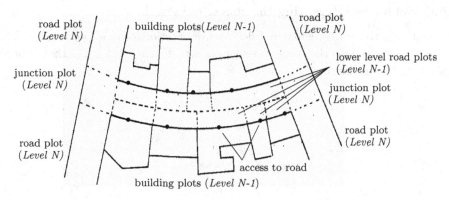

Fig. 7.70   Roads as plots of land at *Level N-1* and *Level N*

Figure 7.70 shows a land area divided into contiguous pieces called *plots*. At *Level N-1* nine of them are *building plots* and fifteen are *road plots*, or pieces of road. At *Level N* there are two *junction* or *road intersection* plots and five road plots. Of these the *Level N* horizontal road segment between the junctions is aggregated from N-1 road plots so that each building plot has access to road via a "lower level" road plot, *i.e.* a level below that of the road and the junctions.

Building plots and the lower level road plots are defined to exist at *Level N-1*. Then, as shown in Fig. 7.71 sets of *Level N-1* plots can be aggregated into *zones* at *Level N*. Here the plots are aggregated into the sides of the streets, Z_6^N, Z_7^N, Z_8^N and Z_9^N, the road segments Z_1^N and Z_2^N, and three junctions Z_3^N, Z_4^N and Z_5^N.

If required higher levels zones could be defined with, for example, Z_6^N, Z_1^N, Z_7^N aggregating to a zone around the street Z_1^N, and Z_8^N, Z_2^N, Z_9^N aggregating to a zone around the street Z_2^N at *Level N+1*.

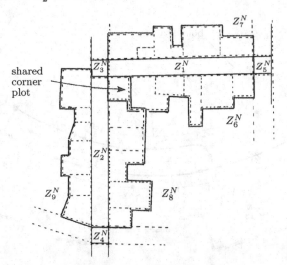

Fig. 7.71   Sets of *Level N-1* plots aggregated into zones at *Level N*

Multilevel nodes integrating the zones, links and routes

The next objective is to create a multilevel representation for roads and routes. This will be integrated into the zone system by a set of *multilevel nodes* that connect zones at the various levels.

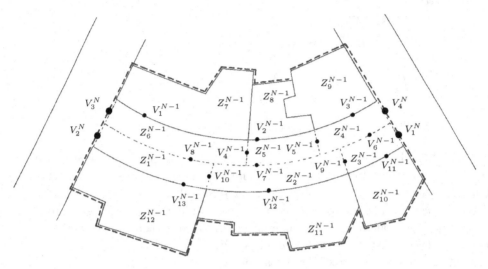

Fig. 7.72 Access between multilevel zones is by multilevel nodes on their shared boundaries

For example, a driver leaving home in zone Z_{12}^{N-1} by V_{13}^{N-1} into Z_1^{N-1} can cross into Z_8^{N-1} via node V_6^{N-1}, and from there drive to the junction via Z_5^{N-1} and Z_4^{N-1}, as illustrated in Fig. 7.73 where vehicles drive on the left of the road.

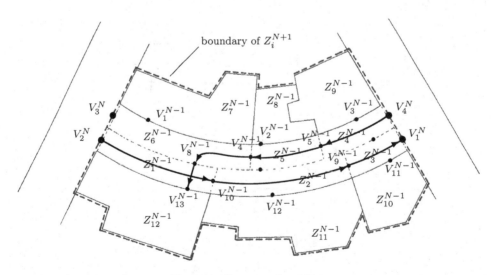

Fig. 7.73 Routes at *Level N-1*

Let a pair of nodes $\langle V_i^{N+k}\rangle$ and $\langle V_j^{N+k}\rangle$ be λ^{N+k}-related if they are nodes on the boundary of a zone Z^{N+k} and it is possible to drive from V_i^{N+k} to V_j^{N+k} in Z^{N+k}. Let the hypersimplex $\langle V_i^{N+k}, V_j^{N+k}; \lambda\rangle$ be called a *link* from V_i^{N+k} to V_j^{N+k}. This will be abbreviated to $\langle V_i^{N+k}, V_j^{N+k}\rangle$. Then

$$R_1^{N-1} = \langle\ \langle V_4^N, V_5^{N-1}\rangle, \langle V_5^{N-1}, V_4^{N-1}\rangle, \langle V_4^{N-1}, V_8^{N-1}\rangle, \langle V_8^{N-1}, V_{13}^{N-1}\rangle; \lambda^{N-1}\rangle.$$

forms a *route* from V_4^N to V_{13}^N at the junction. This can be contrasted with

$$R_2^{N-1} = \langle V_2^N, V_{10}^{N-1}\rangle, \langle V_{10}^{N-1}, V_9^{N-1}\rangle, \langle V_9^{N-1}, V_1^N\rangle; \lambda^{N-1}\rangle.$$

which goes between the two junction nodes V_2^N and V_1^N. Let a route between two nodes V_i^{N+k} and V_j^{N+k} be defined to be an *instantiation* of the link $\langle V_i^{N+k}, V_j^{N+k}; \lambda^{N+k}\rangle$. Thus R_2 is an instantiation of the *Level N* link $\langle V_1^N, V_2^N; \lambda^N\rangle$.

Figure 7.74 shows three routes across the *Level N+1* zone Z_1^{N+1} between the *Level N+1* boundary nodes V_1^{N+1} and V_2^{N+1}. These three routes form a *Level N+1* link between these $N+1$ boundary nodes, as shown in Fig. 7.75.

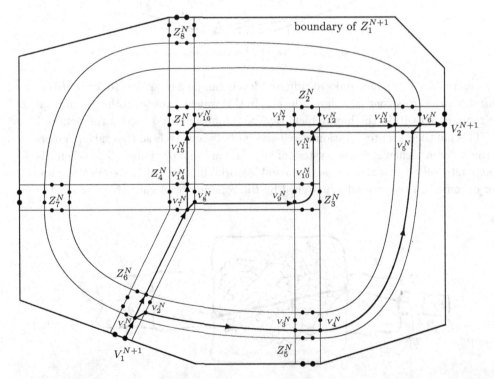

Fig. 7.74 Three *Level N* routes across the zone Z_1^{N+1} between boundary nodes V^{N+1} and V_2^{N+1}

The rules underlying the hierarchical scheme are that

(i) There is a *Level N+k* boundary node wherever a road on the ground crosses the boundary of a *Level N+k* zone.
(ii) *Level N+k* routes are contained entirely within a single *Level N+k+1* zone.
(iii) if there is a *Level N+k* route between the boundary nodes V_1^{N+k+1} and V_2^{N+k+1} of the zone Z_i^{N+k+1} then there exists a *Level N+k+1* link between V_1^{N+k+1} and V_2^{N+k+1} across Z_i^{N+k+1}.

This is illustrated in Fig. 7.75 where the abstract *level N+1* links associated with the routes in Fig. 7.74 traverse Z_1^{N+1}.

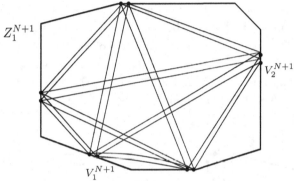

Fig. 7.75 The *Level N+1* links across the zone Z_1^{N+1} between boundary nodes V^{N+1} and V_2^{N+1}

Figure 7.76 shows how links at different levels can be aggregated to form *hierarchical routes* across zone at different levels. In this way any route on the ground can be represented within the hierarchical scheme. [Johnson (1981)] gives the details.

This multilevel representation is *self-similar* between levels so that, given a hierarchical zoning scheme, a road system of any size can be represented. The scheme is *computationally tractable* because the combinatorial increase with area in the number of routes on the ground is matched by the combinatorial gain of the scheme.

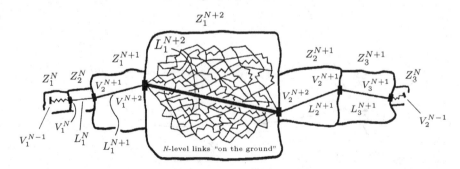

Fig. 7.76 A hierarchical route across hierarchical zones between V_1^{N-1} and V_2^{N-1}

7.21 Example: hospital admission for assault

In Britain, as in many other countries, there are levels of violence between young men that cause concern to policy makers. Aggression and violence are part of our human makeup, and learning how to deal with them are a natural part of growing up for most boys. For example, a study of 1037 boys aged six to fifteen years of age in a high-risk population sample from Montréal identified four percent of boys with relatively high levels of aggression: "Here, we were concerned with identifying characteristics that distinguish the modestly large fraction (28%) of boys who start off displaying high levels of physical aggression but subsequently desist, from the small but prominent group (4%) of boys who continue their physical aggression unabated. Only 2 such characteristics were identified, mother's low educational attainment and teenage onset of childbearing." [Nagel & Tremblay (2001)]. The correlation between the hypersimplex ⟨low educational attainment, teenage onset of childbearing; R_{mother}⟩ and male aggression traffic has obvious policy implications.

Figure 7.77 shows admission statistics for English National Health Service hospitals for assault involving 13 to 14 year olds from the report entitled *Dying to Belong* published in 2009 by the Centre for Social Justice (CSJ) which gives an in-depth analysis of street gangs in Britain [CSJ (2009)].

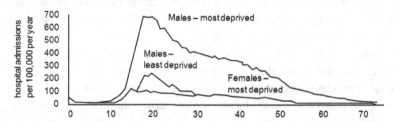

Fig. 7.77 Admissions to English NHS hospitals for assault involving 13 to 14 year olds

These statistics provide an aggregate snapshot of hospital admission traffic on the backcloth ⟨most deprived males, least deprived males, most deprived females⟩ at a relatively high *Level N+k*. Three classes of people are considered at *Level N+k-1*, ⟨males, deprived⟩, ⟨males, not deprived⟩ and ⟨females, deprived⟩.

The statistics can be read as follows. The class ⟨males, deprived⟩, has the highest incidence of hospital admissions peaking around the age of nineteen. Since the class ⟨females, deprived⟩, has much lower frequencies, deprivation alone is not the cause. The statistics show that ⟨males, not deprived⟩ also peaks around the age of nineteen, suggesting that males in general have more assaults. However, if ⟨males, not deprived⟩ represents the baseline of male assaults, after the age of ten the statistics for ⟨males, deprived⟩ are higher by a factor of three.

For policy making purposes, hypersimplices such as ⟨low educational achievement, teenage onset of childbearing; R_{mother}⟩ may provide more practical detail at *Level N+k-2* than the more general *N+k-1* vertex ⟨deprived⟩.

7.22 Example: the London riots of 2011

Between 6th and 10th August in 2011 London and other British cities experienced violent riots, looting, arson, assault and robbery. The fatal shooting by police in London on 4 August 2011 of Mark Duggan, a 29-year old black man, triggered protests in Tottenham on 6th August followed by violent conflict with the police and the destruction of property and looting into the night. Over the next few days many thousands of lawless people took to the streets with unprecedented destruction and looting that spread into other areas of London and other cities.

The media suggested many reasons for this shocking breakdown in law and order. These included welfare dependence, social exclusion, lack of fathers, spending cuts, weak policing, racism, gangsta rap and culture, consumerism, opportunism, technology and social networking [BBC (2011)].

Many believe that the riots reflected underlying tensions between minorities and the police, "Too many black men and women have been treated like criminals when they're not. This is not the cause of these riots, but it's there in the mix". Others thought the causes included, "a perverted social ethos, which elevates personal freedom to an absolute, and denies the underclass the discipline – tough love – which alone might enable some of its members to escape from the swamp of dependency in which they live". Others argued that "it's not just about the underclass – it's about politicians, it's about bankers, it's about footballers ... It's not just about a particular class, it permeates all levels of society".

The media response to the riots provides a rich soup of words and concepts relevant to the social dynamics of the riots and many hypothetical entailments. There is wide agreement that the social system in London was predisposed to extreme behaviours and, by implication, policy interventions earlier may have averted the riots.

In principle, complex systems science can add precision to the vernacular accounts by making systems and their many subsystems well defined and formalising the many behavioural hypotheses for multilevel simulations. For example, the BBC article above on arguments used to explain the riots included the words and expressions paraphrased below:

welfare dependence: social ethos; absolute personal freedom; underclass; discipline; tough love; escape; swamp; dependency; culture of entitlement; underclass; politicians; bankers; footballers; societal levels; all levels of society; leadership;
social exclusion: established community; provide nothing; attack on dignity; repeated humiliation; continuously dispossessed; possession rich society; areas of social deprivation; most vulnerable to riots; excuse;
lack of fathers lack of male role models; jailed youth offenders; stable family setup;
spending cuts mayoral candidate; austerity measures; massive cuts; potential for revolt; cuts to local authority services; impact; next year;
weak policing: water cannon; tear gas; batton rounds; policy during G20 protests;

officers afraid of rioters; legal action; robust policing; rioters enjoyed feeling powerful; political correctness; British justice; large youth prison population; European neighbours;

racism: too many black men killed by police; black people treated like criminals; mix of causes; all people shot in last 3 years were white; Macpherson report; stop-and-search; disproportionate; law abiding young black people; confrontational encounters with police; not true now of other police forces; other forces have similar threats;

gangsta rap and culture: hatred culture around rap music; glorify violence; loathing of authority; police and parents; trashy materialism; raves about drugs; like music but don't agree with lyrics; those susceptible;

consumerism: shopping riots; consumer choices; people with nothing; noses rubbed in; stuff they can never afford; looting is taking what you can; small petty things looted;

opportunism: anonymity in the crowd; unlikely to be caught; opportunism made people act abnormally; feeling of invulnerability; part of something big; doing some thing transgressive; feeling powerful; culture of no power;

technology and social networking: social media; organised criminality; high levels of greed; Deputy Assistant Police Commissioner; Metropolitan police; gangs use technology; mobile phones can counteract criminality;

This soup is partially structured by the underlined headings from the BBC article. It is inconsistent and incomplete. For example, it was claimed that $\langle$social media, organised criminality; $R_{\text{ysed}}\rangle$ played a significant part in the riots. This is part of the story. Certainly the behaviour of many involved in the riots was not orchestrated by social media. Inconsistencies can be seen by the claim of a cause being $\langle$too many black men killed by police$\rangle$, while in reality $\langle$all people shot in last 3 years were white$\rangle$. Similary $\langle$stop-and-search; disproportionate; law abiding young black people; confrontational encounters with police; $R_{\text{cause resentment}}\rangle$ is countered by $\langle$not true now of other police forces, other forces have similar threats; $R_{\text{cause}}\rangle$.

These inconsistencies show that people have opinions and beliefs that are not consistent with the available evidence. To what extent is the hypersimplex $\langle$welfare-dependence, social exclusion, lack of fathers, spending cuts, weak policing, racism, gangsta rap and culture, consumerism, opportunism, technology & social networking; $R_{\text{causes}}\rangle$ a true reflection of the causes of the London riots?

Let $\alpha > \beta > \gamma$ be an ordinal scale for measuring these causes under the mapping μ. Suppose the evidence were that the numbers of welfare dependent rioters was about the same as those not dependent on welfare, so that $\mu(\text{welfare dependence}) = \gamma$; that alienation through social exclusion made people predisposed to riot with $\mu(\text{social inclusion}) = \gamma$; that more rioters had active fathers than those who did not with $\mu(\text{lack of fathers}= \gamma$; that there is no evidence that spending cuts caused the riots with $\mu(\text{spending cuts}) = \gamma$; that the initial police response was weak but became better coordinated with μ(weak policing) $= \alpha$; that μ(racism) $= \beta$ because, although the riots were triggered by the police shooting a black man, the subsequent disorder involved people of all races; that the evidence is that gangs were highly active in the riots with μ(gangsta rap and culture) $= \alpha$; that the looting was driven by material desired with μ(consumerism) $= \alpha$; that few of the rioters would initiate the riots so that μ(opportunism) $= \alpha$, and that technology and social networking played a part in exacerbating the riots with μ(technology & social networking) $= \beta$. Then at the α level there is the hypersimplex

$\langle$social exclusion, weak policing, gangsta rap & culture, consumerism, opportunism; $R\rangle$

If it were correct that this is the hypersimplex of the main causes of the riots how might it be used in policy? The hypersimplex of $\langle$weak policing, opportunism$\rangle$ was important early in the riots, suggesting a missed policy response of stronger policing and removing the opportunity for others to get involved. Gang culture was identified as being important and is discussed in the next section. $\langle$consumerism$\rangle$ is deeply embedded in British culture and there is little any government do to change this in the short term. $\langle$social exclusion$\rangle$ is significant policy problem in Britain as it is in many countries.

Social exclusion is a network phenomenon with individuals becoming disconnected from wider society. However, this kind of network problem must be tackled at all levels including the individual's microlevel networks, or lack of them. For example, the networks of unemployed people are likely to include predominantly other unemployed people, creating a substructure that is disconnected from employment and poorly placed to receive information about possible jobs.

This section scratches the surface of the use of network and hypernetwork theory for modelling multilevel social processes and policy. As illustrated, the vocabulary for representing a social phenomenon such as the London riots can be large and it refers to micro, meso and macro levels. The dynamics of the processes that cause social breakdown need to be understood for policy purposes. The rhetoric of "tough love" and "bleeding hearts" needs to be translated into well defined operational terms that can be tested by policy experiment or pre-policy computer simulation. The multilevel formalism of hypernetworks may bring such multilevel simulations closer, but there is a long way to go.

7.23 Example: street gangs in London

The report *Ending Gang & Youth Violence*, published in November 2011 by the Secretary of State for the Home Department, was the basis for a more general policy response to the London riots. "One thing that the riots in August did do was to bring home to the entire country just how serious a problem gang and youth violence has now become" [HMG (2012)].

This report set out detailed policy plans for the agencies to work together, including providing support to local areas; preventing young people becoming involved in serious violence; pathways out of violence and gang culture; punishment and enforcement to suppress the violence of those refusing to exit violent lifestyles, and partnership working to join up local area responses to gangs and youth violence.

The 2012 CSJ report *Time to Wake Up* questions the effectiveness of the police practice of identifying and removing gang elders: "it seems that an unintended consequence of the arrest of senior gang members has been to heighten tensions and violence ... There was a consensus that the current gangs neither have cohesive leadership, which is resulting in increased chaos, violence and anarchy" [CSJ (2012)].

The reports cited discuss the problem of gangs at the macrolevel of state policy, at mesolevels of local authority policy, and at the microlevel of dealing with individuals. For example, *Ending Gang and Youth Violence* documents the life of "Boy X" from his birth to his seventeen year old crack cocaine addicted mother to life imprisonment for murder at age twenty one, and his many contact points with the social and emergency services. This is summarised in Fig. 7.78.

If this were the story of just one person it would be regrettable but not an issue for policy. Figure 7.78 is a kind of model of the life of this individual at the microlevel and his interactions with the mesolevel social and emergency services. Implicitly it is intended to generalise to classes of individuals at higher mesolevels of aggregation. Policy cannot target individuals at the microlevel but requires models of the behaviour of individuals within classes or higher level aggregate entities, *i.e.* hypernetworks.

Many hypersimplices can be abstracted from Fig. 7.78, for example $\langle$neglected by parents, parental substance abuse, parental violence; $R_{\text{experienced } 0-5}\rangle$ suggests a child at high risk. The hypersimplex $\langle$outbursts of aggression at school, involved in street violence, many visits to A&E; $R_{\text{experienced } 10-15}\rangle$ indicates a child becoming increasingly violent and dangerous, while regularly $\langle$late and truant, school low attainment, excluded from school; $R_{\text{experienced } 10-15}\rangle$ is a likely precursor to $\langle$joined a local gang, selling Class A drugs, early and repeat offending, drug and alcohol abuse; $R_{\text{experienced } 16-21}\rangle$.

In many cases subsequent enquiries discover that neighbours, welfare agencies, schools, the police and even the postman had evidence of tragic events to come, but this evidence was not joined up as a hypersimplex. This is an easy conclusion but it is difficult to rectify the lack of structure to prevent further incidents. How can microlevel individuals join up the information spread across their mesolevel organisations? How can institutional structures such as $\langle$welfare agencies, schools, police, postman, milkman; $R_{\text{share and synthesise information}}\rangle$ be formed and run at politically bearable costs?

244   *Hypernetworks in the Science of Complex Systems*

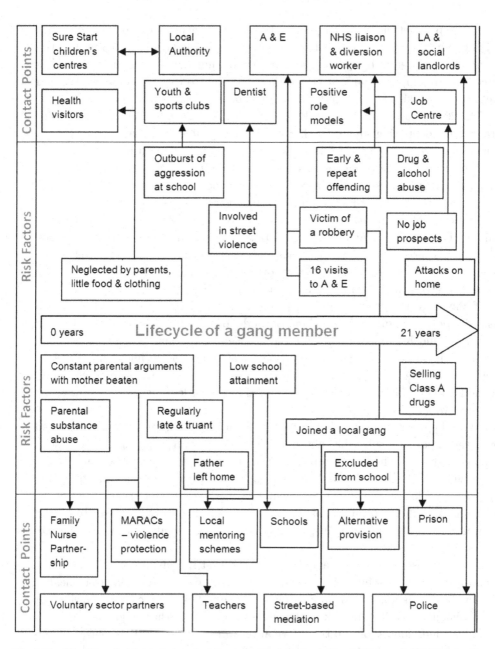

Fig. 7.78   The lifecycle of a gang member (Source: "Ending Gang Youth Violence", [HMG (2012)])

The reports cited above contain propositions on the behaviour of gang members. These include postulating classes of Elders and Youngers and their microlevel interactions, and classes of rival gangs and their interactions at higher mesolevels. Much of this is empirically based and forms a theory of gang behaviour on which to base policy. What can formal modelling with hypernetworks add to this?

Social policy is inevitably expressed in natural language within a legal framework for implementation. In comparison, technical or formal models of systems are stated in their own language which may include mathematics and computation, but they are always embedded in a meta-language such as vernacular English.

The problem with the theories in the reports cited above is that their vernacular models are untestable before implementation. For example, the failure of the 2011 police policy to remove Elders from gang was *predicted* in 2009 [CSJ (2009)] but this was ignored or overlooked by the police. Why? Perhaps due to predictions presented in natural language carrying little more weight than opinions because the outcome of their premises and logic cannot be demonstrated before empirical testing on the street.

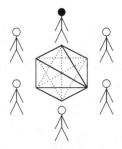 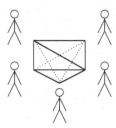

(a) with a head the gang structure is stable   (b) without a head it is unstable

Fig. 7.79   "... an unintended consequence ... has been to heighten tension and violence ..."

Figure 7.79 shows a structure in which the leader of the gang plays a role coordinating the other members. As [CSJ (2009)] suggests, removing the gang leader can disrupt a stable structure making it unstable and leading to an increase in undesirable traffic of violence and lawlessness.

There seems to be a policy gap between leaving such lawless structures to continue, and invoking policies that may have short term costs but remove the entire undesirable structure and replacing it with a more desirable structure. It is a gap of imagination: how can these form productive and satisfying hypersimplices?

Technical models can translate vernacular models into formal theories that can be implemented in computers to generate their logical and empirical consequences. Given the many possible initial conditions such models must be run many times to characterise the space of possible policy outcomes. For policy purposes the output of the computational model will again be evaluated in a vernacular meta-language, but the intermediate step of computation can add a lot in terms of understanding the real social system and its dynamics.

To the best of my knowledge there is no research investigating how deviant young people (*Level N*) can form hypernetwork structures at *Level N+1* (social groups, or benign gangs) and groups of groups (*Level N+2*) that they find satisfying and constructive. This is an opportunity for researchers and policy makers.

Chapter 8

Time, Events, Prediction and Forecasting

8.1 Prediction

The natural sciences have put a high value on prediction, as a way of legitimising their theories and as the basis of applications. This appears in an extreme form in the writings of Laplace in 1814:

> We may regard the present state of the universe as the effect of its past and the cause of its future. An intellect which at any given moment knew all of the forces that animate nature and the mutual positions of the beings that compose it, if this intellect were vast enough to submit the data to analysis, could condense into a single formula the movement of the greatest bodies of the universe and that of the lightest atom; for such an intellect nothing could be uncertain and the future just like the past would be present before its eyes. [Laplace (1814)]

Today this view is not tenable. There is no formula that can capture everything, and even if there were it is now known that it could not give predictions in the way Laplace proposed.

In this chapter it will be explained why this view of prediction is not applicable in the science of complex systems. The reasons include the amplification of tiny but inevitable errors in measurement and interacting feedback loops.

Beyond this it will be shown that there are different kinds of time. Apart from the familiar time of physics there are other kinds of time defined by system events, defined to be the formation and transformation of hypersimplices. Some kinds of prediction can then be seen to be the problem of aligning system events in clock time, but other kinds of prediction are possible and highly relevant to the practical problems of designing, constructing and managing complex systems.

Prediction becomes even more problematic in the context of multilevel systems. Does a prediction mean knowing the future states of every part of the system at some future point in time? Or is there emergent robust predictability at some levels despite highly unpredictable states at lower or higher levels?

Prediction is an essential part of life. At the microlevel, individuals do things because they expect particular outcomes. At the mesolevel, organisations plan future strategies on expected intermediate events and expected outcomes. At the macrolevel, governments set fiscal and social policy on predictions of their outcomes. If prediction cannot be done using formulae alone then how can it be done?

An essential first step in prediction is to understand the dynamics of the system. For example, it is becoming better understood that human behaviour involves network phenomena, and that it can be impossible to understand the behaviour of individuals in the absence of knowledge of relational structure of their networks and hypernetworks. As Atkin showed many years ago, the behaviour of the traffic depends on the topology of the backcloth.

Related to this, it is now understood that some system behaviour *emerges* from the *discrete* interaction of autonomous agents at lower level, as shown in Fig. 8.1. These discrete interactions *generate* continuous patterns of numbers at higher levels. Although the precise behaviour of the individual may be unpredictable, the aggregate behaviour may show regularities.

Generally extreme events cannot be predicted from the high level data, they emerge bottom-up from lower level interactions. Although extreme events may be unpredictable at the macrolevel, they may have *precursors* at lower levels.

As Fig. 8.1 shows, higher level phenomena impose top-down constraints on lower level interactions. For example, macro policy may aim to shift economic activity from public to private sectors at meso levels with individuals at microlevels changing their network behaviour due to redundancy and job seeking.

How can the impact of such policies be predicted if the macrolevel behaviour depends bottom up on the microlevel activity? The best answer produced so far is *agent-based computer simulation* where the interactions between individuals are computed based on assumptions about individual (as vertices) and their interactions (in networks and hypernetworks). In the next chapter this will be put in the context of "policy informatics" where computers routinely make 'predictions' in support of policy, as they have done for many years.

The rest of this section will be devoted to making the concept of prediction better defined and showing that the deterministic point-predictions of physics can at best be replaced by stochastic interval predictions for social systems.

Point-predictions and interval predictions

Let a prediction be defined to be a proposition of the form: "if a system is in state s at time t and an action or intervention a is applied to the system at time t, then the system will be in state s' at some future time within a time interval T. If T is a single point in time, $T = \{t'\}$, the prediction is called a *point-prediction*. If T is an interval $[t_1, t_2]$ the prediction is called an *interval prediction*.

Physics gives point-predictions, *e.g.* if you drop an object from a known height you can calculate the precise point in time at which it will hit the ground.

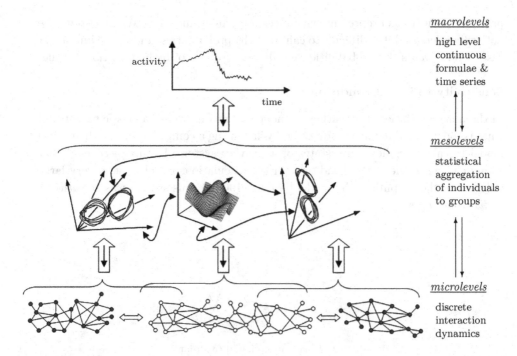

Fig. 8.1 Prediction involves bottom-up and top-down dynamics in complex systems

In contrast, an interval prediction asserts that something will happen within some future interval. For example, "if you are reading this sentence now you will have completed reading it within the next minute" is a prediction in the interval [now, now + one minute]. It will be empirically true for most readers.

The recent interval prediction that: "without public money the global financial system will fail within days", was more instinct and belief than science. It became the prediction that "with (lots of) public money the financial system will not fail within days", which seems to true for the time being.

The interval prediction that, "if greenhouse gas emissions are not capped at the current level the climate will change irreversibly by time t", has been highly contentious. One problem is that t seems to be a long way in the future, in the order of decades. Related to this is the interval prediction that, "irrespective of human activity the climate will change irreversibly between now and t". Since the first prediction was made some decades ago the pattern of extreme weather events, melting glaciers, rising sea levels, and floods has become more immediate to policy makers.

These examples show that predictions can be vague about when things will happen, they can be stated in imprecise ways, and there may be no practical way of testing them. In fact, some predictions are even more vague by asserting the future state is "likely" or "expected but not certain". When systems are well understood probability theory can represent such uncertainty in useful ways. When systems are not well understood it is difficult to calibrate the probabilities, and in the important case of rare events the probabilities are almost zero and of little operational value.

Sensitivity to initial conditions

In the nineteen sixties the weather scientist Lorenz discovered a classic example of what is today called *deterministic chaos*. When using a computer and mathematical model to calculate the future states of a weather system, Lorenz discovered that a tiny change in the initial conditions of the calculations could make a very large different to the outputs of the model [Lorenz (1963)]. This system was *sensitive to initial conditions*.

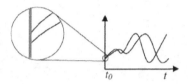

Fig. 8.2   Sensitivity to initial conditions

This is illustrated in Fig. 8.2 where a small difference in the initial conditions at t_0 can result in the system diverging considerably. In the short term the trajectories may be similar and good approximations to each other, but after some time the trajectories are completely different. If the dynamics of a system are sensitive to initial conditions, there is a time horizon beyond which predictions have no value.

Since all measurement has error, even with a perfect model, computing the future states of the system from the "same" initial conditions will result in wide variations when the system is sensitive to initial conditions. Lorenz showed that even deterministic physical systems may not be point-predictable in the conventional sense that the system will be in a particular state at a particular point in future time.

The weather provides an instructive insight into prediction for social purposes. Weather forecasting is extremely important in making public and private plans. In some places the weather is highly predictable, *e.g.* "it will not rain here for a month, then after a month or so there will be a monsoon". In other places the weather is highly unpredictable on a short timescale, *e.g.* "it may rain in the next hour, or it may not" but be highly predictable on a longer timescale, *e.g.* "... but it will definitely rain by the end of the day". Such predictions are useful because they give a range of possible events that can be taken into account when planning activities.

8.2 Feedback, simulation and prediction

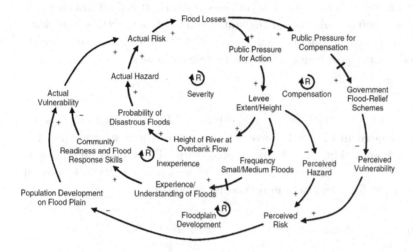

Fig. 8.3 Feedback loops in a causality network (Source: [Wasson (2002)])

Figure 8.3 shows a model of a river flood management system.

> If you look just at the centre of the following diagram, if you take the levees-only policy of dealing with flood containment on rivers, you have flood losses at the top, which are largely economic losses; this produces public pressure for action, which produces levee construction or enlargement of levees in many places, including Australia. Nyngan is the case study for this one. Levee construction actually increases river height when the river goes over-bank, so the flooding actually is worse as a result of the levees a paradoxical result. Probability of disastrous flood goes up; the actual hazard goes up; the actual risk goes up; the flood losses actually go up as a result of levee construction. This is a positive feedback loop. [Wasson (2002)]

There are various other loops in this system with negative arrows that decrease things, for example the Perceived Risk may decrease Population Development on the Flood Plain. In fact the only arrow entering the positive feedback loop at the top centre is the arrow from Actual Vulnerability on the right, which may be decreased by Community Readiness and Flood Response Skills, and a possible decrease in Population Development on the Flood Plain.

Could this model be used to predict the future behaviour of the system? Suppose that values could be assigned to each of the vertices in this diagram, measuring the Flood Losses, Public Pressure for Action, and so on. Suppose also that each of the arrows could map the values on its out vertex to a change value on its in vertex. Then, in principle, initial values could be assigned to each of the vertices and the dynamics of the system could be simulated in time.

When run many times with many combinations of initial conditions this model might suggest that there are fixed points in the system, more or less benign, and it might suggest that some cycles in the system are attractors. It is doubtful that this model could give a point prediction of the states of any of the vertices, especially since it is likely that such a model is sensitive to initial conditions, but it is possible that such a model could give insights into the system dynamics.

System Dynamics

Computing interacting feedback cycles is the basis of Jay W. Forester's System Dynamics as illustrated in Fig. 8.4 [Forrester (1990)]. In the first order case the inventory increases smoothly towards the desired state. The second order case is more complicated with interacting values of the Inventory, Order Rate, Goods on Order, and Receiving Rate resulting in oscillating behaviours.

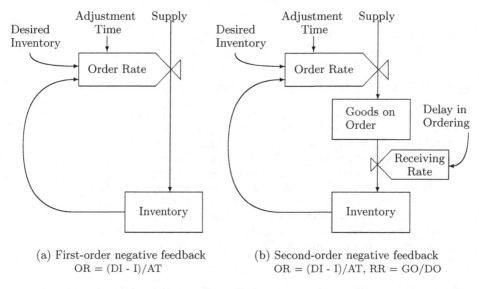

Fig. 8.4 Forrester feedback systems

Soft Systems

Feedback cycles are common in Checkland's Soft Systems methodology [Checkland & Scholes (1990)], typically in the processes of organisations and interactions at different levels: "Soft Systems Methodology was developed in the 1970s. It grew out of the failure of established methods of "systems engineering" when faced with messy complex problem situations ... [it] was developed expressly to cope with the more normal situation in which the people in a problem situation perceive and interpret the world in their own ways and make judgements about it using standards and values which may not be shared by others" [Checkland & Scholes (1990)].

Time, Events, Prediction and Forecasting

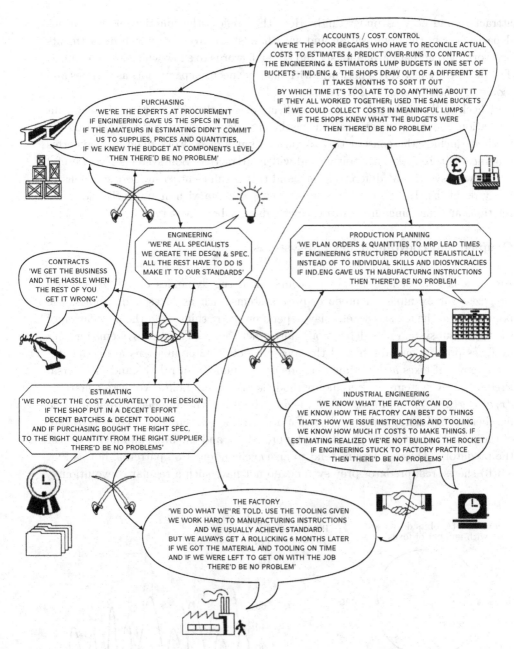

Fig. 8.5 A rich picture of a situation in an engineering company ([Checkland & Scholes (1990)])

Figure 8.5 shows a typical example of a "rich picture" as used in the method, where some of the salient features of an organisation have been abstracted together with some of their relationships. The inclusion of graphic icons makes the pictures attractive and easy to understand. Here they reflect the function of the various departments in the organisation and with handshakes to show which departments are working together harmoniously and crossed swords to show which are in conflict. The picture is effectively a network, with heterogeneous departments as vertices and (some of) their interactions as edges.

Checkland's methodology is very effective for collecting and synthesising information about organisations and has been widely and successfully applied. Its methods include the construction of various kinds of network diagrams, and this graphical nature makes it intuitive and widely attractive. Soft systems give a useful overview of systems at different levels, and this is sufficient to support solutions to many problems. However, when systems get very large with many levels and many relations and mappings more formal methods may be necessary.

Coupled and coevolutionary subsystems

An obvious kind of feedback loop occurs when the dynamics of subsystems are *coupled*. For example, in predator-prey systems such as foxes and rabbits, the population of the fox subsystem may depend on many things and the population of the rabbit subsystem may depend on many other things. These subsystems interact by foxes preying on rabbits and the dynamics of their populations are coupled – the number of foxes at any time depends on the number of rabbits and vice versa. Predator-prey systems are commonly modelled using the Lokta–Volterra equations $dx/dt = x(\alpha - \beta y)$ and $dy/dt = -y(\gamma - \delta x)$ where x is the number of prey, y is the number of predators, and α, β, γ and δ are parameters representing the birthrate of the prey, the number of prey taken by predators, and the birth and death rate of the predators. These equations tend to give cyclic population patterns, but as Fig. 8.6(b) shows, real predator-prey systems do not have such a regular coevolution.

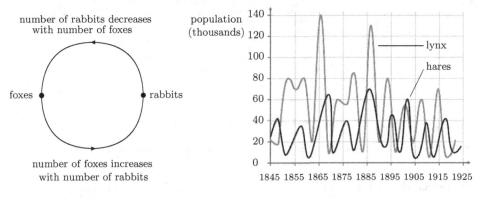

(a) coupled subsystems   (b) coupled populations of snowshoe hares and Canadian lynx

Fig. 8.6 Coupled subsystems have coevolving dynamics

In real predatory-prey systems there are many more factors and variables than accounted for by the Lokta–Volterra equations. Coupled subsystems may make a system unpredictable, but in some systems the dynamics work towards synchronisation, *e.g.* Huygens observed synchronisation between pendulum clocks supported by the same wooden beam.

In hypernetworks, subsystems may be coupled in many ways. For example, the traffic on the system at some level may be coupled by q-transmission, or it may be coupled by aggregation or disaggregation. Whether or not interacting dynamical processes over hypernetworks makes them less or more predictable is a research question to be addressed.

Agent-based systems and prediction

Agent-based systems can also be considered to be feedback systems. For example, consider two chess players, a and b after a has made a move. Although b may have guessed the move, it is possible that an unexpected move is made. Then b makes a move. In this way the game evolves, more or less unpredictably, as an event-driven trajectory as the two players interact. At tournament level no-one can predict which particular trajectory a game may take.

Here a very general interpretation will be given to the term "agent", namely something that either exists on its own or exists in the context of other agents in an environment. An agent-based system is a set of agents with some interacting some of the time.

An agent-based model can be considered to be a way of representing the state of individuals and rules for changing states depending on network connections. The states or rules in any particular case may be "realistic", but even if they are, it does not guarantee predictable outcomes.

The generality is that at the microlevel social systems have network dynamics. The actions of any human agent at any time depends on their previous interactions with other agents. It is now becoming more apparent that individuals *copy* the behaviour of others.

As with chess, the actual trajectory of any agent can be unpredictable, and the microlevel trajectory of the whole system can be unpredictable. At higher levels of aggregation the system traffic may become more regular. Agent-based modelling provides a way of generating the higher level traffic from lower level interactions.

Simulation and prediction

Generally to investigate the outcome of agent interactions requires the use of computers to "simulate" the emergent interaction-generated system behaviour . Taking a broad definition of agent means that any almost any network simulation can be considered to be agent-based, since the vertices can be regarded as agents.

Agent-based models are used in a wide variety of computer simulations. Whether or not agent-based simulations give predictions is the subject of debate. Many agent based simulations are sensitive to initial conditions, so changing things just a little may give very different outcomes. Therefore the best that can be done is to investigate systems is to run simulations many times with an "appropriate" set of initial conditions. In this way it may be possible to understand the space of possible future states of a system, but clearly to claim that an agent based simulation gives meaningful predictions that can be safely used requires justification.

Many simulations make no claim to be models of real things, and although they investigate the behaviour of systems in time, their results are not intended to be predictions of real systems. For example, the classic Schelling model shows that, based on minimal preference assumptions for having neighbours of one colour rather than another, the pixels cluster on computer screens [Schelling (1971)]. This is not a model for ghetto formation in cities. It demonstrates the existence of a system in which slight preferences can cause clustering which might apply to human systems.

Probability and prediction

When point predictions are impossible it may be that one can predict the probability distribution as opposed to the detailed state. However this assumes that the space of possibilities is known, which may not be the case for self-reconfiguring complex systems. Also, there may be rare events with probabilities close to zero.

The limitations of prediction

This short discussion of prediction has shown that in social systems it is very rare that one can make testable point predictions, and that even interval predictions are limited to assertions that some thing might happen in some future time window. The remainder of the chapter will discuss the nature of time in more detail as a necessary precursor to representing the dynamics of complex systems.

8.3 Time and Events

It has long been known that the time of physical systems and the time of social systems are different, *e.g.* in social situations time can fly or time can drag with respect clock time. Given the successes of the physical science it is commonly assumed that clock time is the correct time and any misalignment between clock time and subjective time reflects a fault in the individual or the group.

In the nineteen seventies Atkin suggested that systems have another kind of time measured by multidimensional *events*, a theory developed in his book "Multidimensional Man" [Atkin (1981)]. Atkin's ideas are central to the hypernetwork theory of dynamics.

Clock time

The concept of structural event allows moments of time to be marked by appropriate systems events, such as the tick of the clock for physical systems, and the formation of events in social systems. In his "Principia" [Newton (1687)] defined time as follows:

> Although time, space, place, and motion are very familiar to everyone, it must be noted that these quantities are popularly conceived solely with reference to the objects of sense perception. And this is the source of certain preconceptions; to eliminate them it is useful to distinguish these quantities into absolute and relative, true and apparent, mathematical and common ... Absolute, true, and mathematical time, in and of itself and of its own nature, without reference to anything external, flows uniformly and by another name is called duration. Relative, apparent, and common time is any sensible and external measure (precise or imprecise) of duration by means of motion; such a measure – for example, an hour, a day, a month, a year – is commonly used instead of true time.

Atkin called this *clock time*. To make this operational, Fig. 8.7 shows the observation of a moving pendulum. The observer can mark an instant of time each time the bob is instantaneously stationary as it reverses direction from left to right as Event 1, Event 2, and so on. Let each of these events be called a "tick". For the pendulum the right-to-left reversals could also be observed as "tock" events. In physics it is assumed that the *interval* or *duration* between tick events is the same. In fact it is assumed that time is in one-to-one correspondence with the real numbers under a mapping t where $t(\text{Event } i)$ is the time at which Event i occurs, and $t(\text{Event } i+1) - t(\text{Event } i) = t(\text{Event } j+1) - t(\text{Event } j)$ for all tick events i and j.

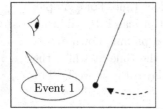 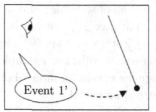 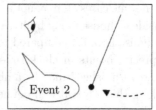

Fig. 8.7 Clock time is measured by physical events

Thus, given an oscillator such as a pendulum or quartz crystal, duration is *defined* to be the time between successive cyclic events such as the ticks. Clock time is usually important in the dynamics of social systems, but there are other important concepts of instantaneous time and time intervals.

Events in system time

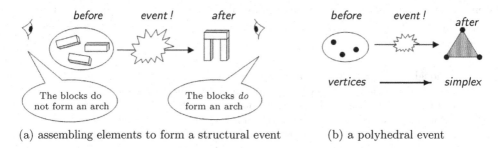

(a) assembling elements to form a structural event (b) a polyhedral event

Fig. 8.8   System time is measured by system events

Figure 8.8 shows how the formation of a hypersimplicex can mark an instant in time. Here it is possible to observe that the blocks are *not* related under R_{Arch} and it is possible to observe that blocks *are* related under R_{Arch} and this provides an operational definition of "before" and "after" for the event of building the arch.

Let $t_{\text{before},i}$ be the observation at clock time t_i that the blocks are not related under R_{Arch} and let $t_{\text{after},j}$ be the observation at clock time t_j that the blocks are related under R_{Arch}. Then the event of forming $\langle b_1, b_2, b_3; R_{\text{Arch}} \rangle$ is sandwiched between t_i and t_j.

Let σ be a hypersimplex, and let $\mathcal{E}_\sigma$ be the event that σ has been formed and $\not{\mathcal{E}}_\sigma$ the event that σ has ceased to exist.

Let $t_{\text{before}}\mathcal{E}_\sigma$ be the latest known clocktime before the event $\mathcal{E}_\sigma$, let $t_{\text{after}}\mathcal{E}_\sigma$ be the earliest known clocktime after the event, and let $t\mathcal{E}_\sigma$ be the clocktime at which $\mathcal{E}_\sigma$ actually occurs. Then $t_{\text{before}}\mathcal{E}_\sigma < t\mathcal{E}_\sigma < t_{\text{after}}\mathcal{E}_\sigma$.

In some cases $\mathcal{E}_\sigma$ may not exist at an observable point in clock time. For example, consider a house being built. At what precise point in clock time can it be said that the house is built, compared to an instant previously? Often pinning down events to precise points in clock time is not useful. For example, the time at which the house is "finished" may be ambiguous within an interval of days or even weeks, but making this interval well defined can be useful.

The lives of most people are marked by significant events such as learning to read, $\langle x, \text{reading}; R_{\text{can}} \rangle$, joining the football team, $\langle x_1, x_2, ..., x_{11}; R_{\text{football\_team}} \rangle$ falling in love, $\langle x, y; R_{\text{loves}} \rangle$, learning to drive a car, $\langle x, \text{driving test}; R_{\text{pass}} \rangle$ becoming qualified $\langle x, \text{certificate}; R_{\text{awarded}} \rangle$, getting married, $\langle x, y; R_{\text{married}} \rangle$, having a child $\langle x, y, z; R_{\text{baby\_born\_to}} \rangle$, getting a job $\langle x, \text{job}; R_{\text{contract}} \rangle$, publishing a paper, $\langle x, \text{paper}; R_{\text{published}} \rangle$, and so on.

The *dimension* of an event is the dimension of its relational hypersimplex, and Atkin defined a *p-event* to be an event with dimension p. For example, the dimension $\langle b_1, b_2, b_3; R_{\text{Arch}} \rangle$ is two, and this is a 2-event.

8.4 Mapping system time to clock time

In real systems clock time may be important, but system events usually constrain the dynamics. For example, infrastructure projects such as building a bridge or a dam may be scheduled to be used on a given day, but the actual day on which they are first used will be the day that they are ready to be used. Generally clock time matters because the costs of the project are related to clock time: people are paid monthly, equipment is hired by the day, interest is paid by clock time, and the income or benefit of the project is experienced in clock time. Thus the planning, design, and management of systems involves scheduling the construction of sequences of p-events in system time in a way that is coherently linked with clock time (Fig. 8.9).

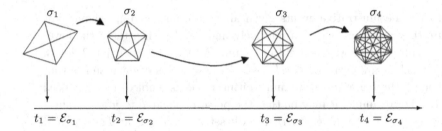

Fig. 8.9 Mapping polyhedral events into clock times

Let $\langle x_1, x_2, ..., x_p; R_{\text{committee}} \rangle$ be a committee and consider the event $\langle x_1, x_2, ..., x_p, y; R_{\text{committee\_agrees}} \rangle$ be the p-event that the committee agrees on y. Depending on y, the committee members may have different initial opinions and spend some time suggesting and discussing its pros and cons, before agreeing one way or the other on accepting it or putting its acceptance to the vote. In general, when committees are small such discussions take relatively short periods of time, but as committees get larger their deliberations seem to take proportionately longer. If each of the p committee members spoke for the same time, t minutes during a discussion then it would take $p \times t$ minutes. Extending the committee by one member involves an extra t minutes. However, suppose that each member of the committee felt they must say something in response to the main pronouncements of the others lasting on average t' minutes. Then the discussion would take $p \times t + p(p-1) \times t'$ minutes. Adding an extra member to the committee now takes $(p+1) \times t + (p+1)p \times t'$ which is an extra $t + [(p+1)p - p(p-1)] \times t' = t + 2p \times t'$. If t = 5 minutes and $t' = 2$ minutes when there are four members of the committee the discussion time is $4 \times 5 + 12 \times 2 = 44$ minutes. Add an extra person and it becomes $5 \times 5 + 20 \times 2 = 65$ minutes. Add an a sixth person and it becomes $6 \times 5 + 30 \times 2$ = 90 minutes. A seventh person would take the discussion time to $7 \times 5 + 42 \times 2$ = 119 minutes. Of course, most people are disciplined in committees and don't exercise their right to comment on what every other speaker says, but here it can be seen that as the dimension of the committee increases its deliberations can take disproportionately longer.

Some committees such as a parliament or the ruling body of a university have hundreds of members. To constrain the time spent debating things there are usually rules constraining who can make a speech, for example, a minister and a shadow minister, and rules that allow the chair to restrict the number of people who can comment on the main speeches. Without these constraints debates could go on for many hours, and even with them committee members may filibuster with the intention of frustrating a vote and a decision.

8.5 Multilevel events

It may seem that the illustrative events given above are all low-dimensional, but this is because they are expressed at a relatively high level with little of the lower level details. Figure 8.10 shows the event of forming a team, $\mathcal{E}\langle$ job-1, job-2, job-3, job-4; $R_{\text{team}}\rangle$ which is a 3-event at *Level N+2*. However, this event assumes that all the members of the team are ready and waiting to be assembled, but this need not be the case. For example, it may be that the person allocated to job-1 requires an extra skill, skill-4, to be added to their skill set as an event at *Level N* and integrated with their other skills as another 3-event at *Level N+1*.

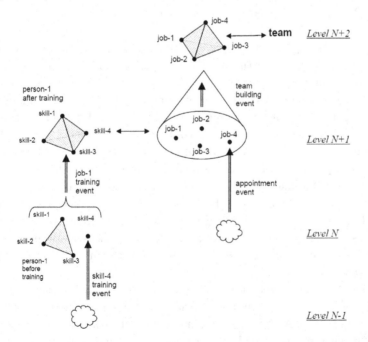

Fig. 8.10 The formation of polyhedra marks multilevel events in system time

In his book "The Sciences of the Artificial", [Simon (1965)] gives the following story in a discussion of the evolution of complex systems:

> There were once two watchmakers, named Hora and Tempus, who manufactured very fine watches. Both of them were highly regarded, and the phones in their workshops rang frequently – new customers were constantly calling them. However, Hora prospered while Tempus became poorer and poorer and finally lost his shop. What was the reason?
>
> The watches the men made consisted of about 1,000 parts each. Tempus had so constructed his that if he had one part assembled and had to put it down – to answer the phone, say – it immediately fell to pieces and had to be reassembled from the elements. The better the customers liked his watches, the more they phoned him and the more difficult it became for him to find enough uninterrupted time to finish a watch.
>
> The watches that Hora made were no less complex than those of Tempus. But he had designed them so that he could put together subassemblies of about ten elements each. Ten of these assemblies, again could be put together into a larger subassembly; and a system of ten of the latter subassemblies constituted the whole watch. Hence when Hora had to put down a partly assembled watch to answer the phone, he lost only a small part of his work, and he assembled his watches in only a fraction of the man-hours it took Tempus.

Simon gives an argument that it takes Tempus about four thousand times as long to make a watch as Hora. This is because, in our terms, for Tempus constructing a watch is a single 999-dimensional event while for Hora it is a sequence of a hundred and eleven 9-dimensional events.

This example makes clear that a divide and conquer strategy, when it can be applied, involves the creation of hierarchical structure. Paradoxically, creating new multilevel structure *reduces* the complexity of the system, in terms of making it more manageable, at the expense of making the representation more complicated.

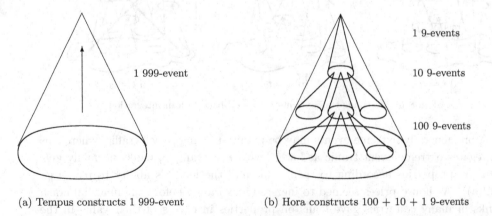

(a) Tempus constructs 1 999-event (b) Hora constructs 100 + 10 + 1 9-events

Fig. 8.11 Tempus requires one 999-event to make a watch while Hora requires 111 nine-events

8.6 Rare and extreme events in multilevel systems

Extreme events are characterised by behaviour outside the normal range of parameters or significant changes in structure. By this definition extreme events are rare and cannot be predicted by extrapolating from normal behaviour. Extreme events include severe earthquakes, financial crashes, extinction of species, and revolutions. Although point-unpredictable, extreme events may have *precursors* enabling interval predictions.

An extreme event in a multilevel system can be a change in the high level traffic of the system, as suggested by the graph at the top of Fig. 8.12. For example, in 1991 the British businessman Gerald Ratner made a speech at the Institute of Directors at which he jokingly denigrated some of the jewellery sold by his company. This was reported in the media and the share value of the company dropped by £500 million. This was a rare and much regretted event with the catastrophic outcome that the company lost value outside anything ever experienced before. In this case the business survived in the short term before changing its name to Signet Group in 1993. By rebranding the backcloth the company has reduced the negative reputation traffic and made the company more disposed to higher volumes of sales traffic.

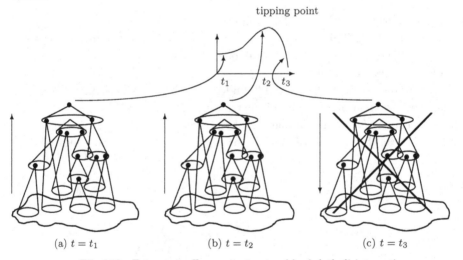

Fig. 8.12 Extreme traffic events structural backcloth disintegration

More generally extreme events may be manifest in high level traffic, where this aggregates extreme event traffic at lower level. For example, banks normally give loans on properties according to their value and the lender's ability to pay (Fig. 8.12(a)). As house prices seemed to increase inexorably, before the financial crash banks in many countries gave loans on properties in excess of their value in the expectation that house prices would continue to increase (Fig. 8.12(b)). Individual loans were bundled up, certificated and traded within the financial institutions at higher levels. Of course, with hindsight it can be seen that the property bubble burst and house prices collapsed. However, not only did the numbers in the high level time series collapse, deep structural changes occurred in the backcloth (Fig 8.12(c)). In some cases financial institutions failed, and many individuals lost their homes (backcloth) and investment in it (traffic). In many countries the housing backcloth has collapsed with hypersimplices being destroyed at many levels.

The perspective of multidimensional events in multilevel systems is important for rare and extreme events. First it raises the question as to what level the "triggering" events happen. Secondly, depending on their dimensions, structural events can take significant clock time to form. This means that there many structural precursors for extreme events with the possibility of monitoring appropriate parts of the system in order to detect them before they happen in clock time.

8.7 Tipping points, nudges, and multilevel cascades

A *tipping point* is an interval during which the some system states change from being possible to impossible or change from being possible to inevitable. For example, there is concern that greenhouse gas induced climate change may irreversibly alter the Gulf Stream which would have profound social and economic impact, but it is not known if the CO_2 level has gone beyond a tipping point for this.

As another example, consider the transmission of diseases or information where the number of people infected or informed may get to a level beyond which nothing can stop transmission to the whole population. In these cases tipping points may be associated with positive feedback cascades. If a system has tipping points, at what levels do they occur in multilevel systems? Although the tipping point may be imagined to be a high level "point of no return" on a graph as in Fig. 8.12, this may be an aggregate effect rather than a disaggregate cause.

[Thaler & Sunstein (2009)] suggested that decision makers can influence system behaviour by small and inexpensive *nudges*, possibly by pushing the system beyond or away from a tipping point, or setting the system in a direction more consistent with policy. Again, in the context of multilevel systems, the level of the nudge can be important, as can unexpected consequences at higher and lower levels.

Although concepts such as tipping points and nudges are useful in applications, their use can be metaphorical rather than technical. In systems that can be managed without explicit data and models these metaphors may be effective. However in more complex systems this will not be the case. For example, where is the tipping point to address the obesity epidemic? What nudges can be applied to create jobs for young people, or to establish long term care for an ageing population?

How can the effect of nudges be predicted? In principle all policy interventions including nudges can be investigated using computer simulations. However, testing nudges made in complex multilevel systems requires the system to be modelled explicitly in the context of "policy informatics" as discussed in the next chapter.

8.8 Time horizons

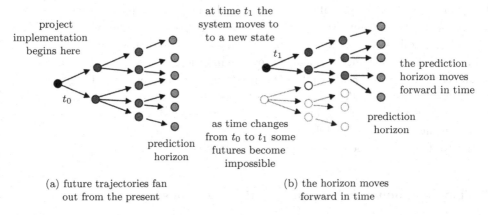

Fig. 8.13 Time horizons in predicting the future states of systems

At any point in clocktime a system may have many possible futures as shown in Fig. 8.13(a). Here it is supposed that the implementation of a project begins at t_0, and that there are two possible future states. As the clock moves on to t_1 one of these states is *realised* and the other state is *released*, along with some of its once-possible but now impossible future states.

In this context prediction moves from saying what *will* happen to what *might* happen, including making the "unknown unknowns" into "known unknowns" which is very important for managing extreme events.

The connectivity of events

Figure 8.13 characterises system states as vertices at the highest level of representation. It should be realised that each of these states is a snapshot of the multilevel system, and that there will be connectivity at lower level. When planning a trajectory of system events it can be useful that some events are highly connected, since this may enable the evolution of the trajectory to be more robust. If one event becomes impossible to realise a highly connected event may be acceptable.

8.9 Time, prediction and events in the science of complex systems

For the reasons discussed in this chapter there is a horizon of systems states beyond which it is impossible to predict anything beyond the possibility of events happening. At each tick of the clock a new system state is realised, many once-possible states are released, and the prediction horizon advances a little further into the future.

This dynamic of predictability is particularly important in the context of policy, which in the next chapter is defined to be *designing the future*. In other words, policy involves making explicit the fans of events from where we are now to where we might want to be – and want not to be – in the near, medium and distant futures.

8.10 Example: Real Time Composition

For some years the world renown choreographer João Fiadeiro has been developing his *Real Time Composition*, researching new formats of encounter, collaboration and exchange. The method suggests that the quality of the decisions we take, individually or in groups, can be improved by excluding, as far as possible, the impact of personal will at the moment of decision. This approach requires individuals to question their habits and patterns of behaviour and increases the potential of social cognition. In the context of bottom-up self-organisation, the method gives deep insights into emergence and the enhanced role that individuals can play in the outcome of group interaction.

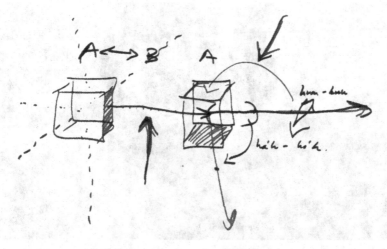

Fig. 8.14 João Fiadeiro's diagram illustrating Real Time Composition

Fiadeiro's research involves trained dancers improvising to create new sequences of collective movement in an evolving environment. For this an "arena" is marked out, typically a rectangle with sides about ten metres. Initially the dancers are outside this space until one or more of them initiate the sequence. For example, a dancer may place a chair in the middle of the space and slowly walk out of the arena. The other dancers are required to contemplate the scene, thinking of ways to develop it. For example, another dancer might place another chair in the arena next to the first. If one or more dancers repeated this the sequence could become boring as the line of chairs grew. However, the method requires the dancers to maintain continuity with what has gone before. This could be done, for example, by a dancer placing a bag instead of a chair, thus maintaining the line while breaking the fixation on chairs. This is illustrated in Fig. 8.14 where the cubes represent the arena at two consecutive times with its faces corresponding to future directions. For example the introduction of the bag takes the development in a new direction represented by the line at right angles from the previous horizontal direction.

> In Real Time Composition there are two things going on: on one hand there is the technique, which is rooted in the idea of 'unpacking decisions' and on the other hand there is what is guiding our decisions. And for that you have different norms to be followed: the norm of "leaving things open for the next person"; the norm of 'not getting obsessed and dragging things into loops'; the norm of 'being part of the community'; the norm of 'leaving as much ego behind as possible'; the norm of 'not getting attached to any particular narrative, past history'; etc. By engaging in an exploration of decision and moral judgment, trough the work of Real Time Composition, you are trying to be fully human. (John Symons in a conversation with João Fiadeiro, http://www.eccs2010.eu/knowbody)

Fig. 8.15 An example of João Fiadeiro's Real Time Composition: *Big Square Example II* (http://www.youtube.com/watch?v=KqFjaQ1PNFs&feature=related)

> João Fiadeiro works exactly with the matter of the 'in-between' and his method is based upon the challenge to produce, by cultivating molecular clarity, a re-assessment of what freedom in improvisation might be, as well as of what the creativity of the artist might be. João has, so to say, woven a whole philosophy of the event applied to dance. His method, encompassed by the dense-light delicacy of simple things, does not deal with anything other than life; it is just a clarification of vital functioning, of the operative dynamics of human relations, of cohabitation. Clarification is by the way a key word to understand his proposal,

which consists in the shared making of an event/action/scene, starting with the inhibition of the I-artist's impulse to act for a 'certain time', enough for the possibilities that the I matches with the action to emerge, and for the very event to ask for the participation of that very I – whose creativity then stops to be associated to the romantic splendour of the outburst-artist, who adds elements to the real or that shapes it according with her own point of view and subjective interpretation, to appear instead as the art of putting oneself at the service of the event, and of choosing with clarity, working on the elasticity of the observable in the name of the weaving of a common experience. The protagonist-artist is no longer to be seen; the only protagonist is the event itself. (Fernanda Eugánio, http://www.eccs2010.eu/knowbody)

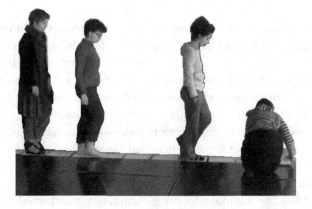

Fig. 8.16 Participant interaction in Fiadeiro's *Real Time Composition*

Fiadeiro's method requires the individual dancers to suppress their own ego in favour of a good outcome for the group. Thus the dancers are required to think of a variety of moves responding to the current state of the arena, rather than blunder in with the first idea that enters their head. Furthermore, a sequence that is developing well can be ruined by an inappropriate action. If the actions are all the same for too long the sequence "loops" and becomes uninteresting. However, an action that is completely disconnected from what went before can break the sequence. Thus the dancers are subject to competing forces, both trying to think of creative actions that will enhance the sequence while knowing that a badly thought through action could spoil the sequence for the whole group.

Although the method does not result in success every time, it can generate some remarkable results. Figure 8.15 show a late stage in a sequence that began with a neat pile of A4 paper in a corner of the arena. One of the dancers began laying out the sheets around the edge, and some of the other dancers followed her, stepping from sheet to sheet. At some point one of the dancers joined in by scrunching up some of the papers as illustrated in Fig. 8.17.

268 Hypernetworks in the Science of Complex Systems

Fig. 8.17   An emergent configuration in João Fiadeiro's *Real Time Composition*

Real time composition is very interesting from the perspective of complex systems science. It can be seen as a method to *generate* desirable emergent patterns in space and time. The 2010 European Conference on Complex Systems in Lisbon was an art-science catalyst bring together Fiadeiro and scientists from the complex systems community. Although the scientists were not trained dancers, it soon became clear that they could participate and apply the method successfully.

Fiadeiro's method *constrains* and therefore guides what can emerge. His "rules" are interpreted by the participants whose personalities can vary considerably. Sometimes individuals cannot adapt to the rules which can result in poor outcomes for the group. However, in my experience the participants were all highly committed and outcomes ranged from the mundane to being completely unexpected, aesthetically outstanding, and very satisfying.

Fiadeiro's method can be viewed as the evolution of connected chains of relational simplices. Like music, dance has its own notation for recording movements through time. However, Real Time Composition is not intended to be re-performed and is generally video-recorded. To illustrate the hypernetwork approach a simple experiment was devised using a computer paint system with two scientists experienced in Fiadeiro's method taking turns in creating an image, adding, changing and deleting graphic objects as they felt appropriate. These are called *moves*.

Figure 8.18 shows the first steps in this process. Initially the arena is defined by a by a square *frame*, denoted f, and the default structure formed from it, denoted $\langle f; R_0 \rangle$. The sequence begins by the introduction of a diagonal line ℓ_1. This line can be characterised by the descriptor simplex $\sigma(\ell_1) = \langle \text{diagonal, long}, \diagdown; D_1 \rangle$.

It could be tempting to add a second diagonal, ℓ'_1, with $\sigma(\ell'_1) = \langle \text{diagonal, long}, \diagup; D_1 \rangle$ which would be 1-near $\sigma(\ell_1)$ with eccentricity 0.33. This maintains the "long diagonal" theme, but his would lead to a possibly uninteresting symmetric cross structure. Instead the next move, Move 2, added a shorter diagonal line, ℓ_2, with descriptor simplex $\sigma(\ell_2) = \langle \text{diagonal, short}, \diagdown; D_1 \rangle$ below the long diagonal to create the structure $\langle f, \ell_1, \ell_2; R_2 \rangle$ (Fig. 8.18(c)). Move 3 adds ℓ_3 and removes the frame to create $\langle \ell_1, \ell_2, \ell_3; R_3 \rangle$,

Move-4, based on structures at *Level N+1*, adds a modified copy of σ_1, $\sigma_2 \stackrel{\text{def}}{=} \langle \ell_4, \ell_5, \ell_6; R_4 \rangle$. This move also adds a new descriptor at *Level N* for the lines, discriminating "solid" lines from "dashed" lines. This gives the descriptor simplices $\sigma(\ell_4) = \sigma(\ell_6) = \langle \text{diagonal}, \text{short}_{D_2}, \diagdown, \text{dashed}; D_2 \rangle$ and $\sigma(\ell_5) = \langle \text{diagonal}, \text{long}_{D_2}, \diagdown, \text{dashed}; D_2 \rangle$. Note that the descriptors "short" and "long" cease to have any absolute interpretation but take their meaning relative to the structure. For example, the "long" line ℓ_5 is shorter than the "short" line ℓ_2. This difference can be maintained by using the notation "short_{D_1}" $\neq$ "short_{D_2}" and "long_{D_1}" $\neq$ "long_{D_2}".

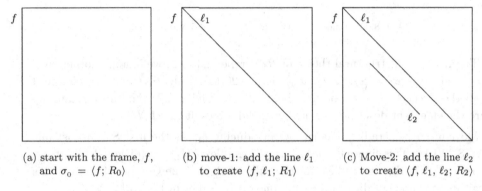

(a) start with the frame, f, and $\sigma_0 = \langle f; R_0 \rangle$

(b) move-1: add the line ℓ_1 to create $\langle f, \ell_1; R_1 \rangle$

(c) Move-2: add the line ℓ_2 to create $\langle f, \ell_1, \ell_2; R_2 \rangle$

Fig. 8.18 Real Time Composition of graphic objects and scenes: Moves 1 & 2

The continuity required by Fiadeiro's Real Time Composition method is given by copying the simplex $\sigma_1 = \langle \ell_1, \ell_2, \ell_3; R_4 \rangle$ to obtain $\sigma_2 = \langle \ell_4, \ell_5, \ell_6; R_4 \rangle$. These are both R_4 structures at *Level N+1*. Also all the lines so far have the descriptors "diagonal" and "$\diagdown$". Since the containing square is now less dominant it is shown by dashed lines.

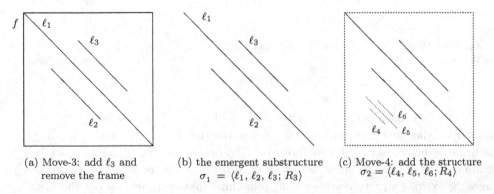

(a) Move-3: add ℓ_3 and remove the frame

(b) the emergent substructure $\sigma_1 = \langle \ell_1, \ell_2, \ell_3; R_3 \rangle$

(c) Move-4: add the structure $\sigma_2 = \langle \ell_4, \ell_5, \ell_6; R_4 \rangle$

Fig. 8.19 Real Time Composition of graphic objects and scenes, Moves 3 & 4

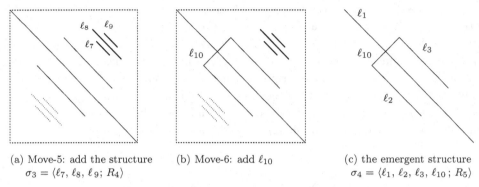

(a) Move-5: add the structure $\sigma_3 = \langle \ell_7, \ell_8, \ell_9 ; R_4 \rangle$

(b) Move-6: add ℓ_{10}

(c) the emergent structure $\sigma_4 = \langle \ell_1, \ell_2, \ell_3, \ell_{10} ; R_5 \rangle$

Fig. 8.20 Emergent structure from Move 5 and Move 6.

The *Level N+1* structural theme of R_4 is repeated in Move 5, and another copy of σ_1 is added as $\sigma_3 = \langle \ell_7, \ell_8, \ell_9 ; R_4 \rangle$ (Fig. 8.20(a)). Thus continuity is reinforced by all the *Level N+1* substructures being determined by R_4, and also σ_1 and σ_3 share the emergent descriptor of having "solid" lines at *Level N*.

Move-6 marks a radical change by introducing ℓ_{10} as the first South-East line (Fig. 8.20(b)). This line has the descriptor simplex $\sigma(\ell_{10}) = \langle \text{diagonal, short}_{D'_1}, \nearrow, \text{red} ; D'_1 \rangle$. Is this too radical for the rules of Real Time Composition, or is this a legitimate change of direction within one of the cubes in Fig. 8.14?

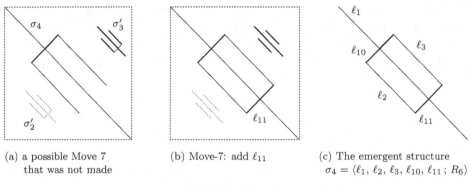

(a) a possible Move 7 that was not made

(b) Move-7: add ℓ_{11}

(c) The emergent structure $\sigma_4 = \langle \ell_1, \ell_2, \ell_3, \ell_{10}, \ell_{11} ; R_6 \rangle$

Fig. 8.21 Move 7

It could be considered to be a legitimate move because, focussing on σ_1 the new South-West diagonal line, ℓ_{10}, strongly groups the the lines ℓ_1, ℓ_2, and ℓ_3 binding them into a new *Level N+1* structure $\sigma_4 = \langle \ell_1, \ell_2, \ell_3, \ell_{10} ; R_5 \rangle$ which completely dominates the original σ_1 substructure (Fig. 8.20(c)). Thus the new line, ℓ_{10}, reinforces the existing structure while adding something that opens up new possibilities. Also, the descriptor $\sigma(\ell_{10})$ shares the vertex $\langle \text{red} \rangle$ with the descriptor simplex of σ_3.

Figure 8.21(a) shows a possible move that creates copies of σ_4 at *Level N+1* by adding lines at *Level N* to σ_2 and σ_3. Let these be denoted as σ_2' and σ_3'. This move would create the *Level N+2* structure $\sigma_{unused} = \langle \sigma_4, \sigma_2', \sigma_3' ; R_{unused} \rangle$ in which the configurations σ_2' and σ_3' have the opposite diagonal orientation to the larger configuration σ_4. This configuration satisfies the requirement that there be continuity with past configurations, but since this move was not made its possible future development remains open.

The actual Move-7 is shown in Fig. 8.21(b). It introduces another North-East line, ℓ_{11} to complete the rectangle, and create the new structure $\sigma_4 = \langle \ell_1, \ell_2, \ell_3, \ell_{10}, \ell_{11} ; R_4 \rangle$ (Fig. 8.21(c)).

Continuity is provided by the new line ℓ_{11} having the same descriptor simplex as ℓ_{10}, $\sigma(\ell_{10}) = \sigma(\ell_{11}) = \langle \text{diagonal, short,} \nearrow; D_3 \rangle$. Also the new simplex $\sigma_4 = \langle \ell_1, \ell_2, \ell_3, \ell_{10}, \ell_{11} ; R_4 \rangle$ has "within it" the simplex $\sigma_3 = \langle \ell_1, \ell_2, \ell_3, \ell_{10}, \ell_{11} ; R_4 \rangle$ as a face, which is that configuration as σ_2 and σ_3.

The R_6 structure shown in Fig. 8.21(c) is very strong, and Move 8 adds two lines ℓ_{12} and ℓ_{13} to make another smaller R_6 structure, $\sigma_5 = \langle \ell_8, \ell_7, \ell_9, \ell_{12}, \ell_{13} ; R_6 \rangle$, as shown in Fig. 8.22(a).

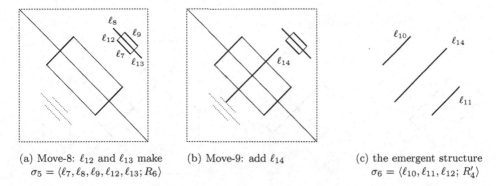

(a) Move-8: ℓ_{12} and ℓ_{13} make $\sigma_5 = \langle \ell_7, \ell_8, \ell_9, \ell_{12}, \ell_{13}; R_6 \rangle$

(b) Move-9: add ℓ_{14}

(c) the emergent structure $\sigma_6 = \langle \ell_{10}, \ell_{11}, \ell_{12}; R_4' \rangle$

Fig. 8.22 The emergence of $\sigma_3 = \langle d, s, s', p, ; R_9 \rangle$ and $\sigma_4 = \langle d, s, s', p, p' ; R_9 \rangle$

Move 9 makes a radical innovation by the addition of the North-East diagonal line ℓ_{14} which cuts across σ_4. Continuity comes from the connectivity of the descriptor simplices $\sigma(\ell_{10}) = \sigma(\ell_{11}) = \langle \text{diagonal, short,} \nearrow; D_5 \rangle$ and $\sigma(\ell_{14}) = \langle \text{diagonal, long,} \nearrow; D_5 \rangle$, and the emergent structure $\sigma_6 = \langle \ell_{10}, \ell_{11}, \ell_{12}; R_4' \rangle$ (Fig. 8.22(c)).

Move 10 shortens the line ℓ_1 to create ℓ_{15} and a very distinct *Level +1* structure, $\sigma_7 = \langle \ell_{15}, \ell_2, \ell_3, \ell_{10}, \ell_{11}; R_7 \rangle$, emerges (Fig. 8.24(b)).

Moves 9 and 10 create an enormous amount of latent emergent structure at all levels. For example four small rectangles, r_1, r_2, r_3, and r_4, and four new lines, ℓ_{16}, ℓ_{17}, ℓ_{18}, and ℓ_{19}, emerge emerge in Fig. 8.22(c). What are the levels of these lines and rectangles? So far all lines have been designated as existing at *Level N*. This suggests that lines ℓ_{16}, ℓ_{17}, ℓ_{18}, and ℓ_{19} be assigned to *Level N* and that the rectangles, r_1, r_2, r_3, and r_4, be assigned to *Level N+1*.

Consider the structure $\langle \ell_{16}, \ell_{17}\,;\, R_8 \rangle$ made of the two North-West diagonal lines in Fig. 8.24(c). This emergent line appears to be the same as ℓ_{15}, and one could write $\ell_{15} \equiv \langle \ell_{16}, \ell_{17}\,;\, R_8 \rangle$. If all lines exist at *Level N* then ℓ_{15} would be at the same level as its parts ℓ_{16} and ℓ_{17}. If ℓ_{15} had been made from *Level N* objects ℓ_{16} and ℓ_{17} it would be a *Level N+1* object, as discussed the end of this section.

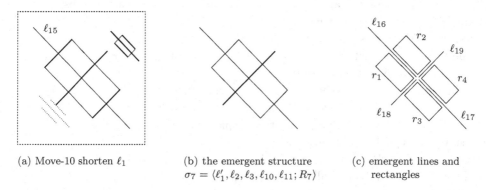

(a) Move-10 shorten ℓ_1

(b) the emergent structure $\sigma_7 = \langle \ell_1', \ell_2, \ell_3, \ell_{10}, \ell_{11}\,;\, R_7 \rangle$

(c) emergent lines and rectangles

Fig. 8.23 Emergent structures and substructures

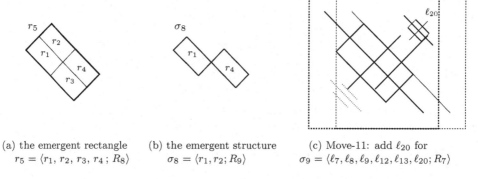

(a) the emergent rectangle $r_5 = \langle r_1, r_2, r_3, r_4\,;\, R_8 \rangle$

(b) the emergent structure $\sigma_8 = \langle r_1, r_2\,;\, R_9 \rangle$

(c) Move-11: add ℓ_{20} for $\sigma_9 = \langle \ell_7, \ell_8, \ell_9, \ell_{12}, \ell_{13}, \ell_{20}\,;\, R_7 \rangle$

Fig. 8.24 The emergence of $\sigma_3 = \langle d, s, s', p,\,;\, R_9 \rangle$ and $\sigma_4 = \langle d, s, s', p, p'\,;\, R_9 \rangle$

Higher level structures can emerge from the *Level N+1* rectangle, e.g. the *Level N+2* rectangle $r_5 = \langle r_1, r_2, r_3, r_4\,;\, R_8 \rangle$ in Fig. 8.24(a) and the structure $\sigma_8 = \langle r_1, r_4\,;\, R_9 \rangle$ in Fig. 8.24(b). There are of course thousands more latent objects that can be created by downwards disaggregation and upwards assembly.

Move 11 creates a smaller copy of σ_7 by adding the diagonal line ℓ_{20} to σ_5 (Fig. 8.24(c)). Move 12 marks a radical departure from what has gone before by making black the rectangle r_1. In Fiadeiro's terms it could be said that the small rectangles are *accidental* consequences of introducing the line ℓ_{14}. The objects created by accidents in Real Time Composition are *latent* and may or may not be integrated into the sequence. In this case the latent emergent rectangle r_1 is *activated* by making it black and henceforth it plays a central role in the composition. The next move recognises the emergent structure σ_8 and the rectangle r_4 within it. Move 12 "decorated" r_1 by making it black. Move 13 "decorates" r_4 by putting in it the structure σ_9. Move 13 also removes the dotted configuration which had become, relatively, increasingly disconnected from the evolving central structure. Move 14 makes black the small emergent rectangle in σ_9 and the participants agreed that this should be the end.

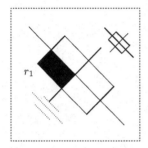 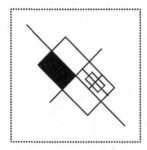 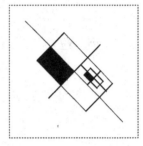

(a) Move-12: the emergent rectangle r, is made black   (b) Move-13 remove σ_2 move σ_9 into σ_z (b) Move 14: make black the the small rectangle

Fig. 8.25 The emergence of $\sigma_3 = \langle d, s, s', p, ; R_9 \rangle$ and $\sigma_4 = \langle d, s, s', p, p' ; R_9 \rangle$

Discussion

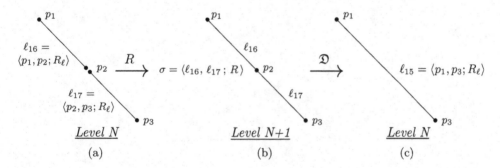

Fig. 8.26 Two lines at *Level N* form a line at a higher level, *Level N+1*

As observed previously, two straight lines ℓ_{16} and ℓ_{17} can be combined under a relation R to form another straight line $\langle \ell_{16}, \ell_{17} ; R \rangle$. In that case this new line appears to coincide exactly with a pre-existing line ℓ_{15}, as shown in Fig. 8.26. This does not mean that the relational simplex $\langle \ell_{16}, \ell_{17} ; R \rangle$ and its elements exist at the same level.

Let $\ell_1 = \langle p_1, p_2; R_\ell \rangle$ and $\ell_2 = \langle p_2, p_3; R_\ell \rangle$ at *Level N* as shown in Fig. 8.26(a). The relation R assembles these two lines into a structure at *Level N+1* denoted $\sigma = \langle \ell_{16}, \ell_{17}, ; R \rangle$ in Fig. 8.26(b). This *Level N+1* structure is certainly not equivalent to ℓ_{15} since it contains information that ℓ_{15} does not, including the point p_2. To obtain the line $\ell_{15} = \langle p_1, p_3; R_\ell \rangle$ requires the *Level N+1* structure σ to be *deconstructed* by an operator $\mathfrak{D}$ which effectively rewrites the available structure into the point data structure at *Level N-1*. Then the relation R_ℓ can assemble the points p_1 and p_3 into the line data structure at *Level N*.

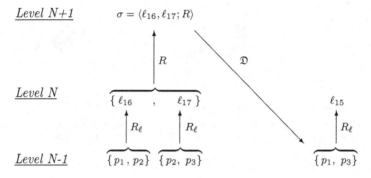

Fig. 8.27 ℓ_{15} reconstructed by R_ℓ from the deconstruction $\mathfrak{D}(\langle \ell_{16}, \ell_{17}; R \rangle)$

Chapter 9

Hypernetworks and Design

It will be argued that design is fundamental in the science of complex systems. By definition, the systems created by human beings are *artificial*. Those created deliberately to achieve specified outcomes are designed [Simon (1965)].

At this moment you are probably surrounded by things intentionally designed by other people. Even if you are in a desert or on wild mountains, you are probably wearing clothes with a map, a compass or a mobile phone. If you are in a city you will be unable to count the designed objects that surround you. Even inside buildings the number of designed things can be astonishing, including the furniture, light fittings, doors and windows, and all the things that you have with you.

Whereas it is obvious that artefacts and infrastructure are designed, it is less obvious that social systems are designed. It will be argued that they are. For example, managers design football teams, generals design wars, entrepreneurs design businesses, bankers design financial systems, artists design exhibitions, academics design universities, mathematicians design proofs, physicians design treatments, criminals design crimes, and governments design policies.

Generally, when social systems are designed those responsible do not consider themselves to be designing. Certainly policy includes the design of infrastructure, usually delegated to specialist designers and engineers, but it also involves the deliberate design of social structures to achieve specific outcomes. If those creating new social systems followed the methods developed by designers, those systems could be more predictable and perform better in the context of their specified desired behaviours.

But the entanglement of design with complex systems science goes beyond the methods of design. Design involves *prediction*. When new systems are being designed, the implementation of the design is an *experiment*. The argument is that design is the experimental laboratory for complex systems science.

9.1 Design

Many areas are recognised as *design disciplines*, for example, electrical, mechanical and civil engineering; architecture and town planning; textiles and fashion; and furniture and furnishings. Although in these cases the systems being designed are very different, many aspects of the design process are common to them all.

Much design is deliberate, expensive and has far-reaching consequences, especially when systems are big and complex: for example, the creation of the new high speed line from London to the Channel Tunnel; developing a new city or suburb; creating new flood defence systems; switching to electric cars; creating educational programmes or the reorganisation of a health system. These are large multilevel hypernetwork systems. They are created by constructing hypersimplices at many levels to create a multilevel backcloth that can support the day to day traffic of the system.

Changing the world by design

The social world is changed by accident or design, in the context of external physical processes that cannot be controlled but can be managed. This book has three entangled theses:

- complex systems have multilevel hypernetwork structure that is essential to understanding their dynamics.
- changing the world is the process of designing, implementing and managing complex multilevel hypernetwork systems, and requires complex system science.
- experiments on complex social systems can only be done by designing and implementing new systems in the context of policy

Design is a *process* that establishes what is wanted, imagines new systems that may deliver what is wanted, evaluates to what extent the imagined systems will deliver what is wanted, and plans the implementation of new systems in order to deliver what is wanted. All of this is pure thought. Professional designers are trained how to think in a *designerly* way [Cross (2007, 2011)]. Design is not an algorithm and the design process is not guaranteed to give good results. Furthermore, the design of products and systems exists in an evolutionary context [Steadman (2008)], driven by the constant desire to make better things and make things better.

Many people who design social and socio-physical systems do not consider themselves to be designers and know little of the large body of knowledge on design and design processes that has accumulated over the last half century. This chapter is intended to introduce some elementary ideas about design and show that it is an essential part of the methodology of complex systems science. The next chapter will put this in the context of policy.

The design process

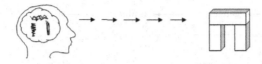

Fig. 9.1 Design begins with thoughts in a person's head, and ends in an artefact

The design process starts with ideas in the designer's head and ends in the production of an artefact (Fig. 9.1). It may involve the creation of documents, from the sketch on a paper napkin, to a fully instantiated document giving all the information necessary to produce an artefact. Here the "final" documentation will be called the *design blueprint*. In simple cases the blueprint may be exactly that, an annotated drawing, but may go far beyond that, to include all documents, drawings, and physical models that are created during the design process. In all cases the complete design blueprint includes an explicit specification of the parts or components, and explicit instructions on how they are to be assembled to form the whole.

Sometimes neither the start nor the end of the design process may be precise points in clock time. The start may formally be the decision to invest resource in a project, but even to get to this stage requires thinking about the project beforehand. Similarly the end of the process might formally be marked by signing off the blueprint, or the production of an artefact, but often design continues alongside manufacture, to correct errors and make improvements during the product or system lifecycle, from which new models, systems and designs evolve [Steadman (2008)].

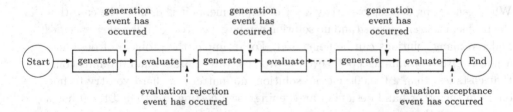

Fig. 9.2 Design as a sequence of generate and evaluate events

Design is a dynamic human process through time. It is a process marked by *design events*. It starts by identifying some needs or *requirements*. These are formally written as a *specification*, which defines the *design problem* of satisfying those requirements. Designers generate potential solutions to design problems. These are *generation events* in the design process. They evaluate their solutions as *accept events* or *reject events*, and if not satisfied generate alternatives. Thus design is characterised by sequences of generation and evaluation events (Fig. 9.2).

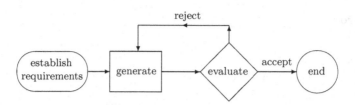

Fig. 9.3   The fundamental generate-evaluate cycle of design

Design is a thought process, a sequence of emerging cognitive events, and it takes clock time. In the beginning the thinking is hazy and the documentation may be just a quickly done sketch drawing that leaves out many details but conveys the main ideas. The process of sketching generates new thoughts and ideas and helps makes things clearer. As time goes on the requirements and possible design solutions become more explicit and better defined.

Design events are characterised by having happened, or not having happened, *e.g.* the proposed solution is ready for evaluation, or it is not ready for evaluation. Following such generation events, a decision will be made whether or not the proposed solution can be accepted as the solution. This decision marks an evaluation event. During the design process, most of these are evaluation-rejection events, as the current candidate solutions are either rejected or subject to more work. The process ends with an evaluation-acceptance event, possibly marked by champagne. During the design process the nature of such design events will become clearer, as will their importance in marking the natural time and rhythm of the design process.

Satisficing

When generating designs to satisfy a set of requirements it is usually the case that the problem is *over-specified* and no solution can be generated, or it is *under-specified* and too many solutions can be generated. For example, the problem of designing a house with a swimming pool, seven bedrooms, three garages and a tennis court that costs less than £100,000 has no solution, no matter how hard you try to find one. On the other hand designing an evening gown costing less than £1,000 has a very large number of solutions. Many design problems are much more complicated than these examples. For example, designing a national health system has many constraints that go beyond cost.

In general designers juggle competing constraints, trading off reduced excellence in some criteria for improvements against others. For example, it might be accepted that the designed object will cost a little more to keep a very desirable feature, such as a larger room in a house, or it may be accepted that price dominates and the room must be smaller than would be ideal. Herbert Simon called this *satisficing* meaning that all the requirements constraints had been more or less satisfied, but some or all constraints have not been optimised [Simon (1965)].

Problem-solution co-evolution in design

In design a remarkable thing often happens: the designer decides that all the satisficing solutions they have generated should be rejected and that in order to make progress the requirements must be changed. For example, a person designing a shaded garden might have complicated constraints on what can be planted to give a pleasing mix of colours throughout the year. After trying many combinations they might give up and decide to cut down a much-loved tree, abandoning the constraint that the tree must stay. As another example, consider a city planner designing a housing project for a hundred families. The original specification might require all the buildings to have at most two storeys on the given site, but all designs satisfying this constraint are undesirable. At some stage the designer may experiment with designs that break the two-storey rule, find a design that is excellent from most other perspectives, and persuade the client to change the specification.

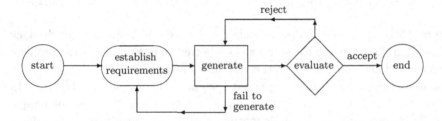

Fig. 9.4 The co-evolutionary generate-evaluate fail-respecify double cycles of design

Changing the requirements and specification can be compared to a student being given the problem of finding 429^{81}, them deciding that this is too difficult to calculate, redefining the problem as that of finding 430×80 and giving the answer as 3420. Here the delivered problem-solution pair is correct but would be rejected because the problem has been changed. Design problems cannot be solved algorithmically. In design it is normal for the requirements and specification to change – it is part of the process. In design the problem and solution *co-evolve* until an acceptable problem-solution pair is found, as shown in Fig. 9.4. Changing the specification for a design is a normal *design event*.

On what authority can the requirements and specification for a design be changed? Ultimately the sanction to change the design is given by the *clients* who will pay for the design and its implementation. Sometimes the designer is their own client, but if not and they want to change the specification they must negotiate and persuade the clients that the revised problem is what they want. Sometimes changes in the requirements or specification are suggested by the client and they have to persuade to designer to accept them.

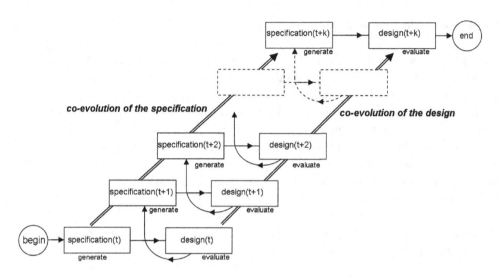

Fig. 9.5 The co-evolution between specification and design through a generate-evaluate spiral

Figure 9.5 shows that the design cycle can be seen as a spiral through clock time. Each iteration through the spiral gives a better understanding of the problem as designers and clients *learn* about the system they are creating.

Design is the process of thinking about systems that do not already exist. In some cases there may be similar systems, but sometimes the system may be unique and completely new. In this case designers and clients must create knowledge about the system, and they are the first scientists of that system. For example, there could be no science of aeronautics before the first aeroplanes were designed. Thus it is argued here that *designers and their clients are the blue-sky scientists at the forefront of the science of new complex socio-physical systems.*

Managing design, its implementation and use of the new system

Although design is marked by design events, in which things are ready when they are ready, design events occurs in a wider social context dominated by clock time. This imposes constraints related to how much resource can be invested in the design process and therefore how long it can take. Also, clock time constrains the feasibility of a design in terms of scheduling the acquisition or creation of multilevel parts and the processes to assemble them into multilevel wholes.

A distinction can be made between the phases of design before and after implementation, and before and after use of the system. Before implementation the main cost is that of the creating the design blueprint. After implementation begins costs can increase steeply until the system enters into use. Apart from the coevolution between specification and design, there is a coevolution between the specification and the implementation of a design, and there is a coevolution between the use of the system and changes that are made to the system in the short and long terms.

Sometimes there are *design errors* in which the assumptions made by the designers and clients are found to be incorrect during implementation. For example, a material may not be strong enough to function as assumed, there may be geometrical errors such as pipe passing through the middle of a room; there may be process errors in which a part of the system cannot be made as assumed; there may be errors in which some part of the design violates statutory regulation, and so on. Generally designers, clients and fabricators find a way around these problem at costs that they are prepared to bear.

A common problem in the design of large complex systems is that the client requests changes during late stages in the design phase or during the implementation phase. This is particularly the case for military systems which may have decades between commissioning the original design and delivery of the first prototype. It can happen, for example, in architecture when the client sees some new opportunity in the emerging building. When specifications change during construction, designers, clients and fabricators again find ways to make the changes at costs that they are prepared to bear, seeking some local redesign that does not cause too much disruption to the whole.

All designed systems are created for a purpose, and when they enter use it can be seen how well they serve this purpose. This can expose design errors in the physical part of a system when things don't work properly, and it can expose design errors in the social part of a system when people do not behave in the way that was assumed. For example, in the postwar twentieth century large tower blocks were built to accommodate large numbers of people, but people did not or could not live in them as had been assumed and many such blocks have now been demolished.

Managing the design, implementation and use of new systems involves the creation of a new multilevel backcloth able to support the desired multilevel traffic of social activity. It involves scheduling the creation of many physical and social hypersimplices at many levels, and assembling these to form the functioning whole.

9.2 Design and the Intermediate Word Problem

When designing things that do not already exist people have an idea of the "the system" at some higher level of representation, and an awareness of many relevant or irrelevant things that we call "The Soup". In the first instance the designer seeks a subset of things that can be selected from the soup and assembled in an appropriate way to make a "System" which has emergent properties satisfying the constraints of the requirements.

In all designed disciplines there is a huge amount of "stuff" in the soup. It includes all the thousands of components that exist in thick catalogues – electronic components for electrical engineers, gears and cogs for mechanical engineers, windows and doors for architects, buttons and fabrics for couturiers, paints and wallpapers for interior designers, woods and veneers for furniture designers, and so

282   Hypernetworks in the Science of Complex Systems

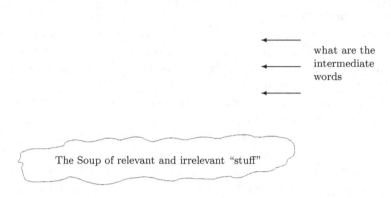

Fig. 9.6   The intermediate word problem of design

on. Part of a designer's professional expertise is an up-to-date knowledge of these thousands of components and the processes that can be used to assemble them into new objects and systems.

In this context it can be seen that designers seek to create emergence and they *predict the emergent behaviour of the whole from the known properties of the parts and their knowledge of the assembly process*. Thus emergence is an everyday thing in the world of design, and the skill of designers is to imagine systems that don't already exist and predict their emergent behaviour when they are assembled from known things.

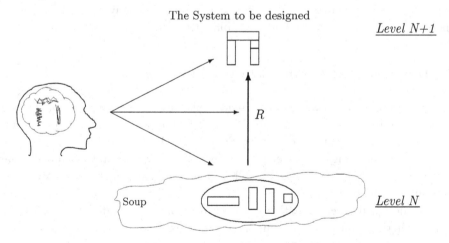

Fig. 9.7   Simple designs can be done entirely in one's head

In the simplest cases, such as building an arch from three blocks, "The System" is the arch and the process of assembling it involves selecting appropriate blocks from the Soup and assembling them appropriately (Fig. 9.7). This requires just two levels of representation making it possible for the designer to "see" in their mind's eye the parts, the whole and the relation that assembles the parts into the whole.

For more complicated systems with many levels of subsystems the designer cannot "just see" the final design. Instead designers work top-down by hypothesising abstract subsystems and bottom-up by identifying components and subsystems that will be assembled into the whole (Fig. 9.9).

Creating explicit representations in design

One of the problems in trying to understand design and designing is that a lot of the process is implicit inside people's heads. Often there is no explicit language for what is being created, especially if it never existed before. As illustrated in Fig. 9.6 the designer faces the intermediate word problem of describing the new object or system at levels between the highest level of "The System" and the soup-level of relevant and irrelevant "stuff" known to or observed by the designer.

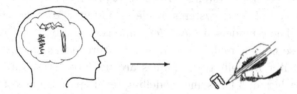

Fig. 9.8   From the implicit idea to an explicit representation of the artefact being designed

At any time the designer may just have a fuzzy and ever-changing abstract vision of the new artefact or system. In order to record their ideas and communicate them to others, designers have to create an explicit representation of the artefact as a drawing, a verbal explanation, a text, or some other external record. For example, Fig. 9.8 shows the designer recording the design as a drawing.

Technical drawing is a means of developing and recording designs that has been developed and used over centuries. It has rules for making assemblies of lines that represent objects, subsystems and the whole. Sometimes the lines are geometric representations of particular objects, and sometimes the drawings are "schematics" representing the interactions between subsystems at higher level. Technical drawing enables a three-dimensional object to be projected onto two dimensions in ways that enable the three dimensional object to be reconstructed.

In the nineteen seventies and eighties technical drawing moved almost entirely from pen, ruler, paper and blue print machine to joystick, screen computer and plotter. Despite resistance from some traditional pen-and-paper designers, computer aided design (CAD) had such enormous advantages that CAD is now almost universal in design. Apart from providing data structures to represent and manipulate objects CAD, gives many other advantages. For example, it enables objects to be displayed in perspective from many viewpoints, allowing them to be rotated, scaled and moved allowing the designer to "see" and even "test" the abstract object before it exists. Other advantages of CAD include the availability of libraries of components with known properties. A great early advantage of CAD was the possibility of automating the creation of lists of parts and costs, removing expensive and error-prone human processes. One of the most important contributions of CAD is that it creates explicit *computer models*.

CAD is now used routinely in the *analysis* of many aspects of systems. Finite element analysis allows parts of physical systems to be approximated by many simple three dimensional geometric elements such as tetrahedra and cuboids. Given the boundary conditions, the physical attributes of the elements such as temperature, stress, and deformation are simple to calculate. These can then be aggregated to give the physical attributes of the whole, which may have a much more complicated geometry than the elements. Finite element analysis allows the dynamics of systems to be investigated before they exist as real objects, providing *predictions* on the future behaviour of the system yet to be constructed.

CAD is instructive for complex systems scientists. It provides a working method of representing complicated physical systems inside a computer as *models* that can be used for *simulation*. These models are *multilevel* and they combine the dynamics at different levels. These models make *predictions* of future states of hypothetical systems. Similar computer modelling techniques are used to investigate chaotic systems such as the weather, and the same modelling techniques are used to create three dimensional animations including the avatars of virtual worlds.

Design as creating multilevel models

At the lower levels, the designer identifies *verifiable* relationships between parts and components, and can experiment by assembling parts into intermediate level components and testing the behaviour of the assemblies created.

At the higher levels of representation the designer hypothesises that various relationships hold between the uninstantiated abstract systems. For example, an architect may hypothesise that a building will have an atrium and that it will have a conference room, and that it will take less than five minutes to walk from one to the other. This makes sense, even though the exact position and dimensions of these spaces may be uninstantiated.

Let the top level of a system be designated *Level $N+h$* and the lowest level be designated *Level N*. The top-down process creates a new level of representation, *Level $N+h-1$* while the first application of the bottom-up process creates a new level of representation, *Level $N+1$*. These top-down and bottom-up processes continue until they meet with *Level $N-k$* = *Level $N+k'$* for some k and k'.

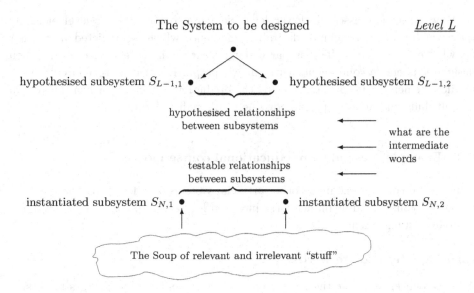

Fig. 9.9 The intermediate word problem of design

Reductionism versus multilevel perception in design

In the science of complex systems one cannot study subsystems in isolation and then simply combine the knowledge of the subsystems to understand the behaviour of the whole system. This *reductionist* approach has worked well in physical science but, as Ross Ashby noted, varying the factors one at a time is fundamentally impossible in complex systems [Ashby (1956)].

The limitations of reductionism are sometimes interpreted as meaning that one should not look at the parts of systems to investigate their properties. The reality is that we unable to stop ourselves seeing the parts of systems. When we encounter a thing at the Gestalt level our brains seem to deconstruct them automatically. Often we can hear the individual instruments in an orchestra or the single string in an harmonious chord. When we see a painting we see the whole but also scan the parts. When we see a building we also see the doors and windows. It seems that our perception is adapted to oscillate between between wholes and their parts, possibly because there is survival value in having a good repertoire of parts for creating new models for predicting the behaviour of unknown new wholes. There is no doubt that most people are able to observe systems instantaneously at many levels of aggregation, and designers are masters of this within their own disciplines.

In design, reducing a system to its lowest level parts is essential. However, this is done with the explicit expectation that those parts will be combined to form hypersimplices, and they too will be assembled to form higher level hypersimplices, and so on until the whole system has been assembled.

When designers assemble lower level elements to form higher level elements they usually do so to achieve desirable emergence which is predicted from their knowledge of the parts. This is particularly clear in the case of electronics when designers assemble components with known properties to construct circuits with predicted emergent properties. Usually the predictions are correct when the system is built, but sometimes chips explode due to unpredicted emergence.

9.3 Example: designing an educational course module

Educational courses, like many other artificial systems, are designed. For example, consider a course module on networks intended for postgraduate students studying complex systems science.

Eliciting intermediate words

Already one knows that the soup contains "course module", "postgraduate students", "networks", and "complex systems science" (Fig 9.10(a)). It is to be expected that many more words will enter the soup as the design process proceeds. Also the requirements, "make a course module on networks intended for postgraduate students studying complex systems science" are not yet precise or well defined. However, in the first instance let the system to be designed be "the course module", renamed as "Network Course" at some level, $N+h$ in Fig. 9.10(b).

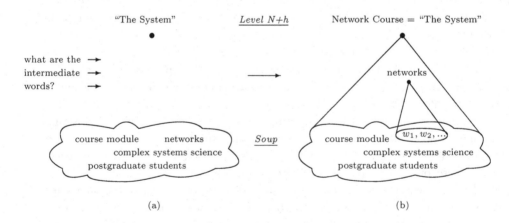

Fig. 9.10 Design begins with the first iteration of the Intermediate Word Elicitation

The designer could proceed in many ways. For example, they could think more about the students. Will they have high or low levels of mathematical ability, or will there be a mixture? In this case the course designer has focussed on networks and *lifted this word out of the soup* to become a formal part of the representation. Also, the designer has prior knowledge of some excellent courses on networks supported by web-based resources, $w_2, w_2, ...$, including slide presentations for lectures and classes given to students at universities in the UK and USA.

In this case the designer used the expression "Network Course" to represent the whole system. In principle the intermediate words could be any symbolic way of representing objects. For example, this course could have been called C101-AX which has no intensional meaning. In contrast the term "Network Course" is rich in intension, and many readers will already have rich implicit or explicit expectation of what a "Network Course" will include (the parts) and how they will be assembled into hypersimplices.

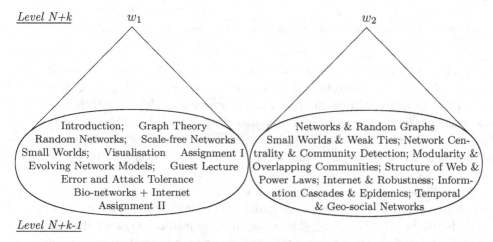

Fig. 9.11 Intermediate words elicited from website w_1 and w_2

Figure 9.11 shows intermediate words elicited from websites w_1 at Northeastern University[1] and w_2 at Cambridge University[2]. These courses will be denoted C_1 and C_2. At *Level N+k-1* they are far from identical but at lower levels they cover a lot of common ideas (Fig. 9.12).

For example, C_1 has the intermediate words or phrases, $w_{1,1}$ = "Graph Theory" and $w_{1,2}$ = "Random Networks" which denote two units of study. In contrast C_2 has one intermediate phrase, namely $w_{2,1}$ = "Networks and Random Graphs" denoting just one unit of study. The intensions of these phrases suggest that they denote very similar things at lower levels in the representation.

Since $w_{1,1}$ and $w_{1,2}$ are consecutive, one would expect $base(w_{1,1})$ to be relatively eccentric with respect to $base(w_{1,2})$. Since the intension of the word $w_{2,1}$ is similar to the intensions of $w_{1,1}$ together with $w_{1,2}$, one would expect $w_{2,1}$ to be less eccentric covering a relatively large portion of concepts. For exam-

---
[1] http://barabasilab.neu.edu/courses/phys5116/, viewed 15-12-12
[2] http://www.cl.cam.ac.uk/~cm542/teaching/2011/stna2011.html, viewed 15-12-12

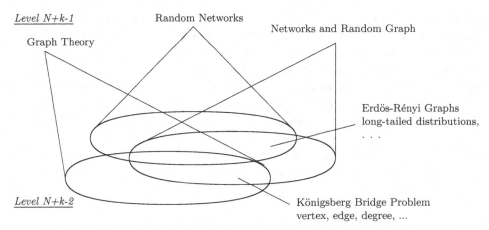

Fig. 9.12 At lower levels the intermediate words for w_1 and w_2 become more similar

ple, in Fig. 9.12 "vertex", "edge" and "degree" appear in $base^{N+k-2}$("Graph Theory") $\cap$ $base^{N+k-2}$('Networks & Random Graphs'), as does "Königsberg Bridge problem". Similarly "Erdös-Rényi Graphs" appears in $base^{N+k-2}$("Random Networks") $\cap$ $base^{N+k-2}$("Networks & Random Graphs"), as does "long-tailed distributions". Thus at the lower level, $N+k-2$ a set of words can be elicited that describe all the concepts common to the first two lessons of the courses, $base^{N+k-2}(w_{1,1}) \cup base^{N+k-2}(w_{1,2}) \cup base^{N+k-2}(w_{2,1})$.

As noted previously, the courses C_1 and C_2 have a great deal of excellent web resource including many slides explaining and illustrating the concepts in W_c. Let these slides be denoted S_1 and S_2. There are bipartite relations between the words W_c and S_1, S_2, and $S_1 \cup S_2$, grouping the slides around the concepts they elucidate. A Q-analysis of this relation can be used for selecting the most appropriate slides to be used in the new course. High dimensional slides cover many words and concepts. Highly connected slides are similar with the possibility of duplication. Highly eccentric slides cover combinations of words and ideas not covered by others.

At this stage the course designer can decide if this set contains all the words required for the new course, or they can add new words to denote new sets and structures. For example, $w_{1,2}$ does not contain the word "hypernetworks". How are such decisions made?

The design cycle: reviewing the requirements and specification

Evaluating the current stage of the design in the design cycle of Fig. 9.4 requires the requirements to be written down as a formal specification. So, should hypernetworks be taught on the course? This depends on the expected *learning outcomes* of the course. Let us suppose that the client for this course is a professional society providing an educational service for its members. Then this client may have already made teaching hypernetworks a requirement. If not the course designer might ask if hypernetworks should be included. It will be supposed that the client says this would be desirable, so "learning about hypernetworks" becomes explicitly added to the specification, and the design problem has changed.

In education the learning outcomes of an activity should be explicit so that the specification can be *tested* in the evaluation stages. How can one test if a student has achieved a stated learning outcome? The conventional way is by tests, quizzes, exercise, essays, *etc.* that are marked (graded) according to an explicit marking guide by a competent marker, ideally with scripts annotated for student feedback.

For simplicity here it will be assumed that the soup contains *questions* and formal procedures for students to answer those questions and formal procedures for the *instructors* to mark those questions.

Instead of creating a new course, why not just give students links to C_1 and C_2 and tell them to study that material? These reason is that these courses were *designed* to be delivered face-to-face with the lecturers giving a lot of extra information beyond that available on the internet. Also, in general, some of the on-line material available from many web-based courses is "private", including the assessment. Let us assume that the requirement for the new networks course is that it is to be delivered over the internet with everything required online as one of the new generation of MOOCs, *Massive Open Online Courses*. These are designed to be studied by large numbers of students worldwide. Neither C_1 nor C_2 are MOOCs and they do not meet this specification.

Notice how the specification is evolving during the design process. Here a potential solution was generated but rejected because it did not meet an implicit requirement. After this the requirements were revisited and the MOOC specification became explicit. Notice too that a lot of new words are entering the soup, *e.g.* learning outcome, questions, scripts, marking, and MOOCs.

Top-down disassembly and bottom-up assembly

The network course being designed will be delivered over time. At the microlevel there is the time it takes a student to read a section of text, look at a picture, watch a video, run a computer exercise, and so on. At the macrolevel there is the time for the whole course. The micro-, meso- and macro- depend a number of things, including the number of concepts and topics in the curriculum, the depth to which they will be studied, the prior knowledge of the students and their ability, and other constraints such as the other course modules being studied on a programme. A main determinant of study time is the expected *learning outcomes* for the course and its parts.

Suppose there is an external constraint that the course must not take more than one hundred hours to complete and the course designer decides that it will last ten weeks and involve ten study hours per week, and deciding to disaggregate the set of all things to be taught into ten *lessons*. Each lesson requires a name so suppose, taking inspiration from C_1 and C_2, the following *soup list* is made:

List of possible lessons

Introduction	Graphs and Networks
Trees and Lattices	Random Networks
Scale-free networks & Power Laws	Small worlds and Weak Ties
Clustering & Community Detection	Network Evolution
Error and Attack Tolerance	Percolation & Cascades
Hypergraphs and hypernetworks	Big Data
Examples: biology, telecoms, transport, business, crime	Assignment & Assessment-1
Assignment & Assessment-2	Hands-on data analysis
Networks of networks - multilevel systems	Visual analytics

In the design process almost everything is iterative and almost no task is completed at the first iteration. Here the task is to structure the course into ten lessons with appropriate names whose intension at *Level N+k-1* matches the intended extension at *Level N+k-2*. A *soup list* is a list of possible intermediate words from which the desired list will be produced.

The list generates new questions regarding the specification, *e.g.* will the course be assessed by one or two assignments? Another question is whether there will be hands-on data analysis? These things are for the client to decide, and the specification is updated yet again. Suppose the client says that hands-on data analysis is highly desirable, but although they definitely have the requirement that the course is rigorously assessed they do not have the expertise to specify exactly how to do it.

Suppose the designer reasons that an assignment half way through the course will give students feedback on their progress and understanding, and that an assignment at the end of the course will test that the learning outcomes have been achieved. The feedback from the first assignment gives students the possibility of changing their study behaviour, perhaps doing less of one thing and more of another, or applying better ideas or principles emphasised in the feedback.

How will the course be graded, *i.e.* what will be the traffic of marks and grades? Will it be pass-fail, as with a driving test? Will there be marks out of a hundred? Will there be a scale such as A, B, C, D, E possibly with plus and minus? These are things for the client to decide. Suppose the client says that the course will be graded as marks out of a hundred and that these will be associated with a scale A, B, C, D, where D means fail, A means passing to very high standard, *etc.* The word "fail" has now entered the soup. Can the student take the assignments or course again if they fail? The client says that students can repeat the course.

Following these deliberations the designer suggests that the middle and last lessons should be dedicated to assignments. The designer also argues that if there are to be hands-on exercises, these should be introduced early in the course so that students can do hands-on exercises as the course progresses. The designer has already agreed with the client that the first lesson will be a general overview. So the following skeleton hypersimplex is emerging: $\langle$Introduction, ... , Hands-on Exercises, ... , Assessment 1, ... , Assessment 2; $R_{\text{lesson-study-order}}\rangle$.

Assembling abstract hypersimplices during the design process

The emerging hypersimplex, $\langle \text{Introduction}, \ldots, \text{Hands-on Exercises}, \ldots, \text{Assessment 1}, \ldots, \text{Assessment 2}; R_{\text{lesson-study-order}} \rangle$, is a typical partly instantiated expression of the design. The symbols ... mean "lesson names to be decided later" as the designer "works up" the hypersimplex into a fully instantiated structure.

According to the specification there are to be ten lessons. Four have been allocated, leaving six to be synthesised from the following list:

List of remaining possible lessons	
Graphs and Networks	Trees and Lattices
Scale-free networks & Power Laws	Random Networks
Small worlds and Weak Ties	Network Evolution
Clustering & Community Detection	Error and Attack Tolerance
Percolation & Cascades	Hypergraphs & hypernetworks
Examples: biology, telecoms, transport, business, crime	Big Data
Networks of networks - multilevel systems	Visual analytics

Abstracting six lesson names from these fourteen possible names involves (i) deconstructing these names at *Level N+k-1* into more detailed words at *Level N+k-2*, (ii) assembling those lower level words into new lessons, and (iii) giving the new *Level N+k-1* assemblies names with intensions that match their *Level N+k-2* extension. Suppose this process results in the following list of lesson names:

1. Introduction
2. Graphs, Networks, Trees and Lattices
3. Hands-on Exercises, Big Data & Visualisation
4. Random Networks, Scale-free Networks & Power Laws
5. Assessment-1
6. Connectivity: Small Worlds, Clustering and Community Detection
7. Percolation, Cascades and Epidemics
8. Hypergraphs, hypernetworks
9. Networks of Networks & Multilevel Systems
10. Assessment-2

This list leaves out the examples. Will these appear in other lessons, or should they be given at the end when students know all the technical details? How practical is this structure? Each of these lessons lasts ten hours. Is it feasible to teach students how to use one or more network computer packages in, say, five hours? Does everything that should be included in the course fit into this structure?

It will be supposed that the client and designer decide that, although it is very desirable, it is not feasible to have hands-on activities in this course with just five hours to download, set up, and learn how to use a computer package. They decide that it is best to have another course focussing on hands-on experimentation using standard computer packages. Note that this is a very big change to the requirements. Including hands-on experimentation is highly desirable, but to include it could compromise the emerging course and, regrettably, this requirement must be dropped.

On making this major change in specification, the following hypersimplex emerges and for the time being this will be the structure of the course:

$\sigma_{\text{course}}^{N+k} = \langle$ 1. Introduction,
2. Graphs, Networks, Trees and Lattices,
3. Random Networks, Scale-free Networks & Power Laws,
4. Connectivity: Small Worlds, Clustering and Community Detection,
5. Percolation, Cascades and Epidemics,
6. Assessment-1,
7. Hypergraphs, hypernetworks,
8. Networks of Networks & Multilevel Systems,
9. Examples, Big Data & Visualisation,
10. Assessment-2; $R_{\text{order-of-lessons}}^{N+k-1} \rangle$

Instantiating lower levels

Having built the abstract hypersimplex $\sigma_{\text{course}}^{N+k}$ it is necessary to instantiate the *Level N+k-2* vertices with *Level N+k-2* objects. For example, what will be in $\langle$1.Introduction$\rangle$. Like almost every other introductory course on graphs and networks, this will include a discussion of the Königsberg Bridge problem and Euler's solution. The final part of the instantiation will be the actual teaching materials at *Level N+k-3* as shown in Fig. 9.13. The intermediate words KB_i identify the actual pieces of teaching materials that will be used to discuss the Königsberg Bridge Problem. At this stage, for example, the text for KB_1 may not be written, but the design specifies that the hypersimplex $\langle KB_1, KB_2, KB_3, KB_4, KB_5, KB_6; R_{KB} \rangle$ will begin with text.

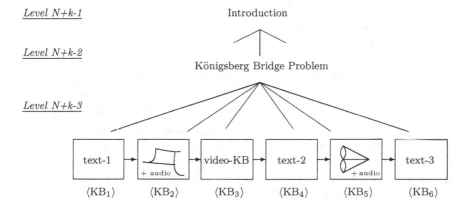

Fig. 9.13 Part of the introduction is fully instantiated a the lowest level

Checking the multilevel coherence of the course

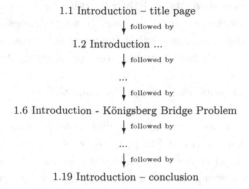

1.1 Introduction – title page
↓ followed by
1.2 Introduction ...
↓ followed by
...
↓ followed by
1.6 Introduction - Königsberg Bridge Problem
↓ followed by
...
↓ followed by
1.19 Introduction – conclusion

Fig. 9.14   The *Level N+k-1* hypersimplex $\langle 1.1, 1.2, 1.3, \ldots, 1.19; R_{\text{introduction}}\rangle$

The *Level N+k-1* hypersimplex $\langle 1.1, 1.2, 1.3, \ldots, 1.19; R_{\text{introduction}}\rangle$ is shown in Fig. 9.14. Part of the relational structure of the course involves the order in which the parts are presented. This must also reflect the lower level order of things that should be studied before others.

Let W_c denote the set of lowest level words that denote the content or the *curriculum* of the new course. This set of words has a quasi-order on it, since for some w_i and w_j is is necessary for the ideas associated with w_i to be understood before the ideas associated with w_j can be understood, $w_i \leq_\tau w_j$. For example, for most words w_i, "vertex" $\leq_\tau w_i$, "edge" $\leq_\tau w_i$, "network" $\leq_\tau w_i$, and "degree" $\leq_\tau w_i$.

As the course is implemented it will be necessary to ensure that the order relation on the lessons respects this pedagogic order relation.

Implementation and use

By now it should be clear that the design of the course involves defining its parts and how they will be assembled as multilevel hypersimplices in the context of emerging requirements and constraints.

At some stage the design will be *implemented*, *i.e.* the abstract design of the system will be instantiated with real things. In the design of simple systems the implementation may be done in parallel to the design. For example, during the design of parts of the course tangible parts may be collected, *e.g.* videos, pieces of text, diagrams and photographs, animations, and so on. The use of these components may become part of the specification, for example the use of a video lecture by a famous network scientist.

Following the design of the course and its construction with all the materials making appropriate hypersimplices bound together as a website, the course must be delivered. Designing the delivery system goes alongside the design of the course itself. This system involves students and their recruitment, management of their records and interactions with the course website. It also includes the design of an assessment system with the requirements that the learning outcomes are adequately tested and that there is a well functioning system of assignment submission and marking.

As it involves student agents interacting with many things including each other, instructor agents, *etc.*, the student-course interaction system is more complex than the course itself. The course designer and their clients will have more or less explicit models of how this system will work. In principle it is possible for these models to be implemented as computer simulations to test the possible behaviours of the system. In very complex systems this would be the only way to try to understand their future behaviour. In this relatively simple case the models are tested *in vivo* with the first cohort of students, with data being collected for analytic purposes that will enable the dynamics of the new system to be better understood. This is analogous to using finite element analysis to predict the behaviour of mechanical parts under stress, or building a physical prototype.

Conclusions from the example

This section has illustrated how the theories of design and hypernetworks apply to the design of a social artefact. For simplicity of presentation the system being designed is not very complex, and usually the design of courses is much more implicit than described here. The main points being illustrated are:

- artificial systems created to fulfil a purpose are designed
- design *always* involves building a multilevel vocabulary to represent the system
- there is a coevolution between specification and design generation and evaluation
- design involves solving the multilevel Intermediate Word problem
- design begins with a multilevel "soup" of things possibly relevant to the design
- design involves deconstructing high level concepts to lower level concepts
- design involves assembling lower level constructs into hypersimplices
- design involves identifying traffic on the emerging backcloth
- design involves instantiating all the abstract system parts with explicit structures
- design involves analysing the interactions between parts of multilevel systems
- design continues through implementation and use as systems adapt and evolve

The most important of these is that design involves bottom-up assembling parts into multilevel hypersimplices with emergent behaviour, hypothesising top-down new and uninstantiated things and possible relationships between them and instantiating, while at the same time reasoning about hypothetical behaviour at high levels and testing actual behaviour at low levels, as illustrated in Fig. 9.15. Alongside this is the importance of understanding that Intermediate Word definition occurs throughout the design process in combination with multilevel assembly and disassembly, and that the evolving representation co-evolves with the specification. This relates to what has been called *ontological uncertainty* [Lane & Maxfield (2003)] in innovation.

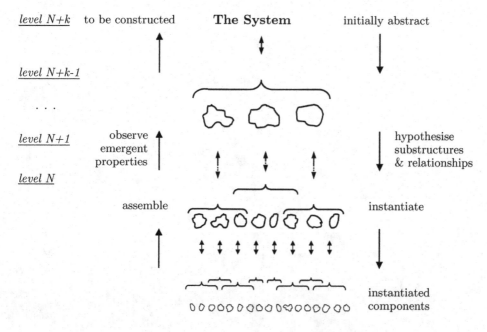

Fig. 9.15 Design as bottom-up assembly, top-down hypothesising, generation & reasoning

Chapter 10

Policy, Design, Planning and Science

10.1 The entanglement of policy, design, planning and science

Social systems are constantly changing and, even if it existed now, a perfect science of society would cease to be perfect in the future. The reason is that human beings deliberately change the rules by which societies work, and in so doing they can change the basic logic of the system.

Social systems are *artificial*. They are designed and planned. The process of designing, planning and implementing the future is called *policy*. Put simply

policy is designing, planning, implementing and managing the future.

Since designers are the first scientists of the systems they create, policy makers are not only engaged in design but they are also engaged in creating the science which will enable them to understand, design, plan, implement and manage the futures they create. Thus complex systems science, policy and design are inextricably entangled.

Fig. 10.1 Complex systems science, policy, planning and design are inextricably entangled

This entanglement is important because policy has the need for new scientific understanding. Policy makers need scientists to help create that understanding in the context of the design and management of complex multilevel systems. Arguably, there can be no science of complex socio-physical systems outside of policy, since complex systems scientists need policy and design as their laboratory. By themselves most scientists do not have the mandate or the money to test their theories.

10.2 Policy and Science

Policy is concerned with artificial systems that are designed, planned, created and managed in terms of implicit or explicit *requirements*. As such it is normative and assumes that, within the constraints of the political system, the policy maker has a mandate to decide what *ought* to be and *how* try to achieve it.

Policy as unrepeatable experiment

It can be argued that most policies are unrepeatable experiments. Although they may aim for specific goals there is generally no certainty that the goal will be achieved as planned. For example, the fisheries policies of Europe over the last quarter century can be seen as an experiment in economics that has resulted in an environmental catastrophe [Booker (2007)], but it is not possible to go back to try policies that might have worked better. Similarly, the recent near collapse of the world financial system is due to failed experiments with new financial instruments and hands-off regulation.

Experimentation by policymakers is usually not conducted in a scientific way. One reason for this is that some policy makers have an unshakable faith that the interventions they make will have the outcomes they predict and they see no need to monitor policy implementation from a scientific perspective so that we can learn if the policy goes wrong. When things do go wrong, as they do from time to time, boards of inquiry try to piece together whatever data exists to find out what happened in order to ensure that it does not happen again.

Related to this is the problem that if systematic data were kept on policy implementation, when things go wrong these data provide an audit trail to policy design errors and policy implementation errors. From a policy maker's perspective there is no need for scientific data collection. If things go as planned collecting data in a systematic way for scientific analysis is an expensive irrelevance, and if things go wrong these data will cruelly expose the shortcomings of the policy, the policymaker, and the implementation.

Some policy data are systematically retained for historical purposes. For example, the Cabinet papers of the UK Government are released after thirty years:

> Cabinet conclusions and memoranda are assessed and (where appropriate) released after 30 years. Some are closed for longer when they contain especially sensitive material, either about people or policy periods ... If you want to see a record that has not been released, you can submit a Freedom of Information request. The National Archives and the relevant government department(s) will examine your request and the document to decide if it can be released. [UK National Archives (2012)]

Confidentiality is essential for effective government and Section 1.6 of *The Cabinet Document Officers' Handbook* states that: "The 'need to know' principle is paramount. No one, other than a member of the Cabinet or the Ministerial Committee concerned, has an automatic right of access to a Cabinet document". [Cabinet Office (2007)]. However, Section 4.12 on disposal of documents states that "Official Committee documents should be destroyed within departments in accordance with normal security procedures".

Thus, although some information is retained and made public after thirty years, other information is routinely destroyed. Also, some documents may not be released even after thirty years if they contain sensitive information.

Here no solution is offered to the conflict between collecting and publishing policy information for scientific purposes, and the need for policy makers to know that they will not be in jeopardy of their inadequacies being routinely exposed at some future date. However, the fact remains that most policy interventions are experiments with unpredictable outcomes.

Policy as scientific experiment

Compared to policymakers, by themselves scientists generally cannot conduct experiments on complex systems such as a city, a national economy or a multinational company. They do not have the mandate and they do not have the money. Usually scientists do not have the moral or legal authority to make interventions on systems, and they do not have millions or billions of dollars necessary to make interventions. Scientists cannot build bridges or shopping malls, they cannot impose new policies on health provision, and they cannot set up new factories, sell banks, or buy large tracts of land. To conduct experiments, scientists must align themselves with policymakers, as consultants or advisors. In such collaborations scientists are usually the junior partner, tolerated as long as they are useful.

Scientists and policymakers can have different ways of looking at the world. For scientists everything is contingent since the scientific method requires that the truths of science must be subject to challenge and refutation. In contrast, many policy makers want certainty. They want to be advised that a particular policy intervention will have a single predictable outcome. Of course there are scientists and policymakers all along the contingency-certainty spectrum.

Another problem with the interface between science and policy is that science is not a single entity, and scientists do not always agree. Not infrequently, when a scientist says one thing another scientist can be found who says the opposite. Understandably policy makers will select scientific advisors whose views match their own.

A particular problem for the policy-science interface is the view of some scientists that science is objective and that the scientist's job is to produce "the science" and for policy to made on the basis of it. Today it is common to hear policy makers speak of *evidence-based policy* implying that their policies are based on incontrovertible scientific evidence.

Evidence-based policy

As an example of evidence-based policy consider the problem of bovine tuberculosis caused, in part, by badgers. In the UK this is an emotive issue because many farmers suffer severe disruption and loss from infections that they passionately believe are carried by wild badgers. The farmers' favoured solution to this problem is to eradicate badgers by culling them. However, "badgers are a protected species and it is illegal to wilfully kill, injure or take one or to interfere with a badger sett. If you are convicted you could face up to six months imprisonment or a substantial fine". [https://www.askthe.police.uk/content/Q31.htm]. To kill badgers legally it is necessary to have a licence.

The UK has a very active animal rights movement with a few individuals going beyond the law causing a security threat. Any proposed badger cull could be expected to result in violent conflict and disorder on a significant scale. No government would sanction a cull unless there were compelling reasons to do so.

In response to intense lobbying the British Government commissioned a study on the effectiveness of a badger cull for controlling bovine tuberculosis. The *Randomised Badger Culling Trial* ran from 1998 to 2007 and was overseen by the Independent Scientific Group on Cattle TB [http://www.defra.gov.uk/animal-diseases/a-z/bovinetb/research/]. A letter to the minister prefacing the final report gave the summary that:

> The [Independent Scientific Group's] work – most of which has already been published in peer-reviewed scientific journals – has reached two key conclusions. First, while badgers are clearly a source of cattle TB, careful evaluation of our own and others' data indicates that badger culling can make no meaningful contribution to cattle TB control in Britain. Indeed, some policies under consideration are likely to make matters worse rather than better. Second, weaknesses in cattle testing regimes mean that cattle themselves contribute significantly to the persistence and spread of disease in all areas where TB occurs, and in some parts of Britain are likely to be the main source of infection. Scientific findings indicate that the rising incidence of disease can be reversed, and geographical spread contained, by the rigid application of cattle-based control measures alone.
>
> Our Report provides advice on the need for Defra [Department for Environment, Food and Rural Affairs] to develop disease control strategies, based on scientific findings. Implementation of such strategies will require Defra to institute more effective operational structures, and the farming and veterinary communities to accept the scientific findings. If this can be achieved, the ISG is confident that the measures outlined in this Report will greatly improve TB control in Britain. [DEFRA (2007)]

This letter to the minister gives strong policy guidance that a cull will be ineffective, and the then Labour Government implemented a policy that there would be no cull. Of course this policy was not welcomed by farmers who continued to lobby the then Conservative opposition.

As this report explains, the effect of culling badgers is not clear-cut and depends on many things, *e.g.* "At the start of the RBCT badgers lived in territorial social groups, and M. bovis infections were found to be strongly clustered on scales of 1–2 km. However, removing badgers by culling was found to disrupt their social organisation, causing remaining badgers to range more widely both inside and around the outside of culled areas. Probably as a result the proportion of badgers infected with M. bovis rose markedly in response to repeated culling, and infections also became less spatially localised. Hence, although proactive culling reduced badger activity by approximately 70%, reductions in the density of infected badgers were much less marked, and infections became more widely dispersed."[DEFRA (2007)].

After the 2010 general election a Conservative-led coalition formed the Government which became more sympathetic to the farmers' lobby for a badger cull. On 19th July 2011 the then Secretary of State gave a speech in which she said:

> Unless we tackle each and every transmission route, including from badgers to cattle, we are likely to see the situation deteriorate further.
>
> There is great strength of feeling on this issue, and that is why I have carefully considered the scientific evidence and the large number of responses to our public consultation. I know that a large section of the public is opposed to culling and that many people are particularly concerned about whether it will actually be effective in reducing TB in cattle and whether it will be humane. I wish that there were some other practical way of dealing with this matter, but we cannot escape the fact that the evidence supports the case for a controlled reduction of the badger population in areas worst affected by bovine TB. With the problem of TB spreading and no usable vaccine on the horizon, I am strongly minded to allow controlled culling, carried out by groups of farmers and landowners as part of a science-led and carefully managed policy of badger control. [Hansard (2011)]

In the debate that followed her announcement the Secretary of State used the term "peer reviewed" four times when citing evidence to support her position, suggesting an irrefutable gold standard of truth for evidence-based policy. However, the scientific community that produces peer-reviewed publications knows that the process can be flawed, *e.g* [Baxt *et al* (1998)].

One of the most interesting things to emerge from this story is that both sides of the badger cull controversy used the same scientific evidence to support their opposing positions. Also, it is interesting that the Labour Government used evidence to implement a policy of not culling while the Conservative-led coalition used the same evidence to implement a policy of culling.

Although the policy was established in law, the programme of badger culling had to be postponed as explained by the Secretary of State:

> The exceptionally bad weather this summer has put a number of pressures on our farmers and caused significant problems. Protracted legal proceedings and the request of the police to delay the start until after

the Olympics and Paralympics have meant that we have moved beyond the optimal time for delivering an effective cull. We should have begun in the summer. In addition to these problems, the most recent fieldwork has revealed that badger numbers in the two areas are significantly higher than previously thought, which only highlights the scale of the problem we are dealing with.

Today I have received a letter from the president of the NFU [National Farmers Union], on behalf of the companies co-ordinating the culls, explaining why they do not feel that the culls can go ahead this year and requesting that they be postponed until next summer. In these circumstances, it is the right thing to do, and as they are the people who have to deliver this policy on the ground and work within the science, I respect their decision. [Hansard (2013)]

Thus the highly contentious policy of culling badgers was blown off course by apparently unrelated events, the Olympic Games, and the weather, neither of which featured in the scientific evidence.

Policy Makers and Scientists

For simplicity a distinction will be made between scientists and policymakers. The term *policy makers* will mean those who make policy including elected politicians, their political advisors and supporters, and their officers who provide politically neutral administrative and practical support. Here the officers will include domain experts and scientists who can advise on the possible outcome or policy, and even help formulate policy. The difference between these civil service scientists and research scientists is that their focus is not on scientific innovation, but on applying known science as effectively as possible to achieve policy aims. In reality these divisions are blurred. Some politicians are scientists, and some scientists are politicians. Some civil service scientists publish original research, and some research scientists act more like officers, giving expert advice as consultants or even employees.

Rhetoric in policy and science

The collaboration between scientists and policy makers is complicated by their different agendas, motivations and methods. Scientists are motivated by objective discovery while policy makers are motivated by the need to get elected and address practical problems. Policy makers must be able to persuade people to vote for them and them must be able to persuade their constituency that their policies are appropriate and will work. In this respect the logic of politicians includes rhetoric, and this is different to the logic of science. In rhetoric a good joke can be more powerful than a theorem.

Addressing the question of why we need rhetoric, an article in the *Stanford Encyclopaedia of Philosophy* on Aristotle's Rhetoric[1] suggests the following:

---
[1] http://plato.stanford.edu/entries/aristotle-rhetoric

It could still be objected that rhetoric is only useful for those who want to outwit their audience and conceal their real aims, since someone who just wants to communicate the truth could be straightforward and would not need rhetorical tools. This, however, is not Aristotle's point of view: Even those who just try to establish what is just and true need the help of rhetoric when they are faced with a public audience. Aristotle tells us that it is impossible to teach such an audience, even if the speaker had the most exact knowledge of the subject. Obviously he thinks that the audience of a public speech consists of ordinary people who are not able to follow an exact proof based on the principles of a science. Further, such an audience can easily be distracted by factors that do not pertain to the subject at all; sometimes they are receptive to flattery or just try to increase their own advantage. And this situation becomes even worse if the constitution, the laws, and the rhetorical habits in a city are bad. Finally, most of the topics that are usually discussed in public speeches do not allow of exact knowledge, but leave room for doubt; especially in such cases it is important that the speaker seems to be a credible person and that the audience is in a sympathetic mood. For all those reasons, affecting the decisions of juries and assemblies is a matter of persuasiveness, not of knowledge. It is true that some people manage to be persuasive either at random or by habit, but it is rhetoric that gives us a method to discover all means of persuasion on any topic whatsoever.

While the objective of rhetoric is persuasion, the objective of science is to formulate theory that is compatible with observation. It is not acceptable for a scientist to suppress inconvenient evidence and it is not acceptable for scientists to tell lies. However, scientists must persuade their peers and the public that their theories are correct and they too will use rhetorical principles. Complex systems science addresses messy multilevel systems that do not allow of exact knowledge and leave room for doubt. It remains a challenge for this science to present its arguments in persuasive ways that do not violate the principle of absolute honesty and transparency. Figure 10.2 shows possible roles of individuals in the partisan-disinterested spectrum.

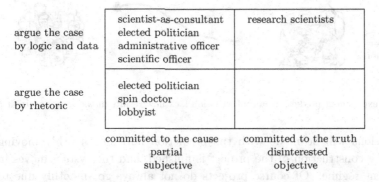

Fig. 10.2 A simplified characterisation of policy and scientific roles

10.3 Policy and Design

Policy as design

The interventions of policy deliberately attempt to change what exists and create artificial systems, and artificial systems are *designed*. As discussed in the previous chapter, the design process begins with the establishment of needs or requirements and proceeds with the generation of possible ways to satisfy those requirements. These are evaluated, and the design solution is either accepted or new and better solutions are generated. This process is open ended and in general there are many possible ways of satisfying the requirements. Some solutions are better than others, but generally the requirements conflict and a compromise has to be reached.

Policy makers must stay in the coevolutionary designing-the-future loop

The iterative generate-evaluate nature of policy and changing requirements underlies a co-evolution between a policy problem and its solution. The design process involves *learning* about the system being created, including its possible futures which may not have been apparent when the process began. It begins by sketching out possibilities in a very general way and proceeds by vague possibilities being instantiated with concrete decisions. The coevolution between the specified requirements and the creation of a satisficing policy requires the policy maker to remain in the designing-the-future loop. Otherwise they will get a future designed by others.

Prediction in policy, design and planning

Design involves identifying appropriate sets of components and specifying how they can be assembled to form structures and, eventually, whole multilevel systems. Planning and implementation of designs to create new systems are dynamic processes that involve *predictions* of how systems will behave.

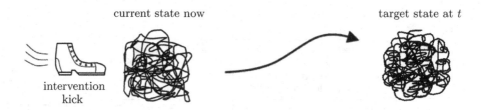

Fig. 10.3 Policy experiment prediction: an intervention kick now results in the target state at t

Ideally decisions will be made with the prediction horizon smoothly moving forward until the construction of the project is finished, and the system moves in to a management regime. Of course projects do not always go smoothly due to unexpected events. Design and implementation trajectories evolve within an ever-unfolding time horizon, where prediction involves hypothesising that certain future states are possible, and that taking a particular action will send the system to one of a set of particular future states. This is shown graphically in Fig. 10.3, where the system is given an *intervention kick* with the expectation (prediction) that it will end up at some target state. From a traditional scientific perspective this becomes a simple experiment that can be used to test the underlying theory: give the system the kick and see if the predicted state emerges, as shown in Fig. 10.3.

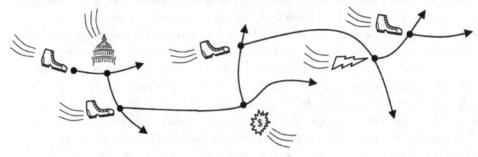

Fig. 10.4 What does prediction mean when the system is continually knocked off target?

Reality is more complicated than this. The decision kick that initiates a project may be taken in ignorance of events that will knock the system off trajectory, *e.g.* Fig. 10.4 shows a project knocked off trajectory by unexpected political decisions, unexpected financial problems, and "Acts of God" such as a lightning strike.

As Fig. 10.4 illustrates, a policy can be characterised as giving the system a kick in the hope that it will reach the target, but with the expectation that unpredictable events will knock it off trajectory before the system can reach the target. What does it mean to make or test a prediction when the final state that characterises the prediction will never be reached? Perhaps the predictions can be tested to the point that the system is first knocked off trajectory by political decisions? After the second restorative kick, perhaps the prediction can be tested at the point that finance knocks it off trajectory? And after the third restorative kick, perhaps the prediction can be tested at the point that lightning strikes?

Prediction in policy and design is more complicated than conventional experiments. How can the correctness of designs be tested statistically in this dynamic environment? What does it mean to make a prediction in a multilevel system? Does it mean predicting some particular system states at micro-, meso- or macro-levels? Or does it mean predicting all system states at all levels, since in complex systems microlevel individuals can have massive effects at meso- and macrolevels.

For systems that are sensitive to initial conditions a single point prediction has almost no information. In general it is necessary to consider distributions of outcomes from the initial conditions. This suggests that statistical tests will have to be multilevel, and that there is a completely new approach to statistical analysis waiting to be discovered and developed.

10.4 Reasoning and logic in policy design

Elements of logic

Complex systems science is based on well defined principles of logical argument established thousands of years ago. A central idea is that one can have propositions, and that these can be true or false. Some propositions are empirical, and some are deduced from others *e.g.* the Aristotelian Darii syllogism gives a way of generating new propositions from existing knowledge: "all men are mortal" and "Socrates is a man", implies that "Socrates is mortal". The logical operators of conjunction (*and*) and disjunction (*or*) allow propositions to be combined into larger propositions, *e.g.* "all men are mortal" is a proposition and "Socrates is a man" is a proposition, while "all men are mortal *and* Socrates is a man" is a compound proposition which is true if both parts are true. Negation flips the *truth value* of a proposition, so that "this house is inhabited" is true, when "this house is not inhabited" is false.

Entailment, that something implies something else, is at the heart of science. One of the simplest rules of entailment is *modus pones* which has the form "if A implies B is true, and A is true then B is true". In policy, *modus pones* can be applied as follows. "The theory tells us that A implies B is true, we want B, so we will take action to make A true". Then since "A implies B is true" and "A is true" it follows that "B is true".

This suggests different roles for the scientist and the policy maker. The scientist's job is to find theory that provides entailments relevant to the objectives of policy makers, such as "A implies B". The policy maker's job is decide whether B is desirable or not, and whether or not it is desirable to take action to induce A and therefore B.

Scientists have no more moral authority than anyone else to assert that either A or B are desirable, and generally they do not have the financial resource or moral authority to induce A. If a scientist wants to test the hypothesis that "A implies B" they will have to convince the policy makers that "A implies B" is true in the context of the policy makers judging that B is desirable and that it is desirable to induce A in order to induce B.

In any logical analysis, inconsistency is fatal. Inconsistency means that the assumptions made allow any proposition to be demonstrated as being both true and false. Part of the scientist's job is to detect incorrect entailments and to resolve logical inconsistencies.

Correctness and proof

Suppose you were making a policy decision about yourself or your family that will greatly improve your life if successful, but have a high cost if unsuccessful. For example, you may be thinking of investing a lot of money on the stock market, buying or selling a house, or doing something else where the stakes are high. Almost certainly you would reason about the possible outcomes of any actions you might take, and almost certainly you will want your arguments to be free of logical flaws. In general policy makers want a convincing argument that their interventions will results in the outcomes they want, and they want those arguments to be be *correct*.

Arguments can be shown to be correct by *formal proof*. To illustrate this, consider Pythagoras' theorem that "in a right angled triangle, the square of the length of the hypotenuse is equal to the sum of the squares of the lengths of the other two two sides", as shown in Fig. 10.5(a).

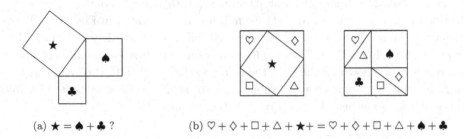

(a) $\bigstar = \spadesuit + \clubsuit$?   (b) $\heartsuit + \diamondsuit + \square + \triangle + \bigstar + = \heartsuit + \diamondsuit + \square + \triangle + \spadesuit + \clubsuit$

Fig. 10.5 Pythagoras' theorem

Figure 10.5(b) gives a graphical argument. The two large squares have equal area, so $\heartsuit + \diamondsuit + \square + \triangle + \bigstar = \heartsuit + \diamondsuit + \square + \triangle + \spadesuit + \clubsuit$. Subtracting $\heartsuit + \diamondsuit + \square + \triangle$ from both sides of this equation leaves $\bigstar = \spadesuit + \clubsuit$ as required.

Although this argument seems convincing, it is not a formal proof and it could be wrong. A *formal system* consists of "meaningless statements", which are finite sequences of abstract symbols and rules for manipulating them. As an example consider the formal system with the symbols =, $\oplus$, $\square$, $\triangle$, $\heartsuit$, $\diamondsuit$, $\bigstar$, $\clubsuit$ and $\spadesuit$. Assume any string of these symbols can be a statement in the formal system.

The *transformation rules* are another part of the formal system. These give rules for transforming one statement into another. Let this system have one rule:

Transformation rule: $a \oplus x = a \oplus y \rightarrow x = y$

where a is any symbol and x and y are statements in the system and the arrow $\rightarrow$ stands for 'is transformed to", or "can be substituted by".

Finally, in order to make a deduction the system must contain *axioms*. For example the system can have the axiom

Axiom: $\heartsuit \oplus \diamondsuit \oplus \square \oplus \triangle \oplus \bigstar = \heartsuit \oplus \diamondsuit \oplus \square \oplus \triangle \oplus \spadesuit \oplus \clubsuit$

A *formal proof* of the theorem $\bigstar = \spadesuit \oplus \clubsuit$ consists of a sequence of transformations (applications of the transformation rule) that lead from the axiom string to the theorem string. For example,

$$\heartsuit \oplus \diamond \oplus \square \oplus \triangle \oplus \star = \heartsuit \oplus \diamond \oplus \square \oplus \triangle \oplus \spadesuit \oplus \clubsuit$$
$$\rightarrow \quad \diamond \oplus \square \oplus \triangle \oplus \star = \diamond \oplus \square \oplus \triangle \oplus \spadesuit \oplus \clubsuit$$
$$\rightarrow \quad \square \oplus \triangle \oplus \star = \square \oplus \triangle \oplus \spadesuit \oplus \clubsuit$$
$$\rightarrow \quad \triangle \oplus \star = \triangle \oplus \spadesuit \oplus \clubsuit$$
$$\rightarrow \quad \star = \spadesuit \oplus \clubsuit$$

is a formal proof that the theorem $\star = \spadesuit \oplus \clubsuit$ is true.

In his book *The Magic Machine* [Dewdney (1990)] tries to convince the reader than he can get a cubic inch of gold for nothing, as illustrated in Fig. 10.6. There are four gold blocks, a, b, c, and d, each an inch thick. Together they are arranged as a square with volume 8" × 8" × 1" = 64 cubic inches. If these blocks are rearranged as a rectangle as shown on the right the volume is 13" × 5" × 1" = 65 cubic inches. Thus, in some mysterious way 64 cubic inches of gold are transformed to 65 cubic inches of gold, so let us set up a formal system to prove it.

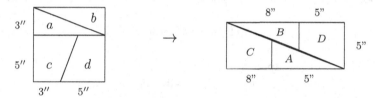

Fig. 10.6 How to get a cubic inch of gold for nothing. (Source: [Dewdney (1990)])

Let the transformation rules be

1	$12 \oplus 12 \oplus 20 \oplus 20 \rightarrow 64$	7	$b \rightarrow B$
2.	$a \rightarrow 12$	8.	$c \rightarrow C$
3.	$b \rightarrow 12$	9.	$d \rightarrow D$
4.	$c \rightarrow 20$	10.	$A \oplus B \oplus C \oplus D \rightarrow 5 \otimes 13$
5.	$d \rightarrow 20$	11.	$5 \otimes 13 \rightarrow 65$
6.	$a \rightarrow A$		

Let the axiom be $a \oplus b \oplus c \oplus d = a \oplus b \oplus c \oplus d$. Then

	Action			Rules used
	$a \oplus b \oplus c \oplus d$	=	$a \oplus b \oplus c \oplus d$	axiom
$\rightarrow$	$a \oplus b \oplus c \oplus d$	=	$A \oplus B \oplus C \oplus D$	6, 7, 8, 9
$\rightarrow$	$12 \oplus 12 \oplus 20 \oplus 20$	=	$A \oplus B \oplus C \oplus D$	2, 3, 4, 5
$\rightarrow$	$12 \oplus 12 \oplus 20 \oplus 20$	=	$5 \otimes 13$	10
$\rightarrow$	64	=	$5 \otimes 13$	1
$\rightarrow$	64	=	65	11

is a formal proof that 64 = 65.

Although the symbols are in principle meaningless, and there is no reason why 64 should not equal 65 in this system, the conclusion that sixty four equals sixty five suggests something has gone wrong and that the argument is flawed somewhere. There is nothing wrong with the axiom that $a \oplus b \oplus c \oplus d = a \oplus b \oplus c \oplus d$ so the fault lies elsewhere. On going through all the transformation rules it can be seen that the fault lies in $A \oplus B \oplus C \oplus D \rightarrow 5 \otimes 13$ since the blocks do not quite fit together.

The point of this example is that it is very easy to make errors when formulating deductive systems. This is why formal logic was invented, since it provides a method of demonstrating whether or not a proof is technically correct. Formal logic knows nothing about the meaning of what is proved. As seen above, formal logic is indifferent to the conclusion that sixty four equals sixty five *within that system*. However, that system is inconsistent with arithmetic, which suggested something is wrong. After this, examination of the assumptions of the argument showed that one of the assumed rules did not reflect empirical reality.

Paraphrasing the question at the beginning of this section, "if a policy maker is taking an important decision should the reasoning behind it be logically and technically correct?". If the answer to this question is affirmative then the argument can be analysed to test its correctness. Furthermore, this testing can be done by computer, provided the argument is well formed.

Surveyabilty

In mathematics it was believed that an individual's mathematical knowledge could be complete, in the sense that every proposition they believed to be true was an axiom or a rule, or could be derived from axioms and rules in a way *surveyed* by that individual. In other words an individual would believe nothing unless they had personally surveyed a proof that it was true. This belief was shaken in 1976 by the computer proof of the *Four Colour Problem* that asks if four colours is sufficient to colour any map so that no two adjacent countries are coloured the same. The proof involved testing a large number of particular cases which took 1,200 hours of computer time and was not surveyable.

The idea of computer proof appalled some mathematics. Some complained that computers make errors, but it turns out that human mathematicians make many more. So the idea that all of mathematical knowledge could be surveyed by a single person was abandoned. At best any individual can know only some locally consistent part, using the results of others as axioms in their system. Then mathematics becomes a kind of social intelligence – just like knowledge in other fields.

In mathematics it is assumed that everything has been surveyed by its own community, and that it is technically correct. However, errors are sometimes found, *e.g.* early but incorrect proofs of the four colour theorem [Wilson (2004)].

Formal systems are particularly well suited to implementation in computers, and automated and semi-automated proof are widely used. Thus the arguments of policy can in principle be computer tested for correctness, and surveyed by all.

10.5 Narratives in policy science

Policy is invariably expressed in vernacular language augmented by tables, diagrams and images. Policies tell stories about how the system is now, how it ought to be, and how it can be changed for the better. As such policies can be considered to *narratives*, stories in which events unfold through time. These are stories that some people believe to be true and others do not. Even scientific theories can be considered to be contingent stories about systems rather than logical-empirical systems that express immutable truths.

There is a view that engineering is based on theories such as electricity and mechanics in physics, with the theory prior to and detached from its applications. Some scientists believe that their theories and models can act as a toolbox for policy makers, rather than expecting policy to create new problems that their existing tools cannot handle. In the sciences of artificial socio-physical systems, scientific knowledge and theory coevolve with applications, *e.g.* the science of aeronautics has developed through the creation and study of aircraft, and the science of medicine has developed through the creation and study of remedies.

At best the existing toolbox for complex systems is incomplete, and tools designed for different systems may do damage. The science of complex systems needs new tools, including a machine-implementable theory of narrative.

Design is a story about systems that don't exist and how they may be brought into being. The narrative of design includes determining requirements and the generation of new systems to satisfy them. The narrative of design is cyclical, or helical through time, with issues being visited and revisited as abstract structures are instantiated with real parts and assembly relations. The narrative includes reasons for rejecting possible solutions and it includes stories about changing the requirements. The evolving narrative of design explicitly includes the creation of the systems of intermediate words required to define policy spaces and their dynamics. Thus designing the future is entangled with policy science, with design being the process of creating possible futures and policy science being the process of testing the dynamics of the future being designed.

10.6 Hypernetwork science for building policy narratives

Since policy is designing futures that do not already exist, like all synthetic systems it can be expected that the new system will have properties and behaviours not seen before in other systems. Then

> *scientists should expect to build new science for new policy problems.*

When scientists do not have an off-the-shelf solution, and their existing tools are not appropriate for a new policy area, hypernetwork theory can help build new system-specific science from first principles.

Narratives and the Intermediate Word Problem

For practical reasons decisions have to be made about what will be included in the analysis for the policy and what will not be included. This is the Intermediate Word Problem – which words in the soup are relevant and which are irrelevant? As shown in Fig. 10.7 establishing the vocabulary will be an iterative process. The dotted lines represent inevitable ambiguities in the *policy*.

The *policy space* for a particular policy will be defined to be the multilevel vocabulary relevant for that policy, and all propositions, models, analysis and simulations constructed within the vocabulary relevant to that particular policy.

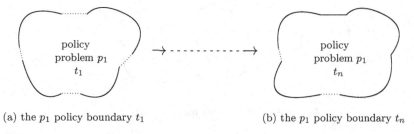

(a) the p_1 policy boundary t_1 (b) the p_1 policy boundary t_n

Fig. 10.7 Iterative definition of the space boundary for a particular policy problem

To illustrate this, consider the policy problem of a city providing infrastructure for electric vehicles. This will involve many agencies including the national government and local government, and cut across many areas including environmental planning, transportation planning, energy, economy, and others. For example, a city might decide to have a shareable electric car scheme like Paris, in which case there may be other stakeholders such as the car manufacturer and those who maintain the charging points, and the companies that generate and distribute the power.

During this process the specification of the policy problem will evolve. For example, it might be decided to make all the public transport vehicles electric, provide special car parks, and so on. It might be decided to work with another city, possibly abroad, to share design and development costs and make bulk purchases.

The boundary of policy problems will be determined by these kinds of consideration, and they illustrate the generality that *no two policy problems are the same*, and that different policy problems will require different combinations of scientific knowledge. This cannot exist *a priori* for all questions.

Multilevel hypersimplices as narrative events

A narrative is a sequence of events in time. As discussed in Chapter 8, events are marked by the creation of hypersimplices, so a *simple narrative* can be defined to be a sequence or trajectory of hypersimplices in time. Generally a *compound narrative* is a collection of parallel trajectories. Even more generally a multilevel narrative is a collection of narratives made of hypersimplex events at different levels.

Question-specific policy spaces

Every policy will have its own policy space, and this policy space must be constructed for each new policy. The collaboration between policy makers and scientists then becomes the policy makers deciding the policy issues and the scientists working with them to build appropriate policy spaces. Of course different policies may have intersecting policy spaces, as shown in Fig. 10.8, and some polices may have exactly the same policy space. But in all cases the policy space has to be built using the methods of science.

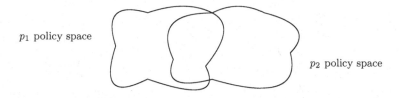

Fig. 10.8 Policy spaces may intersect

10.7 Art in science and policy

If a policy is to be supported by logical argument and evidence, then the argument should be correct. This does not make the premises or the conclusions true, but science provides ways of finding errors in the narrative. This does not guarantee that a policy has examined the best options. The scientific approach provides only one way of viewing the world relevant to policy. Clair O'Farrell writes:

> A scientific practice, in Foucault's account, is a particular set of codified relations between a precisely constructed knower and a precisely constructed object, with strict rules which govern the formation of concepts. Foucault was interested in science for a number of reasons. One of these was that 'science' had set itself up as the ultimate form of rational thought. With the Enlightenment, scientific reason became the privileged way of accessing truth. According to this view for knowledge to acquire value as 'truth', it had to constantly strive to become 'scientific', to construct and organize concepts according to certain rigorous criteria of scientificity. Foucault argues that scientific knowledge is not inherently 'superior' or more 'true' than other forms of knowledge. http://www.michel-foucault.com/concepts/index.html 2007

The argument for surveyable policy does not imply that science is the only form of knowledge. Even within science as defined here there are alternative descriptions of systems. There is the possibility all descriptions and forms of knowledge may be unified at an appropriate metalevel, but this is a speculation to be investigated.

Art provides another way of looking at the world that can inform policy, especially in creating new visions of how the future might be designed.

The science of complex systems must go beyond conventional science which fails to give answers to the many pressing policy issues at local, national and global scales. This involves questioning the methods of traditional science and seeking new ways of understanding. This includes considering the possibility of art making a contribution to the methodology of the emerging science of complex systems.

Surprisingly some scientists believe that art can contribute nothing to science, *e.g.* in a recent radio programme [Lythgo (2007)] the eminent biologist Lewis Wolpert proclaimed with great certainty that: "art has contributed zero to science historically ... There are all sorts of images from science that can give artists something to work on, but it does not go the other way ... The artist couldn't tell us a thing in that particular area". Many would disagree with this.

In the same radio programme Christopher Frayling gave a counter-example from the natural sciences: "Fred Hoyle, [was] beginning to work in Cambridge in the late forties, on his theory of a cyclical cosmology that things don't move in a linear way, they move in circles. He goes to see a film in nineteen forty eight made by Ealing Studios called *Dead of Night*. *Dead of Night* begins with someone pulling up at a country house – it ends with the same scene of someone pulling up at a country house. In between all sorts of things have happened but the entire movie is cyclical. It ends where it begins. It begins where it ends. And he went home and wrote in his diary 'My God! It's a cosmology. Maybe there's something in this cyclical cosmology.' The art had reinforced the idea in Fred Hoyle's mind and off we go with Hoyle's cosmology of the fifties". Frayling's example illustrates art being an inspiration for science.

Art can play a more direct role in science, as illustrated by the work on organisations by Eve Mitleton-Kelly: "During the analysis our resident artist, Julian Burton, will capture some of the themes, dilemmas and underlying assumptions in a picture. This has several advantages: many related aspects that are difficult to think about at the same time, can be captured in one picture; and very sensitive issues that are difficult to talk about, can be presented diagrammatically to workshop participants, before the presentation begins. Once they recognise what is being shown they may laugh and thus break the tension and open the issue(s) to discussion". [Mitleton-Kelly (2003)]

One such drawing shows a tower block with cracks running through it, and people on the top spinning plates. The impression is that this organisation is in danger of falling to pieces while everyone is over-busy attending to immediate tasks.

In the terminology of knowledge engineering, this use of art is a means of knowledge elicitation. The implications are (i) that art can capture information that cannot be captured in more conventional ways such as tape-recording, interview notes, or a questionnaire, (ii) that art can be used in a dialogue that elicits information in new ways, including subjects confirming or correcting the interpretation of previous interviews, (iii) that the work of art can be a way of enabling subjects to see their social situation in new and less threatening ways, and that (iv) that the work of art can be part of a social dynamic, enabling subjects to interact better with the social scientist and therefore being willing to provide information that they might otherwise withhold. In this case art becomes a *scientific instrument*, and is part of the scientific process.

The literature of the world is a huge dataset reflecting human behaviour, both real and imagined. It contains countless narratives of policy interventions and their consequences at micro-, meso- and macrolevels. As complex systems science develops its ability to abstract meaning from these narrative texts, it could contribute greatly to policy creation and policy testing. Similar comments apply to the paintings, sculptures and other artefacts that all tell their own story, no matter how obvious or obscure.

Contemporary art is oriented more towards exploration than traditional aesthetic. As such it provides a means of exploring the search space of possible worlds. We have asserted that policy makers and scientists design the future, creating and predicting the behaviour of new systems that have never existed before. Art can be viewed as the blue sky research laboratory of design, and therefore of science and policy. Art gives glimpses of the unknown unknowns.

Chapter 11

Notes and reflections

It has been argued that complex systems science is inextricably entangled with design and policy and that hypernetworks are central to them all. It is disappointing that so many questions and loose ends remain, and tempting to delay just a little longer to try to tie them all up and provide examples. However, it is time to stop and the rest remains work in progress.

This final chapter looks back on the others to produce a short synthesis. But its purpose is to look forward to a research programme that will address the outstanding problems and put hypernetwork science to work.

11.1 Resumé

Chapter 1 presented some of the fundamental ideas behind this book. The most fundamental is the idea that relations can be defined between more than two things. Some readers will already know about hypergraphs and have used Galois lattices to analyse social data. As argued in Chapters 2 and 3, hypergraphs are a first step towards handling general n-ary relations, but they lack orientation. Simplices overcome this. Chapter 1 gives a brief discussion of simplices in the context of algebraic topology, where topological problems such as finding the holes in a space are mapped to algebraic representations that make possible algorithmic computation. This is discussed in greater detail in Chapter 5. Although simplices are a big step forward, as discussed in detail in Chapter 4, Chapter 1 suggests that the hypersimplex is really what is required and this idea is developed in Chapter 6.

Chapter 1 gives a brief introduction to complexity and complex systems science and in this context introduces the ideas that policy is designing and planning the future, and that the science of complex systems, design and policy are inextricably entangled. These ideas are developed in detail in Chapters 8, 9 and 10.

Chapter 7 on multilevel systems aims to unify partial theories at micro-, meso- and macrolevels using the hypernetwork formalism to represent the multilevel dynamics of heterogeneous systems of systems.

11.2 Networks and Hypernetworks

Many readers will be looking for the new insights that hypernetworks can give to their existing knowledge of network theory, and for new ways of analysing data that adds something extra the techniques they already use. Although networks and hypergraphs have not been the focus here, it must be stressed that with hypernetworks they form a coherent whole. Although similar they are different, and working together they can form a general theory of multilevel relational dynamics.

Hypersimplices

Many examples have been given of hypersimplices and hopefully their existence and importance is beyond dispute.

Those who already use hypergraphs and the Galois lattice may ask what is new and useful in hypersimplices. First, it can be noted that most applications of hypergraphs implicitly order the vertices and implicitly know the relation that binds the vertices together. In other words they are using hypernetworks by another name. What then does hypernetwork theory provide that is not already known implicitly and currently used?

As networks of networks receive more attention the algebra of relations in the hypersimplices becomes more important. The conjunction and disjunction of relations have been known for many years but they are not widely used in network theory. The hypernetwork notation $\langle v_0, v_1, ..., v_p; R_1 \rangle \wedge \langle v_0, v_1, ..., v_p; R_2 \rangle = \langle v_0, v_1, ..., v_p; R_1 \wedge R_2 \rangle$ is a clear invitation to investigate more closely the operations between relations in systems of systems. Let the symbol $\circledast$ represent a relation between relations. Then $\langle R_0, R_1, ..., R_n; \circledast \rangle$ is a "higher metalevel" hypersimplex with vertices relations R_i and "meta-relation" $\circledast$, which suggests studying hypersimplices of the form: $\langle v_0, v_1, ..., v_p; \langle R_0, R_1, ..., R_n; \circledast \rangle \rangle$ when the vertices $v_0, ..., v_p$ are homogeneous, and $\langle a, b, c, ...; \langle R_0, R_1, ..., R_n; \circledast \rangle \rangle$ when the vertices $a, b, c, ...$ are heterogeneous. Conceptually such structures have been known for many years. The challenge is to implement them as computer programmes able to abstract useful information from the many new forms of data that are emerging. This research will generalise the Fundamental Question of Hypernetworks posed in Chapter 6.

Networks and hypernetwork dynamics

Those looking for new ideas and structures may find the combination of Q-analysis and star-hub analysis useful. The computational demands of Q-analysis are less than those of star-hub analysis. It was suggested in Chapter 4 that Q-analysis may make the computation of Galois pairs more tractable in the analysis of Galois lattices, and provide other interesting structures.

Those interested in percolation may find Q-analysis and the dynamics of q-transmission interesting for dynamic backcloth-traffic interdependencies. The idea of q-transmission mechanism can be relaxed with q-nearness being less likely to transmit than p-nearness for $q < p$. This relates to the idea of fuzzy q-nearness defined in the study of road accidents in Chapter 6.

Atkin's idea of p-event and the associated notion of structural time versus clock time introduced in Chapter 8 are essential to understand the dynamics of backcloth change. Creating new theories for the evolution of hypersimplices as structural trajectories in clock time is a big research challenge.

Topology

Hypernetwork theory is based on Atkin's hypothesis that the topology of the backcloth constrains the dynamics of the traffic. Chapter 6 gave a brief overview of this to provide a context for the development of hypernetworks. The power of algebraic topology comes from mapping continuous spaces into discrete algebraic representations suitable for algorithms and computation. The role played by holes and pseudo-holes in hypernetwork dynamics is an open research question.

11.3 Design

Chapter 9 shows how hypernetworks can model the design process, from large complex constructions such as space stations and stadia to softer system such as clothes, gardens and works of art. This view is much broader than *engineering design* with its emphasis on optimisation and control theory.

The argument that policy is designing the future does not require that the future can be uniquely specified and predicted in the way that finite element analysis allows the building and testing of virtual machines. A better analogue is designing and planning a city for which there is no optimum and no final product. Nonetheless cities are designed, planned and managed, as is the future.

The main point in this argument is that the designers and planners of any social system have a *vision* of what the future might be which is created from a sense of dissatisfaction with the current system or an unattractive vision of the what the system may become if no action is taken. As explained in Chapter 9, design is a heuristic coevolutionary process that is more or less the same across many domains of application. For example, design involves identifying parts and assembling them into wholes using appropriate processes. Thus designers must ask themselves "what is the desired whole", "what are the constituent parts", and "what processes might be used to assemble the parts into wholes". These seem to be important questions for policy makers to ask themselves as they make interventions with the explicit intention of making changes.

Generally designers are masters of complexity. They need to know the underlying scientific principles of the systems they are designing, they need to understand the environment for the system being designed such as regulation and fashion, they need to deal with clients who don't know what they want or what is possible within their constraints, and often they need to work in teams as a collective intelligence.

Policy makers also have to deal with complexity and uncertainty and, since they are designing the future, a better understanding of the design process could help them to make better decisions.

11.4 Policy

It is argued in Chapter 10 that complex system scientists must work with policy makers, but also noted that the cultures and agendas of scientists and policy makers may be different. Also the policy making framework can differ considerably between countries.

There are some significant challenges for scientists in working with policy makers. Recently a policy maker told me that the idea of formalising the decision making process, as suggested in this book, is not feasible, and that policy does not work this way. Another policy maker from a federal government department told me that "we don't use the concept of prediction when we make policy".

The polarised caricatures of scientists and policy makers in Chapter 10 overstate the case. Many policy makers are scientifically competent and engaged, while many scientists have political agendas. The point of this exaggerated dichotomy was to focus ideas so that we scientists can better understand why the policy making process can be so different to what we might think it should be.

Persuading policy makers that they do make implicit predictions or forecasts is a major challenge. Persuading them that it would be useful to make their predictions explicit by using scientific methods is a greater challenge, not least because science is contingent while policy makers want certainty.

11.5 Systems theory and complex systems science

Chapter 8 gave an overview of various approaches to systems modelling. One of the main ideas to emerge in the last century was that of networks of interactions with the possibility of feedback loops. Although this can be presented in a formal mathematical way, many policy makers prefer a pictorial approach such as Checkland's soft systems theory. Checkland's method is highly systematic and effective for eliciting information about unknown social systems. As such it can be seen as a useful step towards more formal modelling as required in multilevel systems of systems of systems when data sets are large and there are many complicated multilevel entailments.

In the science of complex systems there are debates about the trade-off between simple but incomplete or incorrect models and complicated but more complete and more correct models. Science demands the best model but policy demands models that can be used in practice. In some sense, the model used becomes part of the system and the analysis of its efficacy is conducted, if at all, at a metalevel.

But some do not subscribe to the idea of best scientific model, and argue that in the science of complex systems it is normal, or even advantageous, to have many models whose output is synthesised in some way.

A major tenet of this book is that there is no adequate scientific model of a multilevel heterogeneous complex system in any domain involving human beings. The reason is that there is no formalism to integrate micro, meso, and macro models in a coherent way. It is suggested that networks and hypernetworks are necessary if not sufficient for such a formalism.

11.6 Multilevel systems, modelling and computation

Chapter 7 of this book presents a hypernetwork-based approach to modelling complex multilevel systems. This formalism translates easily into data structures for computation within and between different levels of aggregation.

It became apparent when writing the chapter that the goal of this book may be unattainable. It may be impossible to formulate integrated multilevel representations of complex social systems. The alternative seems to be muddling through with opportunistic local models using whatever data is available to create narratives and pictures as the basis of understanding the system for policy purposes.

This may be inevitable but it is not very encouraging. Currently policy made on the basis of ambiguous or implicit theory has a poor record in complex systems such as the environment, climate change, finance, terrorism and drug abuse. In comparison, policy made on the basis of *global* models of road traffic or epidemics has been much more successful.

Throughout the book, examples are given of parts being assembled into wholes which, it is argued, gives an immutable upwards arrow to discriminate levels of intermediate words. This may be sufficient to realise the ambition of well-defined multilevel modelling.

11.7 A research agenda for hypernetworks and their applications

There is much more to be done on hypernetworks, including the research agenda:

- automate the construction of multilevel vocabularies of intermediate words
  - develop a machine vision system that creates its own multilevel constructs
  - abstract structures of meaning from large datasets
  - develop a theory of narrative
- use hypernetworks to model the dynamics of a large multilevel system
  - model a large multilevel landuse-traffic system
  - develop a system for personalised education

- technical and theoretical developments
  - implement relational algebras to analyse large heterogeneous data sets
  - answer the Fundamental Question $\langle x_0, x_1, ...; R_1 \rangle \cap \langle y_0, y_1, ...; R_2 \rangle = ?$
  - develop Q-analysis methods for tractable Galois lattice computation
  - give examples of the evolution of a trajectory of hypersimplices
  - investigate the role played by holes and pseudo-holes
  - investigate q-transmission and coupled dynamics using large datasets
  - address the Grand Challenge for multilevel systems
- work with policy makers to design and shape possible futures

Hypernetworks and the science-policy-design entanglement have been presented to many audiences over the last few years, with a variety of responses. Because of the uncertainties of human behaviour, some people are sceptical that the future can be designed and for them the word "shaping" may be more acceptable. However, some people have embraced the ideas enthusiastically, and the prospect of an international community driving forward hypernetwork science is very exciting.

Over the next few years we will develop and share software to process large data sets using hypernetworks – see www.hypernetworks.info. Interested readers are invited to join us as we pursue this research agenda, especially those who are interested in designing and shaping the future.

Bibliography

Agoston, M., K., *Algebraic Topology*, Marcel Dekker, Inc., (New York and Basel), 1976.
Asada, M., Kitano, H., Kuniyoshi, Y., Matsubara, H., Noda, I, Osawa, E., RoboCup: a challenge problem of AI, *AI Magazine*, **18**(1), p. 73, Spring 1997.
Ashby, W. R., *An Introduction to Cybernetics*, Chapman & Hall (London), 1956.
Alexander, C., A city is not a tree, *Architectural Forum* **122**(2) April 1965, 58–62, 1965.
Alexiou, K., Johnson, J., Zamenopoulos, T. (eds) *Embracing Complexity in Design*, Routledge (London), 2009.
Atkin, R. H., Bray, R. W., Cook, I. T., A mathematical approach towards a social science, *Essex Review*, University of Essex, Autumn 1968, No. 2, 3 – 5, 1968.
Atkin, R, H., Johnson, J. H., Mancini, V., An analysis of urban structure using concepts of algebraic topology, *Urban Studies*, **8**, 221–242, 1971.
Atkin, R. H., From cohomology in physics to Q-connectivity in social science, *I. J. Man-Machine studies*, **4**(2), 139–167, 1972.
Atkin, R. H., *Mathematical Structure in Human Affairs*, Heinemann Educational Books (London), 1974.
Atkin, R. H., *Combinatorial Connectivities in Social Systems*, Birkhäuser (Basel), 1977.
Atkin, R. H., *Multidimensional Man*, Penguin Books (Harmondsworth, Middlesex), 1981.
Atkin, R. H., *Mathematical Physics*, Arima Publishing, (Bury St Edmunds), 2010.
Barabási, A-L, *Linked*, PLUME (New York), 2003.
Barabási, A.-L., Albert, R., Emergence of scaling in random networks. *Science* **286** (5439): 509–512, 1999.
Baxt, W. G., Waeckerle, J. F., Berlin, J. A., Callaham, M. L., Who reviews the reviewers? Feasibility of using a fictitious manuscript to evaluate peer reviewer performance. *Annals of Emergency Medicine* **32**(3 Pt 1), 310317, September 1998.
B.B.C., The competing arguments used to explain the riots, *BBC News Magazine*, 11 August 2011. http://www.bbc.co.uk/news/magazine-14483149
B.B.C., Programme schedule, 2-Jan-2013.
http://www.bbc.co.uk/bbcone/programmes/schedules/london/2013/01/02
Bellis, M.A., Hughes, K.,Wood, S.,Wyke, S., Perkins, C., National five-year examination of inequalities and trends in emergency hospital admission for violence across England. *Injury Prevention*, **17**, 319–325, 2011.
Berge, C., *Hypergraphs*, Elsevier Science Publishers B.V. (Amsterdam), 1989.
Booker, C., Fishing quotas are an ecological catastrophe, *The Telegraph* (London), 25 Nov 2007. http://www.telegraph.co.uk/news/uknews/1570439/Fishing-quotas-are-an-ecological-catastrophe.html

Bourgine, P., Johnson, J., Chavalarias, D., (eds), The CSS Roadmap for Complex Systems Science and its Applications 2012-2020, The Complex Systems Society, March 2012.

Buchannan, M., Meltdown Modeling, *Nature*, **460**(6), August 2009.

Castle, S., Europe suffers worst blackout for three decades, *The Independent* (London), 6 Nov 2006. http://www.independent.co.uk/news/world/europe/europe-suffers-worst-blackout-for-three-decades-423144.html

Centre for Social Justice, Dying to belong, London, (February 2009), http://www.centreforsocialjustice.org.uk/client/downloads/DyingtoBelongFullReport.pdf

Centre for Social Justice, Time to wake up. Tackling gangs one year after the riot., CSJ, London (Oct, 2012). http://www.centreforsocialjustice.org.uk/client/images/Gangs%20Report.pdf

Checkland, P., Scholes, J., *Soft Systems Methodology in Action*, John Wiley & Sons (Chichester), 1990.

Cross, N., *Designerly Ways of Knowing*, Birkhuser (Basel), 2007.

Cross, N., *Design Thinking: Understanding How Designers Think and Work*, Berg (Oxford), 2011.

Davies, A., Gardner, B. B., Gardner, M. R., *Deep South*, Chicago University Press, Chicago, 1941.

Dewdney, A. K., *The Magic Machine*, Freeman & Co (New York), 1990.

Dowker, C. H., Homology Groups of Relations. *Annals of Mathematics*, **56**(1), 84–95, Jul., 1952.

Edmonds, B., Syntactic Measures of Complexity, Ph. D. thesis, University of Manchester, England, http://cfpm.org/pub/users/bruce/thesis/all.pdf, 1999.

European Commission, *EU industrial structure 2011: Trends and Performance*, Publications Office of the European Union (Luxembourg), ISBN: 978-92-79-20733-4, 2011. http://bookshop.europa.eu/en/eu-industrial-structure-2011-pbNBBL11001/

Freeman, L., C., Finding social groups: A meta-analysis of the southern women data. Dynamic Social Network Modeling and Analysis. The National Academies, 39–97, 2003.

Forrester, J. W., *Principles of Systems*, Systems Dynamics Series, Pegasus Communincations Inc. (Waltham, MA), (Originally published in 1971 by Wright-Allen Press Inc.), 1990.

Goldratt, E. M., *It's not luck*, North River Press (Great Barrington, MA), 1994.

Gould, P., Letting the data speak for themselves, *Annals of the Association of American Geographers*, **71**(2), 166–176, 1981.

Gould, P., Johnson, J., Chapman, J., *The Structure of Television*, Pion Books Ltd (London), 1984.

Hägerstrand, T., What about people in regional science, *Papers in regional Science*, **24**(1), December 1970.

P. J. Hilton, P. J., Wylie, S., *Homology Theory*, Cambridge University Press (Cambridge), 1965.

Horgan, J., From Complexity to Perplexity, *Scientific American*, **272**(6), 104–109, 1995.

Johnson, J. H., The Q-analysis of road traffic systems, *Environment and Planning B*, **8**(2), 141–189, 1981.

Johnson, J. H., Links, Arrows and Networks: Fundamental Metaphors in Human Thought, *Network in Action*, Eds: D. Batten and J. Casti and R. Thord, 25–48, Springer-Verlag, 1995.

Johnson, J. H., The Future of the Social Sciences and Humanities in the Science of Complex Systems, *Innovation - The European Journal of Social Sciences*, **23**(2), 520-536, June 2010.

Johnson, J. H., Iravani, P., The multilevel network dynamics of complex systems of robot soccer agents, *ACM Transactions on Autonomous and Adaptive Systems*, **2**(2), June 2007.

Katz, P., *Gestalt Psychology*, Methuen & Co. Ltd (London), 1951.

Kelly, G. A., *A Theory of Person Constructs*, W. W. Norton & Co, Inc, (New York), 1963.

Laplace, P-S., *Essai philosophique sur Les Probabilitiés*, MME V^{E} Courcier, Imprimeur-Libraire pour les mathématiques, quai des Augustins, n° 57 (Paris), 1814.

Lane, D., Maxfield, R., Ontological uncertainty and Inovation. SFI Working Paper # 03-09-050, (Santa Fe, NM), Sept, 2003. http://www.santafe.edu/research/working-papers/abstract/e02036fbbce5cbec64b8cac4c7421cc4/

Lloyd, S., Measures of complexity: a non-exhaustive list, "d'Arbeloff Laboratory for Information Systems and Technology, Department of Mechanical Engineering, MIT, 2001. http://web.mit.edu/esd.83/www/notebook/Complexity.PDF

Lorenz, E. N., Deterministic Nonperiodic Flow. *Journal of the Atmospheric Sciences*, **20**, 130–141, 1963.

Lythgo, M., *The Two Cultures*, BBC Radio 4, 25/4/2007. http://www.bbc.co.uk/radio4/science/thenewtwocultures.shtml

McPherson, M., Smith-Lovin, L., Cook, J. M., Birds of a feather: homophily in social networks, *Annual Review of Sociology*, **27**, 41544, 2001.

Mitleton-Kelly, E., 'Complexity Research – Approaches and Methods: The LSE Complexity Group Integrated Methodology, in Keskinen A, Aaltonen M, Mitleton-Kelly E (eds) *Organisational Complexity*. Scientific Papers 1/2003, TUTU Publications, Finland Futures Research Centre, Helsinki, 2003. http://www.psych.lse.ac.uk/complexity/events/PDFiles/publication/complexity_research_approachesandmethods.pdf

Nagin, D. S., Tremblay, R. E., Parental and Early Childhood Predictors of Persistent Physical Aggression in Boys From Kindergarten to High School, *Archives of General Psychiatry*, **58**(4), 389-394, 2001.

Newton, I., The Principia : Mathematical Principles of Natural Philosophy, 1687. Translation by B. Cohen, A. Whitman and J. Budenz, Universty of California Press (Berkeley).

de Nooy, W., Mrvar, A., Batagelj, V., *Exploratory Social Network Analysis with Pajek*, Cambridge University Press, (New York), 2005.

Popielarz, P. A., (In) voluntary association: A Multilevel Analysis of Gender Segregation in Voluntary Organizations, *Gender and Society*, **13**(2), 234-250, April 1999.

Rawlinson, G. E., The significance of letter position in word recognition, PhD Thesis, Psychology Department, University of Nottingham, Nottingham UK, 1976.

Russell, B., *Impact of Science on Society*, Unwin-Hyman (London), 1952.

Schelling, T., C., Dynamic models of segregation, J. of Mathematical Sociology, 1, 143–186,1971.

Simon, H., *The Sciences of the Artificial*, MIT Press (Cambridge), 1965.

Steadman, P., *The Evolution of Designs: Biological Analogy in Architecture and the Applied Arts*, Routledge (Abbingdon), 2008.

Stevens, S. S., On the Theory of Scales of Measurement. *Science*, **103**(2684), 677680, 1946.

Thaler, R. H., Sunstein, C. R., *Nudge: Improving Decisions About Health, Wealth and Happiness*, Penguin Books (London), 2009.

UK Cabinet Office, Cabinet Document Officers' Handbook, The Cabinet Office (London), 2007. http://www.cabinetoffice.gov.uk/media/cabinetoffice/secretariats/assets/cdo_handbook_0712.pdf

UK Department for Culture, Media and Sport, BROADCASTING: Copy of Royal Charter for the continuance of the British Broadcasting Corporation, The Stationary Office (London), 2006. http://www.bbc.co.uk/bbctrust/assets/files/pdf/about/how_we_govern/charter.pdf

UK DEFRA, Bovine TB: The Scientific Evidence. A Science Base for a Sustainable Policy to Control TB in Cattle. An Epidemiological Investigation into Bovine Tuberculosis, Report of the Independent Scientific Group on Cattle TB, Department for theEnvironment, Food and Rural affairs (London), 2007. http://archive.defra.gov.uk/foodfarm/farmanimal/diseases/atoz/tb/isg/report/final_report.pdf

UK H. M. Government, Ending Gang and Youth Violence. A cross-Government Report', Government Command Paper 8211, H. M. Stationary Office, London, 2011. http://www.homeoffice.gov.uk/crime/knife-gungang-youth-violence/

UK Hansard, Parliamentary Debate on Bovine TB, 19 July 2011. http://www.publications.parliament.uk/pa/cm201011/cmhansrd/cm110719/debtext/110719-0002.htm

UK Hansard, Statement on Bovine TB and Badger Control, 23 Oct 2012. http://www.publications.parliament.uk/pa/cm201213/cmhansrd/cm121023/debtext/121023-0001.htm

UK National Archives, The Cabinet Papers 1915-1982, Meetings and papers, The National Archives (Kew, London), viewed 15 Dec 2012. http://www.nationalarchives.gov.uk/cabinetpapers/cabinet-gov/meetings-papers.htm

Urry, J., The complexity turn, *Theory, Culture & Society*, **22**(1), 2005.

Vidal, J., Stratton, A., Goldenberg, S. Low targets, goals dropped: Copenhagen ends in failure, *The Guardian Newspaper*, (London), 18th December 2009. http://www.guardian.co.uk/environment/2009/dec/18/copenhagen-deal

Wasson, R., Integrated Systems: Water, Science at the Shine Dome Canberra, 1–3 May 2002. http://science.org.au/events/sats/sats2002/wasson.html

Wikipedia, Mereology, 2009.

Wilson, R., *Four Colours Suffice: How the Map Problem was Solved*, Princeton University Press, 2004.

Winston, M., and Chaffin, R., and Herrmann, D., A Taxonomy of Part-Whole Relations, *Cognitve Science*, bf 11(2), 417–444, 1987.

Index

2^A, the power set of A, 16
A^C, the complement of A, 16
Diamond, prism operator, 66
α-aggregation, 183
β-aggregation, 183
$\cap$, set intersection, 14
$\cup$, set union, 14
$\emptyset$, empty set, 15
$\in$, belongs to, 14
$\mathbb{A}$, alphabet as a scale, 106
$\mathbb{N}$, integers, 106
$\mathbb{Q}$, rational numbers, 106
$\mathbb{R}$, real numbers, 18, 106
$\mathbb{Z}$, the integers, 135
$\mathcal{P}(A)$, the power set of A, 16
$\prod_{i=1}^{n} A_i$, the product of the sets A_i, 17
$\sigma \cap \sigma'$, 47
$\sigma \lesssim \sigma'$, σ as a face of σ', 46
$\subset$, proper subset, 14
$\subseteq$, subset, 14
$\supset$, proper superset, 14
$\supseteq$, superset, 14

abelian group, 132
abstract p-simplex, 42
address, 217
agenda, research, 319
agent-based systems, 255
aggression, 239
algebraic topology, 2, 130
AND-aggregation, 183
antisymmetric relation, 27
antisymmetry, 107
antivertex, 69
apex, of a hierarchical cone, 181
arson, 240
artificial system, 275

Asada, Minouri, 228
Ashby, W. R., 8
assault, 239
associativity, 132
asymmetric relation, 27
asymmetry, 107
Atkin, R.H., 1, 140
augmentation, of list simplex, 186
augmented phenomenology, 8

backcloth, 4, 101
  backcloth allows, 104
  backcloth does not require, 104
  backcloth forbids, 104
  backcloth topology, 129
  backcloth-traffic design heuristics, 149
Barabási, L., 5
base, of a hierarchical cone, 181
Berge, C., 31
Big Data, 10
bipartite relation, 22, 49
blueprint, 277
boundary of simplex, 133
boundary operator, 133
bounding cycle, 135
Bourgine, P., 8
Boy-X, 243

café wall illusion, 157
canonical faces, 66
Cartesian product, 18
cascade, 263
Centre for Social Justice, 239
chain groups, 132
chain of connection, 51
chains of simplices, 131

changing requirements, 279
chemical isomers, 154
choreography, 265
classes, 14
client, of designer, 279
clock time, 256
closure, of group, 132
clustering, 196
cocycle law, 49, 139
coevolution of problem-solution pair, 279
coevolutionary subsystems, 254
coevolving subsystems, 254
cohomology, 139
combinatorial explosion, 231
commutative group, 132
comparable elements of a relation, 27
complement, 16
complex systems methodology, 8
complex systems science, 7
complexity, 6, 10
composition of mappings, 18
computer simulation, 10
conjugate complex, 49
conjunction, 22
connectivity, 4, 46, 50
consumerism, 241
coupled subsystems, 254
coupling in multilevel systems, 214
cover, of sets, 15, 16
cumulative extension, 186, 187
cumulative intension, 186
cup – handle, 204
cycle, bounding, 135
cycle, non-bounding, 135
cycles as chains, 132

De Morgans' Laws, 17
defender's dilemma, 228
density, road traffic, 233
deprived female, 239
deprived males, 239
descriptor simplex, 69, 116
descriptor simplices, 69, 197
design, 10, 275, 317
 prediction in design, 304
 the design cycle, 288
 the design process, 277
designing the backcloth, 149
designing the future, 276, 297
dichotomy corollary, 198

difference prism, 67
difference set operation, 14
digraph, 151
$\dim(\sigma)$, dimension of σ, 42
dimension, 42
dimension, of a simplex, 42
disjoint sets, 14
disjunction, 22
Dowker, C. H., 1, 140
dual hypergraph, 31
dynamics, 110
 type-one dynamics, 110
 type-two dynamics, 110

eccentricity, 58
empty set, ∅, 13, 15
Ending Gang & Youth Violence, 242
entanglement, complexity, policy, &
 design, 297
equivalence relation, 24, 25
Escher, 73
Essex Review, 48
Euler circles, 14
events, 247, 256, 264
 p-event, 5, 258
 connectivity of events, 264
 extreme events, 262
 mapping p-events to clocktime, 259
 multilevel events, 260
 rare events, 262
evidence-based policy, 300
extension, 14, 287
 cumulative extension, 186, 187
 extension in set definition, 14

face, 46
 face of a relational hypersimplex, 165
 face of a simplex, 44, 46
face operator, 141
family, simplicial, 49
fathers, lack of, 240
feedback, 251, 252, 254, 318
 feedback loop, 252, 254
Fiadeiro, João, 265
flood management, 251
flow-density relationship, 233
folded hierarchies, 232
formal systems, 307
Full Monty, the, 44
function, 18

Index

Fundamental Question of Hypernetworks, 163, 316
fuzzy q–nearness, 160
fuzzy q-near, 160
fuzzy conjunction, 160
fuzzy q-nearness, 317

Galois connection, 36, 65
Galois families, 65
Galois hypergraphs, 36
Galois lattice, 38, 316
Galois pair, 36, 159
Galois prism, 68
gangsta rap, 240
generate-evaluate cycle, 277
geometric realisation, 42
Gestalt, 45, 198
goalkeeper's dilemma, 228
Grand Challenge of multilevel systems, 214
Gregory, Richard, 157
greyscale, 216
group operator, 132

Hägerstrand, Torsten, 102
handle – cup, 204
Hasse diagram, 38
heterogeneous transmission, 127
hierarchical cone, 181
   apex of a hierarchical cone, 181
   base of a hierarchical cone, 181
hierarchical traffic aggregation, 211
hierarchies, 232
   folded hierarchies, 232
   interleaved hierarchies, 232
HITSOCS, 11
holes, in topology, 2
homology, 135
   homology groups, 135
   homology of cylinder, 136
   homology of Klein bottle, 138
   homology of Möbius band, 137
   homology of relations, 140
   homology of torus, 137
homophily, 196
horizontal integrative science, 7
hospital admissions, 239
hospital admissions for assault, 239
hub, 64
hyper-face, 173

hypergraph, 31, 151
   h-connected, 33
   h-neighbour, 33
   dual hypergraph, 31
   Galois hypergraphs, 36
   hypergraph edges, 31
   hypergraph vertices, 31
   neighbour, in hypergraph, 33
   strong connectivity, in a hypergraph, 39
   weak connectivity in a hypergraph, 39
hypernetwork, 2, 5, 151
hypernetworks, 316
hypersimplex, 2, 5, 151
   list hypersimplex, 185, 186, 217
hypersimplices, 316

I.Q. tests, 155
identity, of group, 132
image segmentation, 162
incidence matrix, 21, 51, 60
incomparable elements of a relation, 27
infimum, of a lattice, 38
initial conditions – sensitivity to, 250
instantiation, 217, 292
intension, 14, 287
   cumulative intension, 186
   intension in set definition, 14
interleaved hierarchies, 232
intermediate word problem, 179, 198
intermediate word problem, in design, 281
intersection of simplices, 47
intersection set operation, ∩, 14
intersections of hypersimplices, 163
interval prediction, 248
interval scale, 106
inverse of a mapping, 18
inverse, in group, 132
IQ tests, 155
isomer, in chemistry, 154

Kelly, George, 198
Kitano, Hiroake, 228
Klein bottle, 138
knight fork, 153, 228

Laplace, 247
links, multilevel, 230
list hypersimplex, 185, 186, 217
list simplex, 186
logic, 17

logical operations on hypersimplices, 165
London, 242
loops and q-holes, 144
looting, 240

Möbius band, homology of, 137
machine vision, 161, 216
macro-level, 211
magic roundabout, 90
mapping system time to clock time, 259
mappings, 18
   composition of mappings, 18
   inverse of a mapping, 18
matrices, 112
matrices and relations, 21
measurement scales, 106
   interval scale, 106
   nominal scale, 106
   ordinal scale, 106
   ratio scale, 106
memory, 217
mereological axiom, 181
mereology, 204
meronym, 204
meso-level, 211
methodology – complex systems, 8
micro-level, 211
monopolar construct, 199
MOOC, 289
mulltilevel systems, 319
multidimensional connectivity, 50
multidimensional descriptor spaces, 197
multilevel links, 230
multilevel models, 284
multilevel road traffic, 233
multilevel routes, 230

N+k, in multilevel representation, 180
n-ary relations, 168
n-tuple, 17
naming mappings, 181
neighbour, in hypergraph, 33
network, 151
networks, 71
nilpotent operator, 134
nodes, multilevel in road systems, 236
noisy images, 161
nominal scale, 106
non-bounding cycle, 135
nudge, 263

observation, 170
obstruction, 145
opportunism, 241
OR-aggregation, 183
ordinal scale, 106
orientation of simplex, 130
ought, 297

partial order, 27, 28
partition, 15, 24
path dependence, 6
paying – shopping, 208
Person Construct Theory, 198
personal constructs, 198
pixel, 216
pixels to features, 216
pixels, contiguous, 216
plots, as zones, 235
point prediction, 248
pointer, 217
police, 240
policy, 10, 297, 318
   polcy maker, 297
   policy and scientists, 302
   policy as designing the future, 297
   policy makers, 302
   policy problem, 242
   policy space, 311
   prediction in policy making, 304, 318
polyhedra, 42
polyhedron, 42
power set, 16
predator-prey systems, 254
prediction, 247, 264
   interval prediction, 248
   point prediction, 248
   prediction horizon, 264
   prediction in design, 304
   prediction in policy making, 304
preferential attachment, 5
prism, 66
prism operator, 66
probabilistic q-transmission, 126
problem-solution coevolution, 279
product, Cartesian, 18
products of sets, 17
proof, 307
proper subset, 14
proposition, 17

Index

pseudo homotopy, shomotopy, 143

Q-analysis, 48, 52
q-complex, 146
q-connected, 52
q-face, 46
q-graph, 121
q-holes, 144, 145
q-loops, 144
q-near, 47
q-percolation, 120
q-transmission, 119
q-transmission front, 120
q-transmission property, 119
quasi order, 27, 107

racism, 241
range of convenience, 199
ratio scale, 106
real numbers, $\mathbb{R}$, 18
Real Time Composition, 265
rectangle illusion, 158
reductioonism, 285
reflective relation, 25
reflexivity, 107
relations, 13, 20
requirements, 277
research agenda, 319
rhetoric, 10, 302
rich picture, 252
riots in London, 2011, 240
risk, perceived, 251
road accidents, 159
road intersections, 84
road systems, 230
RoboCup, 228
robot football, 228
routes, multilevel, 230
run hypersimplex, 216
run of pixels, 216

satisfice, 10, 297
satisficing, 278
scientists, 302
segmentation, of image, 162
self-similar backcloth-traffic aggregation, 214
set, 13
set membership, 14
set products, 17

sets, 13
shared face, 47
shomotopy bottle, 145
shomotopy, pseudo homotpy, 143
shootings, 241
shopping – paying, 208
simplex, 42
simplex orientation, 130
simplicial complex, 2, 49, 151
simplicial family, 49, 151
simplicial prisms, 66
Simpson's finger, 26, 210
simulation, 10, 255
Sky and Water – Escher, 73
skyscraper diagram, 59
social exclusion, 240
social networking, 241
soft systems, 252, 318
southern women, 60
southern women data set, 21
space mean speed, 233
space-time backcloth, 102
space-time prism, 102
spaghetti sunction, 88
specification, 280
spending cuts, 240
standard ordering, 66
star, 64
star-hub pairs, 68
Stevens, Stanley Smith, 106
street gangs, 239, 242
strict order, 107
strong connectivity, in a hypergraph, 39
structure vector, 56
structuring space, 103
subset, 14
subset – proper, 14
sun illusion, 158
superset, 14
supertraffic, 128
supremum, of a lattice, 38
surveyability, 309
symmetric difference set operation, 14
symmetric relation, 25
system dynamics, 252
systems of systems of systems, 177
systems theory, 318

taxonomic aggregation, 184
television programmes, 128

time, 247, 256, 264
time horizon, 264
time mean speed, 233
time series, 101
tipping point, 263
topology, 317
topology of backcloth, 129
torus, homology of, 137
total comparability, 107
total order, 27
traffic, 4, 101, 105
　traffic can destroy the backcloth, 111
transitive closure, 26
transitive relation, 25
transitivity, 107
transpose matrix, 60
travel time, road traffic, 233
trellis, 185
triangulation, 2
triangulation of Klein bottle, 138
trichotomy law, 27
truth value, 17
type-one dynamics, 110

type-two dynamics, 110

union set operation, $\cup$, 14
universal set, 16
unrepeatable experiments, 298

vanilla ice cream, 45
vector, 18
vernacular models, 245
vertex order, 152
vertex, of a simplex, 42
vertical in-depth science, 7
vertically adjacent, 216
vertices, 42
virtual contour, 158

weak connectivity, in a hypergraph, 39
weak policing, 240
welfare dependence, 240

zebra fish, 161
zones, multilevel, 235